THE INTERNET OF DRONES

AI Applications for Smart Solutions

THE INTERNET OF DRONES

AI Applications for Smart Solutions

Edited by
Arun Solanki, PhD
Sandhya Tarar, PhD
Simar Preet Singh, PhD
Akash Tayal, PhD

APPLE
ACADEMIC
PRESS

First edition published 2023

Apple Academic Press Inc.
1265 Goldenrod Circle, NE,
Palm Bay, FL 32905 USA

760 Laurentian Drive, Unit 19,
Burlington, ON L7N 0A4, CANADA

CRC Press
6000 Broken Sound Parkway NW,
Suite 300, Boca Raton, FL 33487-2742 USA

4 Park Square, Milton Park,
Abingdon, Oxon, OX14 4RN UK

© 2023 by Apple Academic Press, Inc.

Apple Academic Press exclusively co-publishes with CRC Press, an imprint of Taylor & Francis Group, LLC

Library and Archives Canada Cataloguing in Publication

Title: The Internet of drones : AI applications for smart solutions / edited by Arun Solanki, PhD, Sandhya Tarar, PhD, Simar Preet Singh, PhD, Akash Tayal, PhD.

Names: Solanki, Arun, 1985- editor. | Tarar, Sandhya, editor. | Singh, Simar Preet, editor. | Tayal, Akash, 1973- editor.

Description: First edition. | Includes bibliographical references and index.

Identifiers: Canadiana (print) 20220222657 | Canadiana (ebook) 20220222681 | ISBN 9781774639856 (hardcover) | ISBN 9781774639863 (softcover) | ISBN 9781003277491 (ebook)

Subjects: LCSH: Drone aircraft. | LCSH: Internet of things. | LCSH: Artificial intelligence.

Classification: LCC TL685.35 .I58 2022 | DDC 629.133/39—dc23

Library of Congress Cataloging-in-Publication Data

Names: Solanki, Arun, 1985- editor. | Tarar, Sandhya, editor. | Singh, Simar Preet, editor. | Tayal, Akash, 1973- editor.

Title: The internet of drones : AI applications for smart solutions / edited by Arun Solanki, PhD, Sandhya Tarar, PhD, Simar Preet Singh, PhD, Akash Tayal, PhD.

Description: First edition. | Palm Bay, FL : Apple Academic Press, Inc. ; Boca Raton, FL . CRC Press, 2023. | Includes bibliographical references and index. | Summary: "In recent years, drones have been integrated with the Internet of Things to offer a variety of exciting new applications in different sectors. This volume provides a detailed exploration of adapting and implementing Internet of Drones (IoD) technologies into real-world applications, with the emphasis on solutions to architectural challenges, providing a clear overview of standardization and regulation, implementation plans, and privacy concerns. The Internet of Drones : AI Applications for Smart Solutions discusses the architectures and protocols for drone communications, implementing and deploying of 5G-drone setups, security issues with drone technology, deep learning techniques applied on real-time footage, and more. It also explores some of the varied applications of IoD, such as for use in monitoring and analysis of troposphere pollutants, providing services and communications in smart cities (such as for weather forecasting, healthcare, communications, transport, agriculture, safety and protection, environmental reduction, service delivery, and e-disposal), for disaster relief management (such as for scanning the affected areas for radiation intensity in cases of nuclear disaster, gathering the location of hotspots, looking for the victims, assessing damage), and more. The authors cover package delivery, movement of traffic, crop monitoring, and mass detection. The problems and challenges associated with IoD in air traffic monitoring, communication between drones, optimum route discovery, and security are also addressed. This detailed exploration of adapting and implementing IoD technologies into real-world applications in this volume will be valuable for graduate students in computer science and especially drone technology, as well as researchers and professionals"-- Provided by publisher.

Identifiers: LCCN 2022018166 (print) | LCCN 2022018167 (ebook) | ISBN 9781774639856 (hbk) | ISBN 9781774639863 (pbk) | ISBN 9781003277491 (ebk)

Subjects: LCSH: Internet of things. | Drone aircraft. | Cooperating objects (Computer systems) | Smart cities. | Ad hoc networks (Computer networks)

Classification: LCC TK5105.8857 .I546 2023 (print) | LCC TK5105.8857 (ebook) | DDC 006.3--dc23/eng/20220705

LC record available at https://lccn.loc.gov/2022018166

LC ebook record available at https://lccn.loc.gov/2022018167

ISBN: 978-1-77463-985-6 (hbk)
ISBN: 978-1-77463-986-3 (pbk)
ISBN: 978-1-00327-749-1 (ebk)

About the Editors

Arun Solanki, PhD
Assistant Professor, Department of Computer Science and Engineering, Gautam Buddha University, Greater Noida, India

Arun Solanki, PhD, is working as an Assistant Professor in the Department of Computer Science and Engineering, Gautam Buddha University, Greater Noida, India, where he has been working since 2009. He has worked as in many roles, including timetable coordinator and member of the examination, admission, sports council, digital information cell, and other university teams from time to time. He has supervised more than 60 MTech dissertations. His research interests span expert systems, machine learning (ML), and search engines. He has published many research articles in SCI/Scopus-indexed international journals/conferences, including those published by IEEE, Elsevier, Springer, etc. He has participated in many international conferences as a technical and advisory committee member. He has organized several FDP, conferences, workshops, and seminars and has chaired many sessions at international conferences. Dr. Solanki is working as an Associate Editor of the *International Journal of Web-Based Learning and Teaching Technologies* (IJWLTT), IGI Global. He has been working as Guest Editor for special issues of *Recent Patents on Computer Science*, Bentham Science Publishers. He has also edited many books with reputed publishers, like IGI Global, CRC, and AAP. In addition, he is working as a reviewer for Springer, IGI Global, Elsevier, and other reputed publisher journals.

Dr. Solanki has received his MTech in Computer Engineering from YMCA University, Faridabad, Haryana, India. He has received his PhD in Computer Science and Engineering from Gautam Buddha University in 2014.

Sandhya Tarar, PhD
Assistant Professor Department of Computer Science and Engineering;
Convener. Centre of Excellence on Artificial Intelligence and Cyber
Security, Gautam Buddha University, Greater Noida, India

Sandhya Tarar, PhD, is working as an Assistant Professor in the Department of Computer Science and Engineering and as a Convener of the Center of Excellence on Artificial Intelligence and Cyber Security in Gautam Buddha University, India. She has 14 years of technical teaching and research experience at various technical institutes and universities. She has published papers in national and international journals and conferences. Her area of research is artificial intelligence, biometrics, Internet of Things (IoT), and digital image processing. She has two patents in her credit: "Smart Medicine Dispenser" and "A Device for Contamination Detection in Fuel Supply into a Vehicle."

She has received many awards, including a Young Scientist Award, Bharat Vikas Award, "Best Teacher Award, Best Guru Award, Young Researcher Award, etc., by organizations such as EET CRS Academic Brilliance, Institute of Self Reliance and Rashtriya Shaikshik Mahasangh, GISR Foundation. She received a Distinguished Educator Award in International Business and Academic Excellence Awards (ABAE-2019) at the American College of Dubai.

She had served as a member of the Board of Studies of Mahamaya Technical University and Amity University, India, and is an active Board of Studies member of the School of ICT, Gautam Buddha University, India. She has supervised more than 60 dissertations and 18 BTech minor and major projects. She is on the reviewer panels of many international and national conferences and journals. She is a Technical Program Committee member of many technical events, including seminars, conferences, and symposiums in India and abroad. She is serving as a reviewer of IEEE/ACM Transactions on Networking, Elsevier, and many more. She has designed course curricula for various courses in computer science and engineering and artificial intelligence and has established laboratories of software engineering and software testing at her university. She is associated with Global Public Visibility Committee of IEEE and has published multiple articles for the same. She is a member of various professional societies, such as IEEE, ACM, CSI, ISTE, IACSIT, IAENG, and UACEE.

She obtained her BTech, MTech, and PhD degrees in Computer Science and Engineering.

Simar Preet Singh, PhD
Assistant Professor, Bennett University, Greater Noida, Uttar Pradesh, India

Simar Preet Singh, PhD, is as an Assistant Professor at the School of Computer Science Engineering and Technology (SCSET), Bennett University, Greater Noida, Uttar Pradesh, India. He was previously affiliated with GNA University, Phagwara, and Chandigarh Engineering College, Ajitgarh, India. He also previously worked with Infosys Limited and DAV University, Jalandhar, India, and has worked on international projects. He has published papers in SCI/SCIE/Scopus-indexed journals and has presented many research papers at various national and international conferences in India and abroad. His areas of interests include cloud computing, fog computing, IoT, big data, and machine learning (ML). He earned his doctoral degree at Thapar Institute of Engineering and Technology, Patiala, India, and also holds several specialized certifications, including Microsoft Certified System Engineer (MCSE), Microsoft Certified Technology Specialist (MCTS), and Core Java. He had also undergone a training program for VB.Net and Cisco Certified Network Associates (CCNA).

Akash Tayal, PhD
Associate Professor, Department of Electronics and Communication,
Indira Gandhi Delhi Technical University for Women, Delhi, India

Akash Tayal, PhD, is as an Associate Professor of the Department of Electronics and Communication at the Indira Gandhi Delhi Technical University for Women, Delhi, India, and holds a PhD from IIT Delhi. He has more than 15 years of experience in academics with 465 research citation and a Google Scholar index of (h-index score 12 and i-10 index 13). He has published research articles in SCI/Scopus-indexed international journals and has participated at many international conferences. He has been a technical and advisory committee member of many conferences and has chaired and organized many sessions at international conference, workshops, and seminars.

His current research interests include deep learning, machine learning (ML), data science and data analytics, image processing, stochastic dynamic facility layout problem, optimization, meta-heuristic, and decision science.

He has published articles in various international journals and conference proceedings, such as Sustainable Cities and Society, Annals of Operation Research, IJCS, COI, IEEE, GJFSM, IJOR, IJBSR. He is a life member of GLOGIFT Society and IIIE.

Contents

Contributors

Mehtab Alam
Jamia Hamdard, New Delhi, India, E-mail: mahiealam@gmail.com

Soubhagya Sankar Barpanda
School of Computer Science and Engineering, VIT-AP University, Amravati, Maharashtra, India

Sonalika Bhandari
Department of Electronics and Communication Engineering, Maharaja Agrasen Institute of Technology, New Delhi, India

Arpit Bhardwaj
Department of Computer Science and Engineering, BML Munjal University, Gurugram, Haryana

Harshit Bhardwaj
Department of Computer Science and Engineering, University School of Information and Communication Technology, Gautam Buddha University, Greater Noida – 201310, Gautam Budh Nagar, Uttar Pradesh, India

Namisha Bhasin
Gautam Buddha University, Greater Noida, Uttar Pradesh, India

Korhan Cengiz
Trakya University, Edirne – 22030, Turkey

Akshay Chamoli
Jamia Hamdard, New Delhi, India

Ramakanta Choudhury
Department of Electronics and Communication Engineering, Maharaja Agrasen Institute of Technology, New Delhi, India, E-mail: rkchoudhury1@gmail.com

Jagjit Singh Dhatterwal
Department of Artificial Intelligence & Data Science, Koneru Lakshmaiah Education Foundation, Vaddeswaram, AP, India, E-mail: jagjits247@gmail.com

Nitin S. Goje
Department of Computer Science, Webster University, Tashkent, Uzbekistan

Mahendra Kumar Gourisaria
School of Computer Engineering, KIIT Deemed to be University, Bhubaneswar, Odisha, India, E-mail: mkgourisaria2010@gmail.com

Somya Goyal
Manipal University Jaipur, Jaipur – 303007, Rajasthan, India, E-mail: somyagoyal1988@gmail.com

G. M. Harshvardhan
School of Computer Engineering, KIIT Deemed to be University, Bhubaneswar, Odisha, India

Nabeela Hasan
Jamia Millia Islamia, New Delhi, India

Vivek Jaglan
Department of CSE, Graphic Era Hill University, Dehradun, Uttarakhand, India,
E-mail: jaglanvivek@gmail.com

R. Jayalakshmi
Department of Computer Science, St. Claret College, Bangalore, Karnataka, India,
E-mail: jayabinoy2020@gmail.com

Noor Zaman Jhanjhi
Department of Computing and IT, Taylor's University, Malaysia

Jaideep Kala
Department of Electronics and Communication Engineering,
Maharaja Argrasen Institute of Technology, New Delhi, India

Kuldeep Singh Kaswan
School of Computing Science and Engineering, Galgotias University, Greater Noida,
Uttar Pradesh, India, E-mail: kaswankuldeep@gmail.com

Garima Kulshreshtha
Department of Electronics and Communication Engineering, IILM Academy of Higher Learning,
College of Engineering and Technology, Greater Noida, Uttar Pradesh, India,
E-mail: garima.kulshreshtha.gk@gmail.com

P. Praveen Kumar
Department of Computer Science and Engineering, IFET College of Engineering, Tamil Nadu, India

T. Ananth Kumar
Department of Computer Science and Engineering, IFET College of Engineering, Tamil Nadu, India,
E-mail: ananth.eec@gmail.com

Roshan Lal
Department of Information and Communication Technology, Gautam Buddha University,
Uttar Pradesh, India, E-mail: rchhokar@amity.edu

Chuan-Ming Liu
Department of Computer Science and Information Engineering, National Taipei University of
Technology (Taipei Tech), Taipei – 106, Taiwan

Awadhesh Kumar Maurya
Department of Electronics and Communication Engineering, IILM Academy of Higher Learning,
College of Engineering and Technology, Greater Noida, Uttar Pradesh, India,
E-mail: maurya.akm@gmail.com

Sachi Nandan Mohanty
Department of Computer Science and Technology, IcfaiTech, ICFAI Foundation for Higher Education,
Hyderabad, Telangana, India

Pavithra Muthu
Department of Computer Science and Engineering, IFET College of Engineering, Tamil Nadu, India

Amit Pandey
Computer Science Department, College of Informatics, Bule Hora University, Bule Hora, Ethiopia,
E-mail: amit.pandey@live.com

Sheng-Lung Peng
Department of Creative Technologies and Product Design, National Taipei University of Business,
Taiwan, E-mail: slpeng@ntub.edu.tw

Rajmohan Rajendirane
Department of Computer Science and Engineering, IFET College of Engineering, Tamil Nadu, India

Aditi Sakalle
Department of Computer Science and Engineering, University School of Information and
Communication Technology, Gautam Buddha University, Greater Noida – 201310,
Gautam Budh Nagar, Uttar Pradesh, India

Chandrakanta Samal
Department of Computer Science, Acharya Narendradev College, Delhi University, India

R. Dinesh Jackson Samuel
Faculty of Technology, Design, and Environment, Oxford Brookes University, Oxford, United Kingdom

Uttam Sharma
Department of Computer Science and Engineering, University School of Information and
Communication Technology, Gautam Buddha University, Greater Noida – 201310,
Gautam Budh Nagar, Uttar Pradesh, India

Bharati Singh
Department of Transportation, School of Business, UPES, Dehradun, Uttarakhand – 248007, India,
E-mail: bharti.singh012@gmail.com

Naveen Chilamkurti Smieee
Department of Computer Science and IT, La Trobe University, Melbourne, Australia,
E-mail: n.chilamkurti@latrobe.edu.au

Sandhya Tarar
Department of Information and Communication Technology, Gautam Buddha University,
Greater Noida, Uttar Pradesh, India, E-mail: tarar.sandhya@gmail.com

Saurabh Tiwari
Department of Transportation, School of Business, UPES, Dehradun, Uttarakhand – 248007, India,
E-mail: stiwari@ddn.upes.ac.in

Pradeep Tomar
Department of Computer Science and Engineering, University School of Information and
Communication Technology, Gautam Buddha University, Greater Noida – 201310,
Gautam Budh Nagar, Uttar Pradesh, India

Aanchal Vij
School of Computing Science and Engineering, Galgotias University, Greater Noida, Uttar Pradesh,
India, E-mail: aanchal.vij@galgotiasuniversity.edu.in

Navneet Yadav
Department of Electronics and Communication Engineering, Maharaja Argrasen Institute of
Technology, New Delhi, India

Abbreviations

ADC	analog to digital converter
ADS-B	automatic dependent surveillance-broadcast
AI	artificial intelligence
AMRP	average minimum reachability power
ANN	artificial neural network
A-NSSA	aerial network slice subnet
AoA	angle-of-arrival
AODV	ad-hoc on-demand distance vector
AQI	air quality index
AR	aerial-relays
AR	augmented reality
ARTCC	Air Route Traffic Control Center
ATCSCC	Air Traffic Control System Command Center
ATCT	Air Traffic Control Tower
ATF	air turbine fuel
ATN	air transport network
AUW	all-up weight
BDD	bad data detector
BS	base station
CAIDA	Center for Applied Internet Data Analysis
CAN	control area network
CARPN	class-aware region proposal network
CBC	cipher block chaining
CDL	communication data-link
CNN	convolutional neural network
CPE	complex event processing
CRCs	cyclic redundancy code
CU	central unit
D2BS	drone-to-base station
D2D	drone-to-drone
D2N	drone-to-network
DBM	deep Boltzmann machine
DBN	dynamic Bayesian network
DC	dual networking

DDOS	distributed denial of service
DF	digital forensic
DFIs	digital forensic investigations
DL	deep learning
DO	doppler shift
DoS	denial of service
DSA	digital signature algorithm
DSDV	destination-sequenced distance vector
DSR	dynamic source routing
DTW	dynamic time warping
ECB	electronic code book
EM	expectative-maximization
ERT	emergency response team
ESB	enterprise service bus
F.S.S.	flight service station
FAA	federal aviation administration
FANET	flying ad-hoc network
FDNPS	Fukushima Daiichi Nuclear Power Station
GAN	generative adversarial network
GCS	ground control station
GD	gradient descent
GIT	greedy incremental tree
GPS	global positioning system
GPUs	graphics processing units
GSM	global system for mobile
HALE	high altitude, long endurance
ICOs	internet-connected objects
ICT	information and communication technology
IoD	internet of drones
IoDT	internet of drone things
IoE	internet of everything
IoT	internet of things
IoU	intersection over union
IRT	infrared thermography
KNN	K-nearest neighbors
LALE	low altitude, long endurance
LASE	low altitude, short endurance
LPWAN	low power wide area network
LSTM	long short-term memory

MALE	medium altitude, long endurance
MANET	mobile ad-hoc network
mAP	mean average precision
MAV	micro air vehicle
MeNB	master-eNB
MgNB	master-gNB
ML	machine learning
MLP	multi-layer perceptron
MRO	maintenance repair and overhaul
MTSO	mobile telecommunications switching office
NAV	nano air vehicles
NDT	non-destructive testing
NEF	network exposure feature
NR	new radio
NSaaS	network slice as-a-service
NSIs	network slice managed instances
NSSIs	nets slice subnet instances
NTC	negative thermistor coefficient
OHEM	online hard example mining
OLSR	optimized link state routing
PCA	principal component analysis
PDAs	personal digital assistants
PID	proportional, integral, derivative
PR	pseudo range
QoS	quality of service
RBM	restricted Boltzmann machine
R-CNN	region-convolutional neural network
RCS	regional connectivity scheme
RF	radio frequency
RFID	radio frequency identification
RLC	radio connection control
RLSM	recursive least squares method
RNN	recurrent neural networks
ROIs	region-of-interests
RPN	region proposal network
RSS	received signal strength
SAT	secure aggregation tree
SC	surveillance center
SCA	sudden cardiac arrest

SCDSA	shortened complex digital signature algorithms
SD	slice differentiator
SDAP	service data adaptation protocol
SDN	software defined network
SGD	stochastic gradient descent
SIT	secure IoT
SL	supervised learning
SNR	signal-to-noise ratio
SOAP	simple object access protocol
SOM	self-organized map
SQ	service quality
SSD	single shot detection
SVM	support vector machine
SVN	satellite vehicle number
SWIPT	simultaneous wireless information and power transfer
TCS	task control system
ToA	time-of-arrival
TORA	temporarily ordered routing algorithm
TRACON	terminal radar approach control
TSODR	time-slotted on-demand routing
UAANET	UAV ad hoc network
UAS	unmanned aerial system
UAV	unmanned aerial vehicle
UGV	unmanned ground vehicle
VLOS	visual sight line
Vo	voltage
VTOL	vertical take-off and landing
WSN	wireless sensor network
YOLO	you only look once
ZRP	zone routing protocol

Preface

In recent years, the use of drones has been widely investigated as a solution to various sectors. The sectors are mainly categorized into forces, governmental, and non-governmental models. Drones are primarily used in wars and disasters in recent times. However, in recent years, drones are being integrated with IoT, known as the Internet of Drones (IoD). The main aim of the book to discuss the applications of IoD in different sectors. The other sectors cover package delivery, movement of traffic, crop monitoring, and mass detection.

There are some problems associated with the IoD, including air traffic monitoring, communication between drones, optimum route discovery, and security. This book explores the different structures and frameworks for meeting specific IoD requirements. According to application requirements, to meet user-specific, value-added, and on-demand services, IoDs are integrated with IoT devices such as sensors, motors, cameras, etc. The goal of IoD is to provide safe and timely delivery. To obtain this goal, the company Sun Rising Technologies played a significant role. IoD faces many open challenges such as standardization and regulation, implementation planning, and privacy. Therefore, this book provides a clear overview of IoD, its necessity, working, and implementation.

The main objective is to provide a detailed exploration of adapting and implementing IoD technologies into real-world applications. It can be used at the elementary and intermediate levels for graduate students in computer science, researchers, and professionals. Architectural solutions in implementing IoD are the core of this book. This book contains 13 chapters as follows:

> **Chapter 1: The Internet of Things (IoT) Architectures and Protocols for Drone Communications**
> The Internet of Things (IoT) is a computing, mechanical, and digital device, that can transfer data through an internet-based network without human interaction. The internet-controlled drones are named the Internet of Drones (IoD). IoD offers conventional types of assistance for different automaton applications, such as bundle conveyance, traffic reconnaissance, search, and salvage, and that is

just the beginning. This chapter presents the architecture of how such a design can be sorted out and indicates the highlights that an IoD framework based on our design should execute. In this chapter, we also discuss IoD protocols including the latest applications of IoD systems.

> **Chapter 2: Approaching Internet Renovation of Imperceptible Computers to Facilitate Internet of Drones**
> This chapter identifies the knowledge, performance, and challenges of IoT and IoDT (Internet of Drone Things). The chapter provides a brief overview of UbiComp (ubiquitous computing), IoT, and IoDT. The chapter also describes the main technological tools behind UbiComp, the Internet of Objects, and the Internet of Drones. This chapter describes the relationship between cloud computing, IoT, and IoD with voice data. The chapter examines the major capabilities, controversies, and futures of UbiComp and the importance of the forthcoming internet renewal of invisible computers to activate the Internet of Drones. This chapter also focuses on the most important opportunities and challenges for drones. In addition to these features, this chapter also provides an overview of intelligent community forecasting using Internet of Drone.

> **Chapter 3: Implementation and Deployment of 5G-Drone Setups**
> This chapter discusses both the MED and WID cases under practical limits of 5G at the start of 5G Phase I completion (Rel-15). We also address possible solutions to challenges highlighted, either by implementing existing guidelines or presenting recommendations for further changes. While drones can easily be inserted into cell phone networks, they tend to use 4G LTE-A or 5G Rel-15 substantial achievements constructing main structures. Nevertheless, it is essential to fine-tune potential launches by looking at current MED and WID aspects to fill the gaps with new technologies.

> **Chapter 4: Security Issues on the Internet of Drones (IoD)**
> Drones are the small flying vehicles that have changed life in many ways by providing information transmission, surveillance of a places where it was difficult to travel earlier by humans, becoming aware of hazards, and forwarding that information instantly to relevant stakeholders. These days people want to connect without wasting any time, which is challenging but has become a necessity. The

medium used to communicate with each other is through a network medium that can be easily hacked and exploited; hence security is a significant concern in this network. Except for a physical attack, the attackers can hack the system. To make the system secure, various approaches are followed and many more are required.

➤ **Chapter 5: Real-Time Monitoring and Analysis of Troposphere Pollutants Using a Multipurpose Surveillance Drone**
This chapter presents a multi-tasking quadcopter drone capable of carrying scientific payloads such as an air quality index (AQI) determining IoT system. This can be mounted over the quadcopter to measure tropospheric pollutants such as PM2.5, PM10, CO, SO_2 along with temperature and humidity with a high level of accuracy and a flight time of over 20 minutes. This quadcopter can carry 1 kg of additional payload with a restricted flight time to perform other operations such as holding a thermal imaging camera, advanced sensor equipment or other non-scientific payloads.

➤ **Chapter 6: Advanced Object Detection Methods for Drone Vision**
This chapter describes the potential of deep learning techniques applied on real-time footage gathered by on-board cameras on UAVs. Machine learning algorithms have found immense applications in almost every area of modern research. A specialized branch of machine learning is deep learning, which mainly deals with neural networks. These neural networks are designed to mimic the human brain and form many advanced artificial intelligence applications, such as smart computer vision.

➤ **Chapter 7: Security Analysis of UAV Communication Protocols: Solutions, Prospects and Encounters**
Unmanned aircraft can be piloted independently or remotely by ground crews and can save lives by moving people out of harm and unsafe conditions. Typical UAV applications include monitoring and detection; collaborative search, detection, and tracking areas; consistent household monitoring; radio tracking; and forest fire surveillance. Despite their enormous utility scope, UAV technology's research and development have grown exponentially in the last decade. UAVs are typically remote-controlled, which, in effect, opens paths for cyber attacks. Popular methods for attacks, minor findings, and related vulnerabilities are discussed in this chapter.

> **Chapter 8: Challenges and Opportunities of Machine Learning and Deep Learning Techniques for the Internet of Drones**
> This study focuses on significant research in the area of smart drones. This work finds the various algorithms of machine learning, deepfake, generative adversarial network (GAN), and deep learning. The authors also compare the performance of various algorithms in terms of computational performance, making more smart drones. Hence, the, findings of this study are helpful for multiple application in the field of drones.

> **Chapter 9: Machine Learning and Deep Learning Algorithms for IoD**
> This chapter explores extensively the latest recorded uses and implementations of profound learning and machine learning for UAVs, along with their efficiency and limitations. A thorough description of the core methods of deep learning (DL) and machine learning (ML) is also used for realistic and efficient approaches. By concentrating on supervised and reinforcing learning strategies and, lastly, by describing the main challenges for deep learning in UAV based applications, we address why, where, and which ML approaches help construct U-RANs.

> **Chapter 10: Smart Cities and Internet of Drones**
> Unmanned aerial vehicles (UAVs), or drones, are certainly going to be the future. They operate in the air and keep track of the occurrences in their vicinities. They do not need any human to control them or to operate them. They gather data into the form of images and sounds with various sensors, cameras, and mics built in them and transmit the gathered information to the gateway where data can be processed and analyzed. They will soon be working as aerial and portable base stations (BS) for users as well as for governments, acting like gateways, collecting data from their own sensors and various other sensors deployed on roads, bridges, traffic lights, vehicles, buildings, and various other places. They will help us gather big data for analysis and make smart cities even smarter.

> **Chapter 11: Internet of Drone for Enhancing Service Quality in Smart Cities**
> The Internet of Drones (IoD) can play an essential role in various smart urban applications, for example; weather forecasting,

healthcare, communications, transport, agriculture, safety and protection, environmental reduction, service delivery, e-disposal, etc. The Internet of Drones can alter the surveillance idea because of its mobility and can fundamentally change the method and viewpoint of data collection and input to governments, businesses, and people. This chapter discusses the application of the Internet of Drones (IoD) to make a smart city.

➤ **Chapter 12: Internet of Drones Applications in Aviation MRO Business Services**
This research aims to identify the aviation maintenance repair and overhaul (MRO) business service provider's optimized outcomes to use drones in their daily operation effectively. This study implies that MRO business services require research and investment in technologies like drones, machine learning, and cloud applications to support the maintenance and repair process. This will help in more effective maintenance and component defect identification, optimized technician scheduling for each work plan, and improved turnaround times for aircraft.

➤ **Chapter 13: Deploying Unmanned Aerial Vehicle (UAV) for Disaster Relief Management**
In this chapter, the use of UAVs for relief operations in cases of disasters is proposed. Life-saving task forces work to save lives and minimize damage. For this, accurate location information is desirable to carry out rescue operations effectively. The more we know the affected areas, the better relief operations can be carried out. In such a scenario, drones find applications to scan the affected areas for generating the radiation intensity in case of nuclear disaster, gather hotspots' locations, look for victims, and assess the intensity of damage incurred. The drones can play an essential role in disaster relief management.

PART I

INTRODUCTION TO DRONES

CHAPTER 1

The Internet of Things (IoT) Architectures and Protocols for Drone Communications

GARIMA KULSHRESHTHA,[1] AWADHESH KUMAR MAURYA,[1] and SHENG-LUNG PENG[2]

[1]*Department of Electronics and Communication Engineering, IILM Academy of Higher Learning, College of Engineering and Technology, Greater Noida, Uttar Pradesh, India, E-mails: garima.kulshreshtha.gk@gmail.com (G. Kulshreshtha), maurya.akm@gmail.com (A. K. Maurya)*

[2]*Department of Creative Technologies and Product Design, National Taipei University of Business, Taiwan, E-mail: slpeng@ntub.edu.tw*

ABSTRACT

The internet of things (IoT) is a computing, mechanical, and digital device, which can transfer data through an internet-based network without human interaction. IoT has many applications in various fields as embedded systems, smart-phones, smart-homes, smart-cities, machine learning (ML), agriculture, environmental monitoring, wireless sensor networks (WSNs), control systems, home automation, building automation, industry, healthcare, and robotics. The popular application is the robotics sector because it reduces human efforts and is based on automation. The attractive and most popular application of robotics is a drone. The internet-controlled drones are named the internet of drones (IoD). The IoD is a layered system control engineering structured for the most part for organizing the entrance of automated elevated vehicles to controlled airspace, and giving route administrations between

The Internet of Drones: AI Applications for Smart Solutions. Arun Solanki, PhD, Sandhya Tarar, PhD, Simar Preet Singh, PhD & Akash Tayal, PhD (Eds.)

areas alluded to as hubs. The IoD offers conventional types of assistance for different automaton applications, such as bundle conveyance, traffic reconnaissance, search, and salvage, and that is just the beginning. This chapter presents the architecture of how such a design can be sorted out and indicates the highlights that an IoD framework based on our design should execute. In this chapter, we also discuss the IoD protocols including the latest applications of IoD systems.

1.1 INTRODUCTION

The IoD is a smart Unnamed Aerial Vehicle (UAV) or Aircraft System based on the technology of IoT. Drone technology was developed in the 20th century. Presently this technology is most popular. This technology is related to flexible, dynamic, cost-effective, and real-time robotic systems. The drone technology is now improving by the size reduction, weight reduction, higher battery life, camera integration, global system for mobile communication (GSM) integration, and global positioning system (GPS) integration [1, 2]. These systems are implemented by artificial intelligence (AI) [3], machine learning (ML) [4, 5], deep learning (DL) [6, 7] and Block-chain algorithms to increase the efficiency of the drone [8–12].

The UAV is a smart aircraft system including the major components like microcontroller, electronic communication system, different types of sensors, camera, and structural part of the drone. The IoT based drone having advanced sensors, smart wireless connections, cloud infrastructure, huge data collection, and analysis [13–15]. The UAVs have a wide range of applications like photography [16–18], package or product delivery [19–22], mapping [23–26], disaster management [27–29], agriculture [30–32], search operation [33–35], rescue operation [29, 36–38], weather forecast [39], environmental monitor [40–43], wildlife monitoring [40, 44, 45], law enforcement [46, 47], entertainment [48, 49], defense [50], surveillance [51–53], traffic management or vehicular network management system [54–56], industry 4.0 [57], cyber security [58], and many others [59–62].

1.2 IoD ARCHITECTURES

The IoD has layered architecture which is used to communicate with UAV with the help of the internet to control and monitor. IoD architecture has several types of architecture. The architecture as mainly related to the cloud-based

[63–65], communication network-based [66–68], UAV structure-based [69], computing-based, etc. [70–72]. These architectures are described as follows:

1.2.1   CLOUD-BASED ARCHITECTURE

UAVs are an important application of IoT. The cloud-based architecture plays a vital role to control and monitor the UAVs using IoT. The cloud-based architectural block diagram is given in Figure 1.1. This architecture has three main layers: (i) UAV layer; (ii) cloud services layer; and (iii) client layer [65]. The functions of all the given layers are described as follows:

1.2.1.1   UAV LAYER

The UAV provides the resources to the end-user. All the hardware controls are dependent on the UAV layer like sensor controlling, actuator controlling, etc. This layer allows the development of high-level software development to control all the hardware related issues. It has two types of network interfaces, web services interfaces, and cloud services interfaces. The web-based services allow clients to access the UAV through it and the cloud-based services allow clients to perform UAV communication through a wireless medium.

FIGURE 1.1   Block diagram of a cloud-based architecture for IoD.

1.2.1.2   CLOUD SERVICES LAYER

The cloud services layer has three main components: storage, computation, and interface. The storage components have all the information related to drones like localization parameters, mission information, transmitted data, received data, sensors data, images, environmental variables, etc. It will help to process data and will identify the threats, unauthorized areas, unauthorized users, and other activities. The cloud-based services are based on real-time processing of the given data. All types of real-time processing or computational activities are performed by the computation component of the cloud. The interface component provides the interfaces between the hardware and software of the drone and allows authorized users to access the data from the cloud.

1.2.1.3 CLIENT LAYER

The client layer is always present on the client-side in the cloud-based architecture and also acts as an application layer. It provides the interfaces between the UAV layer and cloud services layer both. The multiple authorized users can access multiple drones with the help of this layer. It can analyze, monitor, modify, control, or change the action of UAVs. Remotely accessing is also possible with the help of this application layer.

1.2.2 COMMUNICATION NETWORK ARCHITECTURE

The communication network architecture is a collaboration between the UAV network and the wireless sensor network (WSN) and is mainly used for surveillance operations. So, Figure 1.2 presents the communication network architecture of the IoD systems. This is an example based on multi-UAVs communication systems [67]. As per the requirement, the network architecture has three sub-networks. The first sub-network provides communication between UAVs and is known as the Ad-Hoc network. The second sub-network is WSN which collects the physical parameters of the environment with the help of sensors. The last sub-network makes the communication between UAV and the base station (BS), and WSN and BS both. This sub-network is known as a mobile communication network.

This network also communicates with ground control station (GCS) and UAVs. The GCS controls all the operations of the UAVs. This also acts as a supervisory control of the UAVs which can give the command to UAVs, monitor the UAVs, and also can change the mission of UAVs. The GCS can be used for different purposes also like autopilot mode using their application control station software. The surveillance system has the following links for the communication (a) between UAVs (b) between BS and UAV (c) between WSN and BS (d) between WSNs (e) between BSs and (f) between GCS and BS. All the information must be traveled from UAV to GCS regularly to communicate with UAVs. Therefore, it has bidirectional links for real-time processing. The GSC collected all the data of drones like the location of the drone and video information, etc. In another way, GCS sends all the controlling command to control the UAVs. Therefore, it is a bi-directional communication process. The main unit or central processing unit is the GCS controller which performs the entire task. Therefore, it is also known as a task manager and integrated with the UAVs.

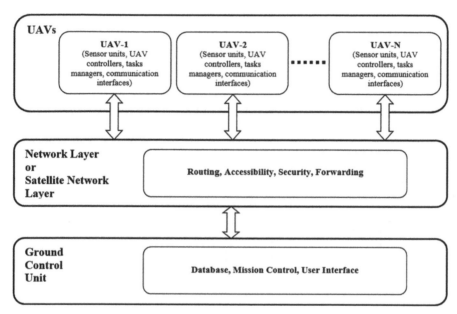

FIGURE 1.2 UAVs processing architecture.

1.2.3 UAV ARCHITECTURE

The autonomous UAV must have decision-making ability in real-time that process on the alternate flight plan. These alternations are dependent on the following events: data acquisition or processing, health management, and regulations. UAV architecture plays a vital role in performing the above events or operations. The architecture has two types of signal or data processing: (a) onboard processing (b) off-board processing. The onboard processing is related to all data collection or capturing, processing, and decision-making process on the drone. The off-board processing is related to the processing at the server end. The server collects all the data from UAVs and sends it to

UAVs again after processing. The health management of the UAV focuses on the techniques of sensing and monitoring. The sensing data can be collected from the onboard sensors of the UAVs. The legal regulations, safety, and reliability are the main concern in drones.

The UAV architecture is used for remote sensing and self-decision making. UAV architecture is divided into two major parts: (i) basic architecture; and (ii) extended architecture is shown in Figure 1.3 [69]. The proposed architecture allows a huge amount of data for processing, collecting, analyzing, and correlating in real-time. The architecture contains software and hardware components both. The hardware components are the I/O hub and onboard computer. The software components are enterprise service bus (ESB), complex event processing (CPE) engine, action trigger, and event pattern editor.

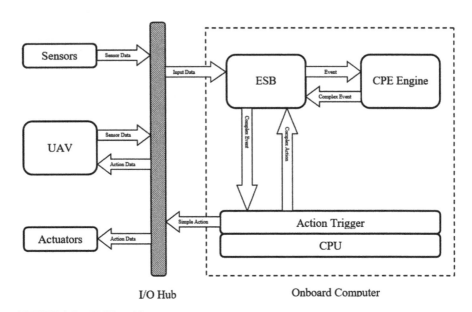

FIGURE 1.3   UAV architecture.

The hardware components are described here. The I/O hub receives all the information from sensors and UAVs like velocity, altitude, humidity, temperature, noise data, air pollution, etc. The I/O hub also transmits these data to the ESB and ESB transforms these data into the event. These events are transmitted to the CEP engine and CEP converts them into the complex event. The complex event is transmitted to the action trigger unit through

ESB. The action trigger unit converts into the simple action and is transferred to the I/O hub. The I/O hub sends this action data to the UAV and the actuators to operate the UAV.

The software components are explained in this segment. The ESB is accountable for all the information coming from different devices. The ESB performs the following actions like data reception through I/O hub, event transformation as per the requirement of the CEP engine, event enrichment, i.e., extra information addition, and domain dynamic generation which is data analyzing process before sending the data to the CEP engine. The CEP engine software matches the event pattern continuously and generates a complex event. The software language for the CEP engine is known as event processing language. The pattern of the events is classified into three categories: data acquisition, health management, and regulations. The categories are also assigned to their priority level like low, medium, and high. The pattern events are solved based on their priorities, and a complex event is generated here. The complex event was transferred to the action trigger through EBS. The action trigger has codes for dealing with the different types of complex events. The complex event can have single action or multiple actions. These actions are triggered by the I/O hub to the specific actuator for their action.

1.2.4 COMPUTING ARCHITECTURE

The IoD's must-have IoT devices. If the storage and services of IoDs are provided and managed by the cloud then it has high latency. The fog is useful to reduce the latency level of the cloud-based IoD systems. Fog is the extended part of the cloud network. It has several benefits: (i) it provides a fast response for multimedia and emergency notifications related applications; (ii) it combines data from various sensors or IoT devices; (iii) it provides data security and protection for sensitive or private information; (iv) it also provides more information about the user including their location and services [70].

Fog computing is also known as edge computing because it spreads the standards of cloud computing to the edge. The fundamentals of fog computing use the networking resources near to the nodes of the IoD systems that generate data. Fog is the high visualization platform to provide computing, storage, and networking services between IoT systems and clouds. Figure 1.4 shows the architecture based on fog computing which using IoT environments with various networks like WSNs, virtual sensor networks, personal area networks, etc. The cloud and fog both provide computing, application, infrastructure, storage, and data resources but have two major differences, accessibility, and proximity. The fog infrastructure is near in proximity to nodes and the cloud is accessible over the internet with the server located anywhere. Fog is a mainly extended version of the traditional cloud to the edge of accessing the network, IoT, WSN, and individual devices.

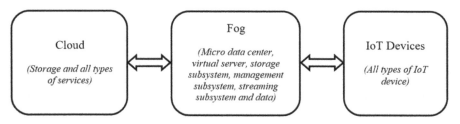

FIGURE 1.4   Block diagram of a computing-based architecture for IoD.

1.3   IoD PROTOCOLS

The IoD network is a combination of WSNs and IoT [73, 74]. The WSN is a self-configured network to monitor physical and atmospheric conditions like pressure, sound, vibration, motion, temperature, etc. The WSN has several protocols for the security of the IoD based networks. The protocol stack is shown in Figure 1.5. The layers of the WSN stack are application, transport, network, data link, and physical [75, 76]. These layers are responsible to perform the power management, mobility management, and task management of the network. These layers have some security requirements for the IoD environment like (a) authentication of sensing devices, users, and gateways, (b) entry under consideration, (c) privacy of communication channel, (d) availability of the network for the authorized users, (e) authorized drone to transmit information to the network, (f) freeness of the information [77].

Apart from the security requirements, the IoD environment has several security challenges [78]. The main challenge is the remote hijacking of the drone. It is possible with the help of some malware programs and the drone can perform unauthorized tasks. The second challenge is privacy. All the information or entities related to the drones are very important and it can be leaked under some condition or situation. The third challenge is mutual authentication. Mutual authentication is compulsory at the user end and drone ends both. The fourth challenge is a replay and man-in-the-middle attack. This challenge is related to the attackers in the communication system which can modify, delete, or inject the messages. The fifth challenge is the impersonation attack. In this challenge, attackers can act as a legal user for the drones. The sixth challenge is a privileged-insider attack. The trusted user of the control centers can act as a privileged-insider attacker. This type of attacker can get the secret information of the authorized users and can misuse those credentials. The seventh challenge is resilience against drone capture attacks. These challenges described that the protection from the physical attack is not possible in the IoD systems. Therefore, it has a change of physical attack. So, attackers can get information that is stored in the drone. Apart from the several given challenge in the IoD systems, it has several security protocols to secure the IoD systems as shown in Figure 1.6 [75] and also described in this section.

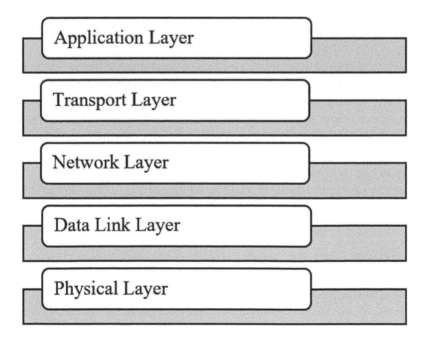

FIGURE 1.5   Wireless sensor network protocol stacks.

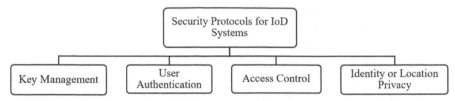

FIGURE 1.6   Security protocols in the IoD systems.

1.3.1   KEY MANAGEMENT

The IoD systems have important security services as key management [79, 80] which is required before the development of the drone. The secret credentials are required for the authentic users know as key rings of the IoD system. These keys are responsible to share the information of neighboring drones. It has two types of key management: (a) probabilistic (b) deterministic. If the key sharing is based on the probability known as probabilistic and if based on their credential then known as deterministic.

1.3.2   USER AUTHORIZATION

The maximum applications related to the IoD systems are based on the real-time systems. The external users are generally accessing the real-time data of the drone from a particular region. This type of action is possible if the user has direct access to the IoD systems and the systems are not connecting with a centralized server. The centralized server has all the information but does not have all the data as real-time data. So, to get real-time data, direct accesses of the authorized users are required for the IoD systems and it will also give the restriction to the unauthorized users. It is helpful to protect the drones from unauthorized user entry and malfunctioning activities.

1.3.3   ACCESS CONTROL

The access control is the development of new nodes in the drones for their deployment. Access control prevents malicious nodes in the IoD systems. The concepts of the access control phenomenon compare the two tasks: drone authentication and key establishment. The access control design is based on the certificate issued by the trusted agencies or authorities or non-certificate based. The non-certificate-based control is mainly used by the personal authorization.

1.3.4   INTRUSION DETECTION AND PREVENTION

Intrusion is the unauthorized access of the give IoD systems. It is divided into two categories: active and passive. The active intrusion is the transmission and interception of the malicious packets and the passive intrusion is information collected from the monitoring networks. The intrusion detection system is the hardware of application software that monitors the network from unauthorized access. The extension of the intrusion prevention systems performs all the operations based on intrusion detection systems. Its design is energy-efficient and best for intrusion detection and prevention from malicious activities.

1.3.5   IDENTITY AND LOCATION PRIVACY

All the geographical places like deployed drones place, working place of industries, traveling paths are very close to each other concerning UAVs. The information of the individual place is the loss of location privacy. The separation of the places is very difficult for the drones but due to the encryption algorithms [66], the IoD systems enhance the location privacy issue and they can easily differentiate the different places of the drone.

1.4   RECENT APPLICATIONS OF IoD

The IoT is a strong innovation and implementation in the field of drones. This technique is known as IoD. The IoD systems have many applications in the real-time implementations of drones [81–85]. The applications related to agriculture, location finding and delivery, healthcare, military, environmental protection, search, and rescue operations are described below:

1.4.1   AGRICULTURE

Today, agriculture is a revolutionary field and few farmers are using electronic systems and drone-based systems for the profusion and monitoring of the crop [86–89]. A smart monitoring system can be implemented with the help of IoD systems, which provides the status of the crops and soil. The formers can use the water and fertilizers as per the status of the crops and soils. The IoD can give information related to threads like animals, and fires. The UAVs were very difficult to utilize in the agriculture field but due to IoD systems, the implementation or utilization of UAVs are very easy [90].

1.4.2 DELIVERY

The industries are using IoD for product delivery [91–95] in terms of the best control and fast response. The drones can transport food items to luxury items in a limited range or controlled range. The controlled range is defined as the drone's coverage area or the range in which the controlling of drones is possible. The e-commerce company AMAZON has already used the IoD based drones in the United States for the package delivery of the product.

1.4.3 HEALTHCARE

The healthcare industries are also using the IoD systems [96–99] for the delivery of medicines in remote areas in very little time. In case of emergency, they can also supply the medical facility at the destination point quickly. The IoD systems can use the flying ambulance for the quick reaching of the patient to the hospital. If the travel time is very high for the remote locating areas, in that case, this drone system is efficient to provide the medical facility on time or quickly.

1.4.4 MILITARY

The military has a wide range of applications for the development of IoD systems. Most of the countries are using drones for different types of applications like border surveillance, early threat detection, soldier, and war items, or weapons safety [100–102]. Smart drones can detect if any thread is crossing the borders. It can reach any destination point within a short time as compared to human time without any humanitarian loss. Drones can also give information from a pinpoint location. The IoD systems can be controlled worldwide if the system is satellite-based. The IoD based monitoring system can give information from the sensitive areas of the borders of war zones.

The smart drone systems also work as intelligence systems and alert the soldiers on the battlefields from the enemies.

1.4.5 ENVIRONMENTAL PROTECTION

Drones are used for managing national parks and agricultural lands, tracking wildlife areas, observing the effects of climate change, and monitoring different ecosystems from rainforests to the oceans. The drones can be used for the recognition and investigation of natural disasters including forest fires, avalanches on mountains, etc. Therefore, drones are also being used for performing environmental actions [34, 39, 40].

1.4.6 SEARCH AND RESCUE

The search and rescue missions are one of the important applications of IoDs. In search and rescue missions, operations are efficient as possible and it is to obtain a rapid overview of the situation. While manned airplanes and helicopters need time to be ready for doing the mission, drones can be put into action immediately without any loss of time. Due to the important role of drones in search and rescue, missions attracted the attention of many researchers; and many drones were designed and fabricated for performing search and rescue operations [37, 38].

1.5 CONCLUSION AND FUTURE SCOPE OF IoD

The IoD is the growing field at this time. It is observed that the application of drones is increasing day by day. This chapter is based on the different types of architectural analysis and network protocols of the IoDs. The cloud-based architecture, network-based architecture, and UAV structure-based architecture are described here. These architectures have many advantages like centralized data storage, controlling, monitoring, large data storage, real-time operations, etc. The IoD systems have network protocols as key management, user authorization, access control, intrusion detection or prevention, and location, or identity privacy to protect the drones. The IoD systems have many applications in the fields of agriculture, military, product delivery, healthcare, and many others.

In recent years, the drone can be transformed in every field. Presently, it is used in several applications but the limited sectors or geographical areas. In the near future, it will be worldwide. The smart industries can use for monitoring purposes. Traffic monitoring and controlling can be performed with the help of UAVs. The weather can be monitored with the help of drones also. These and more other applications of UAVs can be fully implemented in the future.

KEYWORDS

- architecture
- artificial intelligence
- drone
- internet of drones
- internet of things
- protocols
- unmanned aerial vehicle

REFERENCES

1. Alharthi, M., Taha, A. E. M., & Hassanein, H. S., (2019). An architecture for software defined drone networks. In: *IEEE International Conference on Communications* (Vol. 2019).
2. Sekander, S., Tabassum, H., & Hossain, E., (2018). Multi-tier drone architecture for 5G/B5G cellular networks: Challenges, trends, and prospects. *IEEE Commun. Mag., 56*(3), 104–111.
3. Jat, D. S., & Singh, C., (2020). Artificial intelligence-enabled robotic drones for COVID-19 outbreak. In: *Springer Briefs in Applied Sciences and Technology* (pp. 37–46). Springer.
4. Anwar, M. Z., Kaleem, Z., & Jamalipour, A., (2019). Machine learning inspired sound-based amateur drone detection for public safety applications. *IEEE Trans. Veh. Technol., 68*(3), 2526–2534.
5. Shan, L., Miura, R., Kagawa, T., Ono, F., Li, H. B., & Kojima, F., (2019). Machine learning-based field data analysis and modeling for drone communications. *IEEE Access, 7*, 79127–79135.
6. Wu, M., Xie, W., Shi, X., Shao, P., & Shi, Z., (2018). Real-time drone detection using deep learning approach. In: *Lecture Notes of the Institute for Computer Sciences, Social-Informatics and Telecommunications Engineering, LNICST* (Vol. 251, pp. 22–32).

7. Chen, H., Wang, Z., & Zhang, L., (2020). Collaborative spectrum sensing for illegal drone detection: A deep learning-based image classification perspective. *China Commun., 17*(2), 81–92.

8. Alsamhi, S. H., Ma, O., Ansari, M. S., & Almalki, F. A., (2019). Survey on collaborative smart drones and internet of things for improving smartness of smart cities. *IEEE Access, 7*, 128125–128152. Institute of Electrical and Electronics Engineers Inc.

9. Batth, R. S., Nayyar, A., & Nagpal, A., (2019). Internet of robotic things: Driving intelligent robotics of future - concept, architecture, applications and technologies. In: *Proceedings - 4th International Conference on Computing Sciences, ICCS 2018* (pp. 151–160).

10. Singh, M., Aujla, G. S., & Bali, R. S., (2020). A deep learning-based blockchain mechanism for secure internet of drones environment. *IEEE Trans. Intell. Transp. Syst.,* 1–10.

11. Bera, B., Das, A. K., & Sutrala, A. K., (2021). Private blockchain-based access control mechanism for unauthorized UAV detection and mitigation in internet of drones environment. *Comput. Commun., 166*, 91–109.

12. Yazdinejad, A., Parizi, R. M., Dehghantanha, A., Karimipour, H., Srivastava, G., & Aledhari, M., (2020). Enabling drones in the internet of things with decentralized blockchain-based security. *IEEE Internet Things J.*, 6406–6415.

13. Aggarwal, S., Shojafar, M., Kumar, N., & Conti, M., (2019). A new secure data dissemination model in internet of drones. In: *IEEE International Conference on Communications* (Vol. 2019).

14. Yao, J., & Ansari, N., (2019). QoS-aware power control in internet of drones for data collection service. *IEEE Trans. Veh. Technol., 68*(7), 6649–6656.

15. Yao, J., & Ansari, N., (2020). Power control in internet of drones by deep reinforcement learning. In: *IEEE International Conference on Communications* (Vol. 2020).

16. Jiang, B., Yang, J., & Song, H., (2020). Protecting privacy from aerial photography: State of the art, opportunities, and challenges. In: *IEEE INFOCOM 2020 - IEEE Conference on Computer Communications Workshops, INFOCOM WKSHPS 2020* (pp. 799–804).

17. Cheng, E., (2015). *Aerial Photography and Videography Using Drones*. Peachpit Press.

18. Germen, M., (2016). *Alternative Cityscape Visualisation: Drone shooting as a New Dimension in Urban Photography*. Conference name: Electronic Visualisation and the Arts Conference location: London, UK. 150–157. DOI: 10.14236/ewic/EVA2016.31.

19. Tian, Y., Yuan, J., & Song, H., (2019). Efficient privacy-preserving authentication framework for edge-assisted internet of drones. *J. Inf. Secur. Appl.*, 48, 102354.

20. Kornatowski, P. M., Bhaskaran, A., Heitz, G. M., Mintchev, S., & Floreano, D., (2018). Last-centimeter personal drone delivery: Field deployment and user interaction. *IEEE Robot. Autom. Lett., 3*(4), 3813–3820.

21. Zhu, X., Pasch, T. J., & Bergstrom, A., (2020). Understanding the structure of risk belief systems concerning drone delivery: A network analysis. *Technol. Soc., 62*, 101262.

22. Koiwanit, J., (2018). Analysis of environmental impacts of drone delivery on an online shopping system. *Adv. Clim. Chang. Res., 9*(3), 201–207.

23. Paneque-Gálvez, J., Vargas-Ramírez, N., Napoletano, B., & Cummings, A., (2017). Grassroots innovation using drones for indigenous mapping and monitoring. *Land, 6*(4), 86.

24. Shafi, U., et al., (2020). A multi-modal approach for crop health mapping using low altitude remote sensing, internet of things (IoT) and machine learning. *IEEE Access, 8,* 112708–112724.

25. Coeckelbergh, M., (2013). Drones, information technology, and distance: Mapping the moral epistemology of remote fighting. *Ethics Inf. Technol., 15*(2), 87–98.

26. Vacca, A., Onishi, H., & Cuccu, F., (2017). Drones: Military weapons, surveillance or mapping tools for environmental monitoring? Advantages and challenges. A legal framework is required. In: *Transportation Research Procedia* (Vol. 25, pp. 51–62).

27. Erdelj, M., Natalizio, E., Chowdhury, K. R., & Akyildiz, I. F., (2017). Help from the sky: Leveraging UAVs for disaster management. *IEEE Pervasive Computing, 16*(124–132). Institute of Electrical and Electronics Engineers Inc.

28. Apvrille, L., Tanzi, T., & Dugelay, J. L., (2014). Autonomous drones for assisting rescue services within the context of natural disasters. In: *2014 31*th *URSI General Assembly and Scientific Symposium, URSI GASS 2014.*

29. Subhedar, S., Gupta, N. K., & Jain, A., (2019). Identification of living human objects from collapsed architecture debris to improve the disaster rescue operations using IoT and augmented reality. In: *Communications in Computer and Information Science* (Vol. 1033, pp. 521–527).

30. Saha, A. K., et al., (2018). IOT-based drone for improvement of crop quality in agricultural field. In: *2018 IEEE 8*th *Annual Computing and Communication Workshop and Conference, CCWC 2018* (Vol. 2018, pp. 612–615).

31. Tripicchio, P., Satler, M., Dabisias, G., Ruffaldi, E., & Avizzano, C. A., (2015). Towards smart farming and sustainable agriculture with drones. In: *Proceedings - 2015 International Conference on Intelligent Environments, IE 2015* (pp. 140–143).

32. Kulbacki, M., et al., (2018). Survey of drones for agriculture automation from planting to harvest. In: *INES 2018 - IEEE 22*nd *International Conference on Intelligent Engineering Systems, Proceedings* (pp. 000353–000358).

33. Besada, J. A., Bernardos, A. M., Bergesio, L., Vaquero, D., Campana, I., & Casar, J. R., (2019). Drones-as-a-service: A management architecture to provide mission planning, resource brokerage and operation support for fleets of drones. In: *2019 IEEE International Conference on Pervasive Computing and Communications Workshops, PerCom Workshops 2019* (pp. 931–936).

34. Silvagni, M., Tonoli, A., Zenerino, E., & Chiaberge, M., (2017). Multipurpose UAV for search and rescue operations in mountain avalanche events. *Geomatics, Nat. Hazards Risk, 8*(1), 18–33.

35. Mishra, B., Garg, D., Narang, P., & Mishra, V., (2020). Drone-surveillance for search and rescue in natural disaster. *Comput. Commun., 156,* 1–10.

36. Chuang, C. C., Rau, J. Y., Lai, M. K., & Shih, C. L., (2019). Combining unmanned aerial vehicles, and internet protocol cameras to reconstruct 3-D disaster scenes during rescue operations. *Prehospital Emerg. Care, 23*(4), 479–484.

37. Kashihara, S., Wicaksono, M. A., Fall, D., & Niswar, M., (2019). Supportive information to find victims from aerial video in search and rescue operation. In: *Proceedings - 2019 IEEE International Conference on Internet of Things and Intelligence System, IoTaIS 2019* (pp. 56–61).

38. Mejia, A., Marcillo, D., Guano, M., & Gualotuna, T., (2020). Serverless based control and monitoring for search and rescue robots. In: *Iberian Conference on Information Systems and Technologies, CISTI* (Vol. 2020).

39. Thibbotuwawa, A., Bocewicz, G., Nielsen, P., & Zbigniew, B., (2019). Planning deliveries with UAV routing under weather forecast and energy consumption constraints. *IFAC-PapersOnLine, 52*(13), 820–825.
40. Wang, E. K., Chen, C. M., Wang, F., Khan, M. K., & Kumari, S., (2020). Joint-learning segmentation in internet of drones (IoD)-based monitor systems. *Comput. Commun., 152*, 54–62.
41. Potter, B., Valentino, G., Yates, L., Benzing, T., & Salman, A., (2019). Environmental monitoring using a drone-enabled wireless sensor network. In: *2019 Systems and Information Engineering Design Symposium, SIEDS 2019.*
42. Gallacher, D., (2016). Drones to manage the urban environment: Risks, rewards, alternatives. *J. Unmanned Veh. Syst., 4*(2), 115–124.
43. Gallacher, D., (2017). Drone applications for environmental management in urban spaces: A review. *Int. J. Sustain. L. Use Urban Plan., 3*(4).
44. Barnas, A. F., Chabot, D., Hodgson, A. J., Johnston, D. W., Bird, D. M., & Ellis-Felege, S. N., (2020). A standardized protocol for reporting methods when using drones for wildlife research. *J. Unmanned Veh. Syst., 8*(2), 89–98.
45. Rebolo-Ifrán, N., Grilli, M. G., & Lambertucci, S. A., (2019). Drones as a threat to wildlife: Youtube complements science in providing evidence about their effect. *Environ. Conserv., 46*(3), 205–210.
46. Pike, G. H., (2017). Legal issues: The internet of drones. *SSRN Electron. J.*
47. Clarke, B., (2013). Arming drones for law enforcement: Challenges and opportunities for the protection of human life. *SSRN Electron. J.*
48. La Bella, L., (2016). *Drones and Entertainment.* The Rosen Publishing Group, Inc.
49. Kim, S. J., Jeong, Y., Park, S., Ryu, K., & Oh, G., (2018). *A Survey of Drone use for Entertainment and AVR (Augmented and Virtual Reality)* (pp. 339–352). Springer, Cham.
50. De Swarte, T., Boufous, O., & Escalle, P., (2019). Artificial intelligence, ethics and human values: The cases of military drones and companion robots. *Artif. Life Robot., 24*(3), 291–296.
51. Long, T., Ozger, M., Cetinkaya, O., & Akan, O. B., (2018). Energy neutral internet of drones. *IEEE Commun. Mag., 56*(1), 22–28.
52. Nikooghadam, M., Amintoosi, H., Islam, S. H., & Moghadam, M. F., (2020). A provably secure and lightweight authentication scheme for internet of drones for smart city surveillance. *J. Syst. Archit.,* 101955.
53. Dilshad, N., Hwang, J., Song, J., & Sung, N., (2020). Applications and challenges in video surveillance via drone: A brief survey. In: *2020 International Conference on Information and Communication Technology Convergence (ICTC)* (pp. 728–732).
54. Hossein, M. N., Taleb, T., & Arouk, O., (2016). Low-altitude unmanned aerial vehicles-based internet of things services: Comprehensive survey and future perspectives. *IEEE Internet of Things Journal* (Vol. 3, No. 6, pp. 899–922). Institute of Electrical and Electronics Engineers Inc.
55. Rumba, R., & Nikitenko, A., (2020). The wild west of drones: A review on autonomous-UAV traffic-management. In: *2020 International Conference on Unmanned Aircraft Systems, ICUAS 2020* (pp. 1317–1322).
56. Koubaa, A., & Qureshi, B., (2018). Dronetrack: Cloud-based real-time object tracking using unmanned aerial vehicles over the internet. *IEEE Access, 6*, 13810–13824.

57. Zhang, P., Wang, C., Qin, Z., & Cao, H., (2020). A multidomain virtual network embedding algorithm based on multiobjective optimization for internet of drones architecture in industry 4.0. *Softw. Pract. Exp.*, spe.2815.
58. Kharchenko, V., & Torianyk, V., (2018). Cybersecurity of the internet of drones: Vulnerabilities analysis and IMECA based assessment. In: *Proceedings of 2018 IEEE 9th International Conference on Dependable Systems, Services and Technologies, DESSERT 2018* (pp. 364–369).
59. Putranto, D. S. C., Aji, A. K., & Wahyudono, B., (2019). Design and implementation of secure transmission on internet of drones. In: *Proceedings of the 2019 IEEE 6th Asian Conference on Defence Technology, ACDT 2019* (pp. 128–135).
60. Gharibi, M., Boutaba, R., & Waslander, S. L., (2016). Internet of drones. *IEEE Access, 4*, 1148–1162.
61. Shi, W., Zhou, H., Li, J., Xu, W., Zhang, N., & Shen, X., (2018). Drone assisted vehicular networks: Architecture, challenges and opportunities. *IEEE Netw., 32*(3), 130–137.
62. Lin, C., He, D., Kumar, N., Choo, K. K. R., Vinel, A., & Huang, X., (2018). Security and Privacy for the Internet of Drones: Challenges and Solutions. *IEEE Commun. Mag., 56*(1), 64–69.
63. Koubaa, A., Qureshi, B., Sriti, M. F., Javed, Y., & Tovar, E., (2017). A service-oriented cloud-based management system for the internet-of-drones. In: *2017 IEEE International Conference on Autonomous Robot Systems and Competitions, ICARSC 2017* (pp. 329–335).
64. Koubâa, A., et al., (2019). Dronemap planner: A service-oriented cloud-based management system for the internet-of-drones. *Ad Hoc Networks, 86*, 46–62.
65. Qureshi, B., Anis, K., Mohamed-Foued, S., Yasir, J., Maram, A., et al., (2016). *Dronemap-A Cloud-based Architecture for the Internet-of-Drones Dronemap-A Cloud-based Architecture for the Dronemap-A Cloud-based Architecture for the Internet-of-Drones Poster: Dronemap-A Cloud-based Architecture for the Internet-of-Drones Mohamed-Foued Sriti*.
66. Chen, Y. J., & Wang, L. C., (2019). Privacy protection for internet of drones: A network coding approach. *IEEE Internet Things J., 6*(2), 1719–1730.
67. Krichen, L., Fourati, M., & Fourati, L. C., (2018). Communication architecture for unmanned aerial vehicle system. In: *Lecture Notes in Computer Science (Including Subseries Lecture Notes in Artificial Intelligence and Lecture Notes in Bioinformatics)* (Vol. 11104 LNCS, pp. 213–225).
68. Yao, J., & Ansari, N., (2020). Online Task allocation and flying control in fog-aided internet of drones. *IEEE Trans. Veh. Technol., 69*(5), 5562–5569.
69. Boubeta-Puig, J., Moguel, E., Sanchez-Figueroa, F., Hernandez, J., & Carlos, P. J., (2018). An autonomous UAV architecture for remote sensing and intelligent decision-making. *IEEE Internet Comput., 22*(3), 6–15.
70. Aazam, M., Zeadally, S., & Harras, K. A., (2018). Fog Computing Architecture, Evaluation, and Future Research Directions. *IEEE Commun. Mag., 56*(5), 46–52.
71. Cheng, N., et al., (2018). Air-ground integrated mobile edge networks: Architecture, challenges, and opportunities. *IEEE Commun. Mag., 56*(8), 26–32.
72. Molina, J., Muelas, D., De Vergara, J. E. L., & Garcia-Aranda, J. J., (2020). Network quality-aware architecture for adaptive video streaming from drones. *IEEE Internet Comput., 24*(1), 5–13.
73. Lv, Z., (2019). The security of internet of drones. *Comput. Commun., 148*, 208–214.

74. Sharma, B., Srivastava, G., & Lin, J. C. W., (2020). A bidirectional congestion control transport protocol for the internet of drones. *Comput. Commun., 153*, 102–116.
75. Wazid, M., Das, A. K., & Lee, J. H., (2018). Authentication protocols for the internet of drones: Taxonomy, analysis and future directions. *J. Ambient Intell. Humaniz. Comput.,* 1–10, Aug.
76. Boccadoro, P., Striccoli, D., & Grieco, L. A., (2020). Internet of Drones: A Survey on Communications, Technologies, Protocols, Architectures and Services. *Ad Hoc Network, 122*, doi:10.1016/j.adhoc.2021.102600, pp. 1–38, Publication: Elsevier.
77. Pathan, A. S. K., Lee, H. W., & Hong, C. S., (2006). Security in Wireless Sensor Networks: Issues and challenges. In: *8th International Conference Advanced Communication Technology, ICACT 2006 – Proceedings* (Vol. 2, pp. 1043–1048).
78. Ilgi, G. S., & Kirsal, E. Y., (2020). Critical analysis of security and privacy challenges for the Internet of drones: A survey. In: *Drones in Smart-Cities* (pp. 207–214). Elsevier.
79. Zhang, Y., He, D., Li, L., & Chen, B., (2020). A lightweight authentication and key agreement scheme for internet of drones. *Comput. Commun., 154*, 455–464.
80. Gope, P., & Sikdar, B., (2020). An efficient privacy-preserving authenticated key agreement scheme for edge-assisted internet of drones. *IEEE Trans. Veh. Technol., 69*(11), 13621–13630.
81. Nayyar, A., Le Nguyen, B., & Nguyen, N. G., (2020). The internet of drone things (Iodt): Future envision of smart drones. In: *Advances in Intelligent Systems and Computing, 1045*, 563–580.
82. Kaleem, Z., et al., (2018). *Amateur Drone Surveillance: Applications, Architectures, Enabling Technologies, and Public Safety Issues: Part 1* (Vol. 56, No. 1, pp. 14, 15). *IEEE Communications Magazine*. Institute of Electrical and Electronics Engineers Inc.
83. Kaleem, Z., et al., (2018). *Amateur Drone Surveillance: Applications, Architectures, Enabling Technologies, and Public Safety Issues: Part 2* (Vol. 56, No. 4, pp. 66, 67). IEEE Communications Magazine. Institute of Electrical and Electronics Engineers Inc.
84. Hassanalian, M., & Abdelkefi, A., (2017). Classifications, applications, and design challenges of drones: A review. *Progress in Aerospace Sciences, 91*, 99–131. Elsevier Ltd.
85. Kugler, L., (2019). Real-world applications for drones. *Commun. ACM, 62*(11), 19–21.
86. Petkovic, S., Petkovic, D., & Petkovic, A., (2017). IoT devices VS. drones for data collection in agriculture. *DAAAM Int. Sci. B., 16*, 63–80.
87. Krishna, K. R., (2018). *Agricultural Drones: A Peaceful Pursuit*. Taylor & Francis.
88. Boursianis, A. D., et al., (2020). Internet of things (IoT) and agricultural unmanned aerial vehicles (UAVs) in smart farming: A comprehensive review. *Internet of Things*, 100187.
89. Ren, Q., Zhang, R., Cai, W., Sun, X., & Cao, L., (2020). Application and development of new drones in agriculture. In: *IOP Conference Series: Earth and Environmental Science* (Vol. 440, No. 5, p. 052041).
90. Vuran, M. C., Salam, A., Wong, R., & Irmak, S., (2018). Internet of underground things in precision agriculture: Architecture and technology aspects. *Ad Hoc Networks* (Vol. 81, pp. 160–173).
91. Vashist, S., & Jain, S., (2019). Location-aware network of drones for consumer applications: Supporting efficient management between multiple drones. *IEEE Consum. Electron. Mag., 8*(3), 68–73.

92. Yu, J., Subramanian, N., Ning, K., & Edwards, D., (2015). Product delivery service provider selection and customer satisfaction in the era of internet of things: A Chinese e-retailers' perspective. *Int. J. Prod. Econ., 159*, 104–116.

93. Kim, J. J., & Hwang, J., (2020). Merging the norm activation model and the theory of planned behavior in the context of drone food delivery services: Does the level of product knowledge really matter?. *J. Hosp. Tour. Manag., 42*, 1–11.

94. Yoo, W., Yu, E., & Jung, J., (2018). Drone delivery: Factors affecting the public's attitude and intention to adopt. *Telemat. Informatics, 35*(6), 1687–1700, Sep.

95. Farris, E., &. McGee, I. I. W. F., (2018). *System and Method for Controlling Drone Delivery or Pick up During a Delivery or Pick up Phase of Drone Operation.* Google Patents.

96. Wulfovich, S., Rivas, H., & Matabuena, P., (2018). *Drones in Healthcare* (pp. 159–168). Springer, Cham.

97. Poljak, M., & Šterbenc, A., (2020). Use of drones in clinical microbiology and infectious diseases: Current status, challenges and barriers. *Clinical Microbiology and Infection, 26*(4), 425–430. Elsevier B.V.

98. Kumar, A., Sharma, K., Singh, H., Naugriya, S. G., Gill, S. S., & Buyya, R., (2021). A drone-based networked system and methods for combating coronavirus disease (COVID-19) pandemic. *Futur. Gener. Comput. Syst., 115*, 1–19.

99. Hiebert, B., Nouvet, E., Jeyabalan, V., & Donelle, L., (2020). The application of drones in healthcare and health-related services in North America: A scoping review. *Drones, 4*(3), 30.

100. Choudhary, G., Sharma, V., & You, I., (2019). Sustainable and secure trajectories for the military internet of drones (IoD) through an efficient medium access control (MAC) protocol. *Comput. Electr. Eng., 74*, 59–73.

101. Hall, R. J., (2016). An internet of drones. *IEEE Internet Comput., 20*(3), 68–73.

102. Fitwi, A. H., Nagothu, D., Chen, Y., & Blasch, E., (2019). A distributed agent-based framework for a constellation of drones in a military operation. In: *Proceedings - Winter Simulation Conference* (Vol. 2019, pp. 2548–2559).

CHAPTER 2

Approaching Internet Renovation of Imperceptible Computers to Facilitate the Internet of Drones

R. JAYALAKSHMI[1] and CHUAN-MING LIU[2]

[1]*Department of Computer Science, St. Claret College, Bangalore, Karnataka, India, E-mail: jayabinoy2020@gmail.com*

[2]*Department of Computer Science and Information Engineering, National Taipei University of Technology (Taipei Tech), Taipei – 106, Taiwan*

ABSTRACT

The world is getting ready to embrace the new lifestyle entrenched in the technological innovations of imperceptible computers. The world of computing is sprouting day by day to function for mankind in the backgrounds to meet their expectations and requirements. If you can aspire for the shelves in the shopping mall placing the orders for particular products from the inventory department or your vehicle finding out a parking lot by itself or the plants in your garden managing watering by themselves, then it is time to greet the next Internet transformation which is primarily ingrained in UbiComp. UbiComp is the budding trend in Information Technology, about implanting microprocessors in various objects around us to converse for making our lives more comfortable. UbiComp is also called ubiquitous or pervasive computing. Pervasive computing is deep-rooted in the internet of things (IoT). The accountability of the IoT is remarkable to create information about various connected objects, to analyze and to make smart decisions based on this analysis. IoT upgrades the performance of pervasive computing via its unique features. The revival of current drone technology

The Internet of Drones: AI Applications for Smart Solutions. Arun Solanki, PhD, Sandhya Tarar, PhD, Simar Preet Singh, PhD & Akash Tayal, PhD (Eds.)

can be initiated by the internet of drone things (IoDT) by augmenting its scope via better collaboration and coordination among the objects in the smart cities. IoDT or internet of drones (IoD) or unmanned aerial vehicle (UAV) can be contemplated as 'internet of things with wings' or 'flying IoT.' When drones take wings on the sky with complete network connectivity resources, you will change your traditional perception about the world to a smart world composed of many smart cities. This indicates that there is a robust association between UbiComp, IoT, and IoD.

In the prospect of this chapter, a study has been conducted to recognize the existing knowledge, findings, and challenges in the background of ubiquitous computing, IoT, and IoDT. The chapter gives a brief idea about UbiComp, IoT, and IoDT. The chapter also explains the underlying major technological tools behind UbiComp, IoT and IoD. The correlation between pervasive computing, IoT, and IoD is depicted in the chapter with sound details. The chapter explores the major potentialities, disputes, and future of UbiComp along with its relevance related to the forthcoming Internet renovation of imperceptible computers to activate the existence of IoD. The chapter also focuses on the key opportunities and challenges in connection with the IoD. In addition to these aspects, the chapter gives an insight into the anticipation in envisioning smart cities with the IoD. The chapter concludes by giving an idea about the significance and future directions of ubicomp, IoT, and IoD in the revitalization of the Internet to make into reality the endurance of a digital world of conscientious civilization and imperceptible computers.

2.1   INTRODUCTION

Ubicomp or ubiquitous computing or pervasive computing is an emerging technology in computer science where computing is easily accessible without any time and space constraints [1]. 'Existing everywhere' is the implication of the words ubiquitous and pervasive. Ubiquitous computing devices are entirely connected and continuously accessible [2]. Contrary to computing traditionally, ubicomp can occur using any object, in any location, and any set-up. In 1988, the term ubiquitous computing was coined by Mark Weiser to portray a prospect in which imperceptible computers, implanted in real-time objects, substitute PCs [3].

Ubiquitous computing devices are very minute-even unseen-gadgets embedded in about various ranges of things conceivable, comprising vehicles, utensils, domestic devices, garments, and a variety of end-user commodities [4]. All ubiquitous computing devices will be communicating constantly through progressively more interconnected networks. Ubicomp deals with providing the computational facility in the objects around you. When these objects with the computational facility are linked to the Internet, there comes the effect of IoT [5]. Therefore, it is quite apparent that concepts related to ubicomp are widespread to IoT in various ways. Internet of drone things (IoDT) or IoD (IoD) is the budding form of IoT devices that envisions a world of invisible computers with flying abilities to ensure automation in every aspect of life [6]. Ubicomp, IoT, and IoD go hand in hand to fulfill the dream of technological advancements for a better mode of living.

This chapter describes the approaching Internet renovation of imperceptible computers for the betterment of mankind by combining the quintessence of ubicomp, IoT, and IoDT in a resourceful manner. Section 2.2 of the chapter gives the outcome of the literature review that has been conducted for the concoction of the chapter. The involvement of invisible smart devices around us in resolving the real-world problems will become very effective with the digital advancements of ubicomp, IoT, and IoD. The vital technological tools behind ubicomp, IoT, and IoD are generally explained in Section 2.3 of the chapter.

Section 2.4 of the chapter depicts the connection of ubiquitous computing with IoT. The existing real-time applications of ubicomp, IoT, and IoDT are highly appreciable. The potentialities of ubicomp are narrated in Section 2.5 with examples. Section 2.6 of the chapter describes the challenges associated with ubicomp. Future of ubicomp is discussed in Section 2.7 of the chapter.

The cognitive processing skills of an intelligent drone demand the potential to automatically identify and track the smart objects as per the requirement in a self-controlled manner. Section 2.8 of the chapter explains the role of ubicomp as a facilitator for the IoD. The prospects and challenges related to IoD are described in Sections 2.9 and 2.10 of the chapter, respectively. The IoD along with IoT and pervasive computing are penetrating to a new paradigm wherein there is a demand of both understanding aptitude and the capability to interact directly with their environments in real-time by the imperceptible smart objects. Enhancing the infrastructure using intelligent solutions to become smart and responsive by improving sustainability, reliability, economic progress, and quality of life can trigger the creation of intelligent cities. The vast perspective of smart cities in opening up the opportunities at various sectors of life for the personal, professional, and societal developments is becoming prevalent. IoD has gained its attention mainly while envisioning automated intelligent cities. Section 2.11 of the chapter expresses the visualization of intelligent cities with flying IoT.

Conclusion and future enhancements are described in Section 2.12 of the chapter. The chapter also contains the references related to various sources in the preparation of its content.

2.2   LITERATURE REVIEW

The pace in which technological advancements happen has become unpredictable in the present world. There have been quite a lot of studies, research enhancements and technological developments in the focus of refining the attributes of life with the help of invisible computers by being in the natural environment. The microprocessors and minute sensors in the ubicomp activated the new revolution of imperceptible computers in the arena of information and communication technology (ICT). Abilities of interconnected invisible devices in gathering, recording, and processing the data obtained from the surroundings to produce meaningful information to share among other devices without the intervention of human beings established the foundation of computing from anywhere at any time. The thermal, audio, visual, and radar sensors in these imperceptible smart devices help in the understanding of the environment and detection of other connected devices as per the requirement [7]. Radio frequency identification (RFID) technique used by the sensors in the smart devices of ubiquitous computing assist them to trace all types of smart objects involved in the communication, automatically [8]. The support that is offered by ubiquitous computing in everyday life can give rise to an infrastructure-based modernization.

IoT has become one of the buzzwords in the arena of Information Technology. The imperceptible ubiquitous devices in the real world can be integrated under a common virtual environment with persistent connectivity with the help of IoT to serve the needs of mankind [9]. The requirements of large organizations have driven the development of IoT to assist to a great extent from the prudence and obviousness provided by the capability to trace all smart objects via the service connections in which they are implanted. IoT facilitates organizations to become more resourceful by speeding up the operations, dropping faults and avoiding thefts. The prospect of computing and communications rely upon the dynamic developments in IoT. We reside in a digitally-driven world that pedals our natural surroundings. The demanding use of wireless interconnectivity supports these digital devices for use basically by people. People prefer mobile devices such as smartphones and tablets over personal computers or laptop computers. This means that the retrieval of information and services will become more feasible at all times and everywhere with the help of IoT.

The social impact of IoT is very impressive. Intelligent electronic gadgets play an important role in the IoT for communication, healthcare system, home automation, automated transport, smarter energy management systems, water supply, the security of the city, and the development of the smart city. Many experts and businesses foresaw an enormous growth of the internet of things (IoT) for future development. The IoT can build a future-oriented society. All devices would be connected via the Internet via unique IPv6 addresses and would therefore be able to be controlled by the owner of any location at any time. The government is working to increase output and capacity, reduce costs and improve the quality of life of its citizens with the help of intelligent surroundings [10]. The IoT will serve government organizations and large businesses to monitor, collect, analyze, and deliver any solution in a short period. Many research communities are working towards a smart corporate vision. The intelligent world would be possible through the involvement of computer engineers, electrical, and electronic engineers, and researchers. Many researchers are working on real-time data engineering, mobile computing, pervasive computing, the WSNs, signal processing and artificial intelligence (AI), and more importantly on cyber-headlines.

The imperceptible smart objects with ubiquitous potentialities in the environment of IoT not only includes the electronic gadgets but also the 'things,' which you do not consider as electronic such as furniture, food, clothing, monuments, plants, animals, human beings, and the list goes on to anything and everything [11]. The worldwide network which permits the communication between objects and objects, individuals, and objects and individuals and individuals by uniquely identifying each constituent in the environment becomes a reality with the help of IoT [12].

To guarantee the safety of human operators and allow new functions, automation, and remote technology have been very helpful. Aerial drones were used to disrupt the atmosphere over the metropolis, issue orders over the land of the opponent, and conduct fire drills for aircrew during the First and Second World Wars. Drone locomotives have been used by railways to support locomotive operators for a while. Railways have been using drone locomotives to support locomotive operators for some time. While UAVs have an extended history of military deployment, their growing use in unarmed tasks also must be recognized. Even though the existing use is narrow while the expertise is in the developmental stage, as they acquire high impending adaptability drones can renovate the way strategic services are delivered. In the attainment of new commercial, communal, ecological, and other aspirations, there is undoubtedly a unique contribution from drones [13].

The potential of drones was put to greater use during the COVID-19 crisis. People have begun to use the complementary nature of skills to alter the deliverance of existing services to get better safety and capacity levels, including the release of face masks and gloves to isolated places. It is vital to note that the uses of UAVs in larger business applications are also raising leading to considerable cost savings and capacity augmentation. The capabilities of IoDs to see large, economical areas from elevation provide an innovative outlook of observation and capacities for data acquirement [14]. Similarly, aerial camera work has penetrated a new phase of progress with operators, huge, and minute, able to give clients new descriptions. Moreover, the recent heave in retail sales of small-scale entertainment and business drones has move forwarded aerial drones into the amusement arena.

IoT with wings or IoD is introduced to the virtual environment for establishing a sustainable and resourceful ecosystem to augment the eminence of existence that can lead towards the visualization about smart cities. The collaborative effort of IoD along with IoT helps to improve the quality of service (QoS), in a smart virtual environment. IoD ensures the betterment of smart cities by improving the techniques of data collection, privacy, and safety. The presence of imperceptible ubiquitous smart devices with wireless connectivity and the miniaturized processors and sensors are the driving forces of IoD to enhance the prominence of lifestyle [15]. Ubiquitous computing, IoT, and IoD play a very crucial task in assisting a cluster of purposes such as agriculture, safety, disaster management, service delivery, weather monitoring, healthcare, and so on in the smart environment [16]. Literature survey also indicates that the idea of ubicomp, IoT and IoD have evolved as auspicious fields by offering opportunities in the domain of networking, sensors, AI, distributed computing, and mobile computing [17].

2.3   VITAL TECHNOLOGICAL TOOLS BEHIND UBICOMP, IoT, AND IoD

The underlying technologies to support ubiquitous computing are circuitry, electricity, sensors, connectivity, localizing, surveillance, machine-machine interface, and human-machine interface [18]. The accurate, as well as the uninterruptable coordination among all these technological tools, are crucial for the apt and error-free performance of ubicomp devices. Creating a semantic model of the real world wherein machines can sense and understand the meaning of spoken sentences and actions done by human beings will always make easy, the functioning of ubicomp devices [19].

IoT tries to overpass the space between the virtual and the real world by creating smart surroundings. The technologies which are parts of ubiquitous computing help in the implementation of IoT as well [20]. IoT has a robust correlation with constant connectivity among the smart objects in the environment [21]. Bluetooth is believed to be as the main solution for the future of invisible computers embedded in the environment to establish communication as per the requirement without many obstacles [22]. IoD can be considered as 'internet of things with wings.' IoD can renovate various sectors of life by breaking the barriers that exist in the current scenario of transactions. The potential of IoD lies in the technologies behind IoT. In addition to the technologies in the rear of ubicomp and IoT, there is a demand for unmanned aerial vehicle (UAV) operators and cloud connectivity for controlling IoD [23].

2.4 CONNECTION OF UBICOMP WITH IoT

There is a strong association between ubicomp and IoT [24]. This well-built association can instigate the forthcoming Internet revolution. In ubiquitous computing, a user mingles with the computer, which can be present in diverse designs, including tablets, smartphones, personal digital assistants (PDAs), laptops, and terminals in real-world objects such as a fridge or a washing machine or grinder or television [25]. IoT is an assemblage of attached devices and utilities that toil jointly to do something functional [26]. In 1985, the term IoT was introduced by Peter Lewis to illustrate how an omnipresent arrangement of sensors joined to the Internet can put across the facts about the actual world unswervingly to business routines without any involvement by human beings [27].

Ubiquitous computing meets on a broad range of research themes, by taking into account distributed computing, mobile computing, etc. On the other hand, when mostly regarding the objects involved, it is also known as real-world computing or haptic computing [28]. The IoT is an advertising catchphrase for this real-world computing for the betterment of existing facilities enjoyed by people around the world [29]. In this prototype, regular devices and objects are understood to be smart computing devices linked to the Internet and competent to have interaction with users or clients or customers and other related devices [30]. IoT is a mainstay of the future Internet. IoT will encompass zillions of internet-connected objects (ICOs) or 'things' that can feel, correspond, compute intelligently. The ideas about ambient, pervasive, and ubiquitous computing are assimilated in IoT. IoT will be a facilitator of ubiquitous sensing [31].

2.5   POTENTIALITIES OF UBICOMP

As a multi-disciplinary theme of research and advancement, the evolution of pervasive computing has occurred systematically. In a wide range of fields of knowledge such as health care, business, skills management, education, logistics, user-friendliness, and gaming, ubicomp has induced its applications. The prospects in connection with Pervasive Computing Technology spread over its various application areas:

2.5.1   u-HEALTH

Scrutinizing the health conditions of a patient in real-time has become possible by combining the developments of sensor technology with u-health services [32, 33]. Ubicomp infrastructure in the health area can be useful in the administration of hospital procedures, monitoring the conditions of patients and to support welfare services (Figure 2.1) [34, 35].

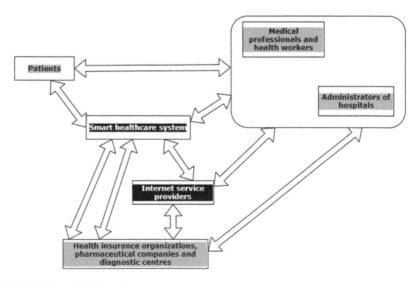

FIGURE 2.1   u-Health environment.

2.5.2 *u-ACCESSIBILITY*

The u-Accessibility is to support different types of people with disabilities. This ensures a scenario which helps people with disabilities to be part of the digital world themselves without any obstacles [36]. The ubiquitous computing environment can increase the accessibility to various resources in a sustainable manner. When we experience an environment where imperceptible computers are everywhere to offer services by automatically identifying the requirements, people with multiple disabilities will be able to experience better accessibility [37]. Ubiquitous accessibility enhances the familiarity of available resources to disabled persons without the assistance of anyone [38].

2.5.3 *u-LEARNING*

The u-Learning allows learning to happen autonomously of location and instance. The u-Learning can be considered as a blend of e-learning and m-learning. Virtual as well as real objects and people, as well as events, can be connected via u-Learning systems to support uninterrupted, appropriate, and relevant knowledge acquirement [39]. With u-Learning, anyone will be able to learn at any place and at any time. The benefits of digital content, natural environment and smart gadgets with pervasive components are used to make possible u-learning. It is an emerging paradigm to accomplish meaningful learning through smart experience. The realization of u-Learning can bring in a drift in the digital teaching-learning environment by widening the opportunities of knowledge acquisition and sharing (Figure 2.2).

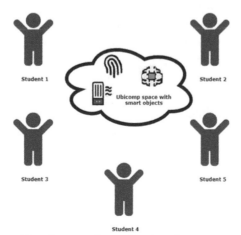

FIGURE 2.2　Enhanced learning experience via u-learning.

2.5.4   u-COMMERCE

The use of networks based on ubicomp is to support tailored and unrelenting interactions and dealings between a business organization and its various collaborators to offer a level of importance ahead of conventional commerce [40]. It is an amalgamation of wireless/mobile, visual, voice, electronic, and traditional commerce (Figure 2.3).

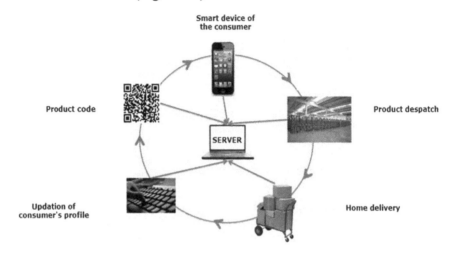

FIGURE 2.3   u-Commerce for better consumer experience.

2.5.5   U-GAMES

It is a multiple player mobile game by the incorporation of wireless technologies. The experience of u-Games gives another dimension to entertainment using technology [41]. Ubiquitous paradigm focuses on both the community and technical aspects. To enjoy interactive gaming experience by being anywhere and anytime, u-games play a major role [42].

Incorporating in the atmosphere and enclosing users with its ubiquity can be considered the essential characteristic of an invasive system. Smartphones may serve as an interface device that links the user to the ubicomp environment. Certification courses in Pervasive Computing, Mobile, and Pervasive Design, Real-Time Pervasive Computer Systems, etc., offered by well-reputed institutes explore a wide range of career opportunities too. New business opportunities specifically in the areas of supervising information overload, building transparent computing environments and enabling mobile and ambient devices to dole out as smart interfaces to our real-world surroundings can be created by pervasive computing environments.

2.6 DISPUTES ASSOCIATED WITH UBICOMP

Forthcoming computers will be a set of connections with a range of specialized devices that "know" what is happening around them, can "talk" to each other and can productively utilize that information. But the challenges that can create are unavoidable to a great extent. There will be most of the time too little trust and lack of reception on behalf of the user when it comes to computing existing everywhere. The implementation of ubicomp also leads to a lack of personal advantages. Another major challenge would be related to privacy. Depending on the application area considered, the relevance of privacy is assessed differently [43]. There are quite a few concerns of security along with ubicomp, and due to which the acceptance ratio of this computing technology may vary from place to place. The high cost of running various technological tools in connection with ubicomp is another challenge [44]. There can be technical obstacles too when it comes to the availability and reliability of ubicomp services. Creating the context awareness to the machines involved in ubicomp is another major challenge of ubiquitous computing. A pervasive computing environment must be advised of its client's situation and surroundings and must get adapted by changing its behavior based on this information.

Ensuring trouble-free, cost-effective, and unimpeded access to services is an indispensable issue in establishing ubicomp infrastructure [45]. Due to the additional material and energy consumption in the production and utilization of ICT for the accomplishment of pervasive computing can give rise to serious environmental risks in terms of clearance of electronic wastes. Being deficient in commercial concepts/business models, the reluctance of customers to spend for experiencing pervasive computing services, unconstructive ecological impact, elevated resource utilization, the inadequacy of legal guidelines, scarcity of regularity, high-level energy management, inadequate human-machine interface, etc., are also the key challenges related to ubicomp [46].

The interface between human and machine, security of data and obstacles in technical support are contemplated by experts as major restrictive factors in general. Environmental suitability, resource utilization and the shortage of legal regulations are portrayed as lesser limiting factors by expert studies [47]. To avoid a digital divide in which society is split between those with and those without access to pervasive computing, there is a need for easy access to ubicomp in technical, financial, and intellectual terms [48]. A few experts also speak of a possibility for a digital divide on the global level due to the introduction of ubicomp surroundings.

2.7   FUTURE OF UBICOMP

Design of an environment which is saturated with computing, communication capability and the graceful integration of human users is considered as the main essence of pervasive computing [49]. In software, we have processing of digital signals and programming using object-oriented concepts and supporting hardware includes mobile devices, sensors, smart pieces of equipment, etc., to make possible the subsistence of pervasive computing everywhere [50]. Even though today all the fundamental technologies are present, to spread the possibilities of ubiquitous computing, the research challenges are unavoidable in the dome of pervasive computing. Researches and advancements in networking hold up ad hoc routing, mobility management and global accessibility to make possible ubiquitous computing. Hence, shortly, we can expect our homes, malls, and workplaces equipped with network smart devices that sustain our information and communication necessities [51].

2.8   UBIQUITOUS COMPUTING AS A FACILITATOR FOR IoD

Ubiquitous computing ensures the presence of imperceptible computers by providing extensibility with ubiquitous experience. Ongoing network connections of minimum power IoT nodes can be facilitated by the demonstration of low power wide area network (LPWAN) technologies [52]. Computing can be made truly ubiquitous with the help of IoT [53]. IoD demands the presence of a virtual world with smart objects for its perseverance in actions [54]. In the virtual world, envisioned by IoD, each real-world object can act as an access point to offer connectivity services [55]. IoD works flawlessly in an environment that has been built by a global network of invisible computers that ensures wide exposure, ubiquitous connection, and substantial access [56]. To implement IoD, it is essential to set up smart objects with the power of local sensing and control with ubiquitous connections. Perfect orchestration of imperceptible computing resources and scheduling of various computing tasks at different levels permits ubiquitous computing facility across the environment created by IoT [57]. This can lead to the progress of aptitude in the virtual environment at a global level [58]. The existing communication network can be extended to areas with a weak connection through the IoD aided communications network. Great proficiency in every aspect of your daily life can be achieved when ubiquitous intelligence acts as a facilitator of IoD [59, 60].

2.9   PROSPECTS RELATED TO IoD

The virtual world of global intelligence will start enjoying the potentialities of IoD, soon. Employing the great mobility and re-programmability, the prospects in connection with IoD mainly spans across pervasive connections, airborne intelligence, sensor powering and consumption and self-management of communications [61]. For an environment based on IoT, ubiquitous connections are indispensable. The network offered by IoD can help to increase the scalability of the atmosphere created by IoT and thereby achieve ubiquitous connection even in flying mode [62]. Aerial intelligence can be activated through IoD, by considering the data which is available in the smart environment and thereby, making use of the AI that is put up locally in the object it operates [63]. Ambient sensors assist in the collection of data from the smart environment. IoD can conduct intelligent perception and decision making for the smart objects which are part of the connected virtual environment [64]. Self-management of communications can be supported by IoD to ensure unblemished communications in the smart environment of IoT [65].

IoD can help to reorganize the lost connections of smart objects which are involved in the communication service [66]. IoD can reduce the risk of spying and misleading attacks on the internet of everything (IoE), to a great extent by acting as a defensive wall [67]. By building dedicated protocols, IoD offers advanced encryption safety measures in the smart environment [68]. The sustainability of smart environments can be improved by way of the IoT with Wings by reprogramming to achieve sensor powering and reprocessing tasks [69]. IoD can reduce the e-waste due to battery draining of smart objects by charging them with energy as per the requisite. Simultaneous wireless information and power transfer (SWIPT) technologies used by IoD help to identify and recharge the nodes in IoE whenever needed [70].

2.10   CHALLENGES OF IoD

Even though IoD ensures a lot of prospects for the sustainability of a smart environment, the challenges are inevitable. In the perception of lively network connection, the inclusion of adaptable network topology and its accurate control for the uninterrupted communication services insist on experiments at different levels [71]. The elementary security and confidentiality concerns related to IoD demand a lot of research with an apt outcome to gain acceptance and popularity in a virtual world with widespread intelligence. Implementation of efficient lightweight intelligent algorithms in IoD for ensuring encryption has a significant role to protect the data [72]. Despite being beneficial, the requirement of significant mobility in IoD is an apprehension for the networks in which sufficient control over aerial vehicles is difficult to achieve. If proper security measures are not taken, a hacker can easily capture and modify the information which is part of the IoD [73]. As a threat to the confidentiality and integrity of the information in IoD, lack of robust security methods can give a chance to the attacker to change the authorization or access permissions by attaining access controls leading to conflicts and system crashes [74]. The performance of smart objects in IoD can be affected by hacking GPS spoofing. Probabilities to take control of the coordination of IoD devices by an intruder can generate collisions in the smart environment [75]. The chances for both physical and information attacks on the IoT with Wings are more compared to a traditional limited smart environment [76]. To spoil the confidentiality and availability of services, the attacker can increase the resource depletion rate and turn off the visual accompaniments, Internet facility, etc., of IoD devices by gaining unauthorized access [77, 78]. If the challenges in connection with IoD are not addressed appropriately, chances for vulnerabilities can exceed.

2.11 VISUALIZATION OF SMART CITIES WITH FLYING IoT

Technological advancements are a key means of facilitating existing resources to meet the growing demand for improved lifestyles. The dream of smart cities with imperceptible computers and ubiquitous connections can be functionally fulfilled through flying IoT [79]. Information is the most essential component of communication. One of the main reasons for the requirement of connectivity among people via smart objects is the necessity to carry out various services via information sharing (Figure 2.4) [80].

The foundation for a smart city can be laid down with the help of IoD because of its ability to gather, analyze, segregate, and deliver information precisely within a smart environment. IoT with Wings will be the prime benefactor in the endeavor of a smart city [81]. In the various phases of enhancement and subsistence of smart cities, IoD has a well-defined role. The flexible nature of IoD in performing the tasks assigned makes them unique in a smart environment [82]. In the visualization of smart cities, IoD can be used for surveillance and safety, perusal, and perception, reviewing, and mapping, transportation, and delivery, traffic management, agricultural activities, crowd management, natural disaster management, resource management, firefighting, etc.

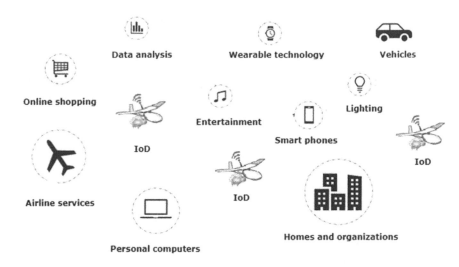

FIGURE 2.4   Flying IoT or the IoDs in the visualization of smart cities.

2.12   CONCLUSION AND FUTURE DIRECTIONS

Advancement of computing technology at reinforced speeds resulted in the evolution of ubiquitous computing. It is the proposal that anything including a real-world object or even human can be implanted with chips to join to a vast network of further human or objects or devices [83, 84]. We can expect that the espousal of ubicomp by the IoD will spawn an analogous influence that the acceptance of the Internet has created in a range of application fields (Table 2.1) [85, 86].

Prosperous source of demanding research queries will be prompted by IoD for several years. Quite a lot of disjoint areas of research such as computer-human interaction, expert systems, software agents, AI and so on can be merged by IoD. IoD can formulate our lives easier via digital surroundings that are adaptive, perceptive, ubiquitous, and alert to the needs of individuals [87, 89]. The time to experience the rich open network space of imperceptible computers around us, where the regulations have yet to be written and the boundaries have yet to be drawn, is approaching the world.

TABLE 2.1   Outcome of the Analysis Carried Out on Ubiquitous Computing, IoT, and IoD

Aspect	Opportunities	Challenges	Scope
Ubiquitous computing	• Development of invisible smart environments • Better socialization • Intelligent decision making. • Creation of dynamic and adaptive surroundings • Quick processing of information. • Enhanced experience • An environment that has been complemented by interconnected digital technologies to deliver improved convergence.	• Insufficient protection of personal information. • Increase in energy usage. • Ethical dilemmas implying user consent. • Maintenance of a sustainable environment.	• Enhanced mobility and ad hoc networking. • Advancement of contextual knowledge and integrated systems.
Internet of things	• New business possibilities. • New capacities for forecasting and acting. • Increased engagement with clients. • Sophisticated services and products. • Emerging sources of revenue. • Improved surveillance of business processes.	• Compatibility • Complexity • Confidentiality • Safety	• Development of smart cities • Establishment of the publish-subscribe paradigm
Internet of drones	• Minimization of visible hazards and health risks. • Comprehensive and more detailed data gathering. • Rapid deployment/launch. • Ability to respond to a majority of inspections. • Comfortable sharing of information • Enhancements in marketing and advertising • Ability to inspect hard-to-reach and dangerous zones. • Save time and effort.	• Legislative uncertainty • Complexity • Privacy • Safety	• Creation of flexible and efficient smart environments with the internet of robotic things

Quite a lot of work has been going on to analyze the role of imperceptible computers for the transition to an intelligent environment through ubiquitous computing, IoT, and IoD. By 2050, a printer may receive the command directly from the smart device on your pen or pencil to produce print outs, the invisible smart device in your shoe might give alert messages to you about your jogging speed, the smart device in the walking stick might show the right path for a disabled person to the destination and so on [90–92]. Imperceptible computers, ubiquitous computing, IoT, and IoD go hand in hand in the visualization of smart cities [93–95]. Even though the studies and researches in these fields can produce a remarkable outcome, shortly, by ensuring a better mode of living, the environmental threats that can be raised are unavoidable [96–98]. Therefore, the flavor of imperceptible computers, ubiquitous computing, IoT, and IoD within the smart cities must adopt the essentials of green computing to safeguard the natural environment to a great extent [99–101]. Let us imagine an era where the IoT with Wings will be competent to sense our situations and foreseeing our needs and adequately acting in our finest welfare, greatly similar to very good human friends or our parents in an environmentally sustainable manner.

KEYWORDS

- **imperceptible**
- **information and communication technology**
- **internet of drones**
- **internet of things**
- **pervasive**
- **UbiComp**
- **ubiquitous computing**

REFERENCES

1. Weiser, M., (1991). The computer for the 21st century. *IEEE Pervasive Computing, 4,* 94–104.
2. Banavar, G., Beck, J., Gluzberg, E., Munson, J., Sussman, J., & Zukowski, D., (2000). Challenges: An application model for pervasive computing. In: *Proceedings of the Sixth Annual ACM/IEEE International Conference on Mobile Computing and Networking (Mobicom2000)* (pp. 266–274). Boston, Massachusetts.
3. Ma Poon, E., Chen, D. J., & Chen, D. T. K., (2004). Extending dynamic web-based multimedia templates to reach users of pervasive computing devices in an e-learning environment. In: *Proceedings of the IEEE Sixth International Symposium on Multimedia Software Engineering* (pp. 110–113). Miami, Florida.

4. Satyanarayanan, M., (2001). Pervasive computing: Vision and challenges. *IEEE Personal Communications, 8,* 10–17.

5. Holland, J. L., & Lee, S., (2019). *Internet of Everything (IoE): Eye Tracking Data Analysis* (pp. 215–245). Harnessing the internet of everything (IoE) for accelerated innovation opportunities, IGI Global.

6. Jiang, J., & Han, G., (2018). *Routing Protocols for Unmanned Aerial Vehicles* (Vol. 56, No. 1, pp. 58–63). IEEE Communications Magazine.

7. Zhang, J., Yan, J., Zhang, P., & Kong, X., (2018). Design and information architectures for an unmanned aerial vehicle cooperative formation tracking controller. *IEEE Access, 6,* 45821–45833.

8. Kharchenko, V., & Torianyk, V., (2018). Cybersecurity of the internet of drones: Vulnerabilities analysis and IMECA based assessment. *Nineth IEEE International Conference on Dependable Systems, Services and Technologies (DESSERT)* (pp. 364–369). IEEE.

9. He, D., Qiao, Y., Chan, S., & Guizani, N., (2018). *Flight Security and Safety of Drones in Airborne Fog Computing Systems* (Vol. 56, No. 5, pp. 66–71). IEEE Communications Magazine.

10. Xu, J., Zeng, Y., & Zhang, R., (2018). UAV-enabled wireless power transfer: Trajectory design and energy optimization. *IEEE Transactions on Wireless Communications, 17*(8), 5092–5106.

11. Duan, Y., Edwards, J. S., & Dwivedi, Y. K., (2019). Artificial intelligence for decision making in the era of big data-evolution, challenges and research agenda. *International Journal of Information Management, 48,* 63–71.

12. Menouar, H., Guvenc, I., Akkaya, K., Uluagac, A. S., Kadri, A., & Tuncer, A., (2017). *UAV-Enabled Intelligent Transportation Systems for the Smart City: Applications and Challenges* (Vol. 55, No. 3, pp. 22–28). IEEE Communications Magazine.

13. Lin, C., He, D., Kumar, N., Choo, K., Vinel, A., & Huang, X., (2018). *Security and Privacy for the Internet of Drones: Challenges and Solutions* (Vol. 56, No. 1, pp. 64–69). IEEE Communications Magazine.

14. Devi, K. R. D., Sherin, J., & Santhiya, G. A., (2018). IoT based smart environment and its applications. *International Journal of Trend in Scientific Research and Development, 2*(4), 2456–6470.

15. Shagufta, R., Swati, K., & Sandhya, P., (2015). Analysis of IoT in a smart environment. *International Journal of Enhanced Research in Management & Computer Applications, 4*(4).

16. Renu, T., Ravi, K. P., Ajit, K. Y., & Akhilesh, S., (2018). IoT for smart environment and integrated ecosystem. *International Journal of Engineering & Technology, 2,* 1218–1221

17. Gokulnath, C., Marietta, J., Deepa, R., Senthil, P. R., Praveen, K. R. M., & Kavitha, B. R., (2017). Survey in IoT based smart city. *International Journal of Computer Trends and Technology, 46*(1), 23–28.

18. Ameyed, D., Miraoui, M., & Tadj, C., (2015). A survey of prediction approach in pervasive computing. *International Journal of Scientific & Engineering Research, 6*(5), 1–11.

19. Bae, C., Yoo, J., Kang, K., Choe, Y., & Lee, J., (2003). Home server for home digital service environments. *IEEE Transactions on Consumer Electronics, 49*(4), 26–34.

20. Evangelos, A. K., Nikolaos, D. T., & Anthony, C. B., (2011). Integrating RFIDs and smart objects into a unified internet of things architecture. *Advances in Internet of Things, 1*, 5–12.

21. Gubbi, J., Buyya, R., Marusic, S., & Palaniswami, M., (2013). Internet of things (IoT): A vision, architectural elements, and future directions. *Future Generation Computer Systems, 29*(7), 1645–1660.

22. Arasteh, H., Hosseinnezhad, V., Loia, V., Tommasetti, A., Troisi, O., Shafie-khah, M., & Siano, P., (2016). IoT-based smart cities: A survey. *IEEE 16th International Conference on Environment and Electrical Engineering* (pp. 1–6). Florence.

23. Trasviña-Moreno, C. A., Blasco, R., Marco, Á., Casas, R., & Trasviña-Castro, A., (2017). Unmanned aerial vehicle based wireless sensor network for marine-coastal environment monitoring. *Sensors, 17*(3), 460–465.

24. Kumar, S., Gupta, G., & Singh, K. R., (2015). 5G: Revolution of future communication technology. *International Conference on Green Computing and Internet of Things (ICGCIoT)*, 143–147.

25. Rigo, S. J., Cambruzzi, W. L., & Barbosa, J. L. V., (2015). Dropout prediction and reduction in distance education courses with the learning analytics multi trail approach. *Journal of Universal Computer Science, 21*(1), 23–47.

26. Shekar, S., Nair, P., & Helal, A. S., (2003). iGrocer: A ubiquitous and pervasive smart grocery. In: *Proceedings of the ACM Symposium on Applied Computing* (pp. 645–652). Melbourne, USA.

27. Chiu, D. K. W., & Leung, H. F., (2005). Towards ubiquitous tourist service coordination and integration: A multi-agent and semantic web approach. In: *Proceedings of the 7th International Conference on Electronic Commerce* (pp. 574–581). New York, USA.

28. Caceres, R., & Friday, A., (2013). Ubicomp systems at 20: Progress opportunities and challenges. *IEEE Pervasive Computing, 11*(1), 14–21.

29. Abowd, G. D., (2012). What next ubicomp? Celebrating an intellectual disappearing act. *Proceedings of the ACM Conference on Ubiquitous Computing* (pp. 31–40).

30. Wagner, A., Barbosa, J. L. V., & Barbosa, D. N. F., (2014). A model for profile management applied to ubiquitous learning environments. *Expert Systems with Applications, 41*(4), 2023–2034.

31. Barbosa, J. L. V., Hahn, R., Barbosa, D. N. F., & Saccol, A., I. C. Z., (2011). An ubiquitous learning model focused on learner integration. *International Journal of Learning Technology, 6*(1), 62–83.

32. Orwat, C., Rashid, A., Holtmann, C., Wolk, M., Scheermesser, M., & Kosow, H., (2010). Adopting pervasive computing for routine use in healthcare. *IEEE Pervasive Computing, 9*(2), 64–71.

33. Hung, K., Zhang, Y. T., & Tai, B., (2004). Wearable medical devices for tele-home healthcare. *IEEE Proceedings of the 26th Annual International Conference of Engineering Medicine and Biology Society* (Vol. 2, pp. 5384–5387).

34. Vergados, D., (2010). Service personalization for assistive living in a mobile ambient healthcare- networked environment. *Personal and Ubiquitous Computing, 14*(6), 575–590.

35. Buttussi, F., & Chittaro, L., (2010). Smarter phones for healthier lifestyles: An adaptive fitness game. *IEEE Pervasive Computing, 9*(4), 51–57.

36. Lin, K. J., Yu, T., & Shih, C. Y., (2005). The design of a personal and intelligent pervasive commerce system architecture. *Proceedings of the 2nd IEEE International Workshop on Mobile Commerce and Services* (pp. 163–173).

37. Roussos, G., & Moussouri, T., (2004). Consumer perceptions of privacy security and trust in ubiquitous commerce. *Personal and Ubiquitous Computing, 8*(6), 416–429.

38. Ubiquitous Accessibility, (n.d.). NOVA. Retrieved from https://www.pbs.org/wgbh/nova/article/people-with-disabilities-use-ai-to-improve-their-lives/ (accessed on 19 November 2021).

39. Lee, S., & Lee, K. C., (2012). Context-prediction performance by a dynamic Bayesian network: Emphasis on location prediction in ubiquitous decision support environment. *Expert Systems with Applications, 39*(5), 4908–4914.

40. Anagnostopoulos, T., Anagnostopoulos, C., Hadjiefthymiades, S., Kyriakakos, M., & Kalousis, A., (2009). Predicting the location of mobile users: A machine learning approach. In: *Proceedings of the International Conference on Pervasive Services (ICPS '09)* (pp. 65–72). London, UK.

41. Eagle, N., & Pentland, A. S., (2006). Reality mining: Sensing complex social systems. *Personal Ubiquitous Computing, 10*(4), 255–268.

42. Ubiquitous gaming, (n.d.). *Video Games Ubiquitous Environment.* Retrieved from https://mashable.com/2013/06/20/twitch-e3/ (accessed on 19 November 2021).

43. Han, G., Shen, J., Liu, L., Qian, A., & Shu, L., (2016). TGM-COT: Energy-efficient continuous object tracking scheme with two-layer grid model in wireless sensor networks. *Personal and Ubiquitous Computing, 20*(3), 349–359.

44. Oliveira, J. L., Souza, R., Geyer, C. F. R., Costa, C. A., Barbosa, J. L. V., Pernas, A., et al., (2014). A middleware architecture for dynamic adaptation in ubiquitous computing. *Journal of Universal Computer Science, 20*(9), 1327–1351.

45. Franco, L. K., Rosa, J. H., Barbosa, J. L. V., Costa, C. A., & Yamin, A. C., (2011). MUCS: A model for ubiquitous commerce support. *Electronic Commerce Research and Applications, 10*(2), 237–246.

46. Barbosa, D. N. F., Barbosa, J. L. V., Bassani, P. B. S., Rosa, J. H., Lewis, M., & Nino, C. P., (2013). Content management in a ubiquitous learning environment. *International Journal of Computer Applications in Technology, 46*(1), 24–35.

47. Atzori, L., Iera, A., & Morabito, G., (2010). The internet of things: A survey. *Computer Networks, 54*(15), 2787–2805.

48. Bardram, J. E., & Christensen, H. B., (2007). Pervasive computing support for hospitals: An overview of the activity-based computing project. *IEEE Pervasive Computing, 6*(1), 44–51.

49. Barbosa, J. L. V., Barbosa, D. N. F., Oliveira, J. M., & Rabello, S. A. J., (2014). A Decentralized infrastructure for ubiquitous learning environments. *Journal of Universal Computer Science, 20*(2), 1649–1669.

50. Hindus, D., & Schmandt, C., (1992). Ubiquitous audio: Capturing spontaneous collaboration. *Proceedings of the ACM Conference on Computer-Supported Cooperative Work (CSCW '92)* (pp. 210–217). Toronto, Canada, ACM Press, New York.

51. Morgenthaler, S., Braun, T., Zhao, Z., Staub, T., & Anwander, M., (2012). UAVNet: A mobile wireless mesh network using unmanned aerial vehicles. *IEEE Globecom Workshops*, 1603–1608.

52. Jiang, T., Geller, J., Ni, D., & Collura, J., (2016). Unmanned aircraft system traffic management: Concept of operation and system architecture. *International Journal of Transportation Science and Technology, 5*(3), 123–135.
53. Morgenthaler, S., Braun, T., Zhao, Z., Staub, T., & Anwander, M., (2012). UAVNet: A mobile wireless mesh network using unmanned aerial vehicles. *IEEE Communications Letters, 16,* 785–788.
54. Rametta, C., & Schembra, G., (2017). Designing a softwarized network deployed on a fleet of drones for rural zone monitoring. *Future Internet, 9,* 1–21.
55. Bor-Yaliniz, I., & Yanikomeroglu, H., (2016). *The New Frontier in RAN Heterogeneity: Multi-Tier Drone- Cells* (Vol. 54, No. 11, pp. 48–55). IEEE Communications Magazine.
56. Cohn, P., Green, A., Langstaff, M., & Roller, M., (2017). Commercial Drones are Here: The Future of Unmanned Aerial Systems. [online] *McKinsey & Company Capital Projects & Infrastructure.* Available at: https://www.mckinsey.com/industries/capital-projects-and-infrastructure/our-insights/commercial-drones-are-here-the-future-of-unmanned-aerial-systems (accessed 10 January 2020).
57. Genc, H., Zu, Y., Chin, T. W., Halpern, M., & Reddi, V. J., (2017). Flying IoT: Toward low-power vision in the sky. *IEEE Micro, 37*(6), 40–51.
58. Bekmezci, I., Sahingoz, O. K., & Temel, Ş., (2013). Flying ad-hoc networks (FANETs): A survey. *Ad Hoc Netw., 11*(3), 1254–1270.
59. Zhang, Z., Wang, H., & Zhao, H., (2018). An SDN framework for UAV backbone network towards knowledge centric networking. *Proc. IEEE Conf. Comput. Commun. Workshops (INFOCOM WKSHPS),* 456–461.
60. Motlagh, N. H., Bagaa, M., & Taleb, T., (2017). *UAV-Based IoT Platform: A Crowd Surveillance Use Case* (Vol. 55, No. 2, pp. 128–134). IEEE Communication Magazine.
61. Motlagh, N. H., Taleb, T., & Arouk, O., (2016). Low-altitude unmanned aerial vehicles-based internet of things services: Comprehensive survey and future perspectives. *IEEE Internet of Things Journal, 3*(6), 899–922.
62. Koulali, S., (2016). *A Green Strategic Activity Scheduling for UAV Networks: A Sub-Modular Game Perspective, 54*(5), 58–64. IEEE Communication Magazine.
63. Tareque, H., Hossain, S., & Atiquzzaman, M., (2015). On the routing in flying ad hoc networks. In: *2015 Federated Conference Computer Science and Information Systems* (Vol. 5, pp. 1–9).
64. Motlagh, N. H., Bagaa, M., & Taleb, T., (2016). UAV selection for a UAV-based integrative IoT platform. *Proceedings IEEE GLOBECOM 2016.* Washington, DC.
65. Schaffers, H., Komninos, N., Pallot, M., Trousse, B., Nilsson, M., & Oliveira, A., (2011). Smart cities and the future internet: Towards cooperation frameworks for open innovation. *The Future Internet Lecture Notes Computer Science, 6656,* 431–446.
66. Hernández-Muñoz, J. M., Vercher, J. B., Muñoz, L., Galache, J. A., Presser, M., Hernández G. L. A., et al., (2011). Smart cities at the forefront of the future internet. *The Future Internet Lecture Notes Computer Science, 6656,* 447–462.
67. Mulligan, C. E. A., & Olsson, M., (2013). *Architectural Implications of Smart City Business Models: An Evolutionary Perspective, 51*(6), 80–85. IEEE Communication Magazine.
68. Walravens, N., & Ballon, P., (2013). *Platform Business Models for Smart Cities: From Control and Value to Governance and Public Value, 51*(6), 72–79. IEEE Communication Magazine.

69. Lynch, J. P., & Kenneth, J. L., (2006). A Summary review of wireless sensors and sensor networks for structural health monitoring. *Shock and Vibration Digest, 38*(2), 91–130.

70. Bonetto, R., Bui, N., Lakkundi, V., Olivereau, A., Serbanati, A., & Rossi, M., (2012). Secure communication for smart IoT objects: Protocol stacks use cases and practical examples. *Proceedings of IEEE IoT-SoS, 1–7.*

71. Chourabi, H., Nam, T., Walker, S., Gil-Garcia, J. R., Mellouli, S., Nahon, K., et al., (2012). Understanding smart cities: An integrative framework. *Proceedings of 45th Hawaii International Conference System Sciences (HICSS), 2289–2297.*

72. Lytras, M. D., & Visvizi, A., (2018). Who uses smart city services and what to make of it: Toward interdisciplinary smart cities research. *Sustainability, 10*(6), 1998.

73. Bibri, S. E., & Krogstie, J., (2017). Smart Sustainable cities of the future: An extensive interdisciplinary literature review. *Sustainable Cities and Society, 31*, 183–212.

74. Guedes, A. L. A., Alvarenga, J. C., Goulart, M. D. S. S., Rodriguez, R. Y. R. M., & Soares, C. A. P., (2018). Smart cities: The main drivers for increasing the intelligence of cities. *Sustainability, 10*(9), 3121.

75. Rani, S., Talwar, R., Malhotra, J., Ahmed, S. H., Sarkar, M., & Song, H., (2015). A novel scheme for an energy efficient internet of things based on wireless sensor networks. *Sensors, 15*(11), 28603–28626.

76. Guvenc, I., Koohifar, F., Singh, S., Sichitiu, M. L., & Matolak, D., (2018). *Detection Tracking and Interdiction for Amateur Drones, 56*(4), 75–81. IEEE Communication Magazine.

77. Menouar, H., Guvenc, I., Akkaya, K., Uluagac, A. S., Kadri, A., & Tuncer, A., (2017). *UAV-Enabled Intelligent Transportation Systems for the Smart City: Applications and Challenges* (Vol. 55, No. 3, pp. 22–28). IEEE Communication Magazine.

78. Alsamhi, S. H., Ma, O., Ansari, M. S., & Gupta, S. K., (2019). Collaboration of drone and internet of public safety things in smart cities: An overview of QoS and network performance optimization. *Drones, 3*(1), 13.

79. Gharaibeh, A., Salahuddin, M. A., Hussini, S. J., Khreishah, A., Khalil, I., Guizani, M., et al., (2017). Smart cities: A survey on data management security and enabling technologies. *IEEE Communications Surveys and Tutorials, 19*(4), 2456–2501.

80. Erman, A. T., Hoesel, L. V., Havinga, P., & Wu, J., (2008). Enabling mobility in heterogeneous wireless sensor networks cooperating with UAVs for mission-critical management. *IEEE Wireless Communications, 15*(6), 38–46.

81. Chaoxing, Y., Lingang, F., Jiankang, Z., & Jingjing, W., (2019). A comprehensive survey on UAV communication channel modeling. *Access IEEE, 7*, 107769–107792.

82. Ahmad, S. H., Noor, S. O., Hazim, S., & Abdallah, K., (2019). Wireless coverage for mobile users in dynamic environments using UAV. *Access IEEE, 7*, 126376–126390.

83. Shakhatreh, H., Khreishah, A., Othman, N. S., & Sawalmeh, A., (2017). maximizing indoor wireless coverage using UAVs equipped with directional antennas. *Proceedings of IEEE 13th Malaysia International Conference on Communications (MICC), 175–180.*

84. Mozaffari, M., Saad, W., Bennis, M., & Debbah, M., (2017). Mobile unmanned aerial vehicles (UAVs) for energy-efficient internet of things communications. *IEEE Transactions on Wireless Communications, 16*(11), 7574–7589.

85. Gharibil, M., Boutaba, R., & Waslander, S., (2016). Internet of drones. *IEEE Access, 4*, 1148–1162.

86. Yanmaz, E., Yahyanejad, S., Rinner, B., Hellwagner, H., & Bettstetter, C., (2017). Drone networks: Communications coordination and sensing. *Ad Hoc Networks, 68*, 1–15.

87. Pierre, D. W., Wendy, E. M., & Rich, G., (1993). "Computer-augmented environments; back to the real world. *Communications of the ACM, 36*(7).

88. Shankar, R. P., Brian, L., Armando, F., Pat, H., & Terry, W., (2001). ICrafter: A service framework for ubiquitous computing environments. *Ubiquitous Computing, 2201*, 56–75.

89. Gregory, A. D., & Elizabeth, M. D., (2000). Charting past, present, and future research in ubiquitous computing. *ACM Transactions on Computer-Human Interaction, 7*(1), 29–58.

90. Edwards, W. K., & Mynatt, E. D., (1997). Timewarp: Techniques for autonomous collaboration. In: Pemberton, S., (ed.), *Proceedings of the ACM Conference on Human Factors in Computing Systems* (pp. 218–225). ACM Press, New York.

91. Henderson, D. A., & Card, S., (1986). Rooms: The use of multiple virtual workspaces to reduce space contention in a window-based graphical user interface. *ACM Transactions on Graphics, 5*(3), 211–243.

92. Minneman, S., Harrison, S., Janssen, B., Kurtenbach, G., Moran, T., Smith, I., & Van, M. B., (1995). A confederation of tools for capturing and accessing collaborative activity. In: Zellweger, P., (ed.), *Proceedings of the 3rd International Conference on Multimedia* (pp. 523–534). ACM Press, New York.

93. Li, S., Zhang, N., Lin, S., Kong, L., Katangur, A., Khan, M. K., Ni, M., & Zhu, G., (2018). Joint admission control and resource allocation in edge computing for internet of things. *IEEE Network, 32*(1), 72–79.

94. Lin, X., Bergman, J., Gunnarsson, F., Liberg, O., Razavi, S. M., Razaghi, H. S., Ryden, H., & Sui, Y., (2017). *Positioning for the Internet of Things: A 3GPP Perspective, 55*(12), 179–185. IEEE Communications Magazine.

95. Ghaderi, M., & Boutaba, R., (2006). Call admission control in mobile cellular networks: A comprehensive survey. *Wireless Communications and Mobile Computing, 6*(1), 69–94.

96. Zeng, Y., Zhang, R., & Lim, T. J., (2016). *Wireless Communications with Unmanned Aerial Vehicles: Opportunities and Challenges, 54*(5), 36–42. IEEE Communications Magazine.

97. He, D., Chan, S., & Guizani, M., (2017). *Drone-Assisted Public Safety Networks: The Security Aspect, 55*(8), 218–223. IEEE Communications Magazine.

98. Koubaa, A., Alajlan, M., & Qureshi, B., (2017). *ROSLink: Bridging ROS with the Internet-of-Things for Cloud Robotics, 2*, 265–283. Cham: Springer International Publishing.

99. Ahlgren, B., Dannewitz, C., Imbrenda, C., Kutscher, D., & Ohlman, B., (2012). *A Survey of Information- Centric Networking, 50*(7), 26–36. IEEE Communications Magazine.

100. Altawy, R., & Youssef, A. M., (2016). Security, privacy, and safety aspects of civilian drones: A survey. *ACM Transactions on Cyber-Physical Systems, 1*(2), 7:1–7:25.

101. Roberts, J., (2009). The clean-slate approach to future internet design: A survey of research initiatives. *Annals of Telecommunications, 64*(5), 271–276.

CHAPTER 3

Implementation and Deployment of 5G-Drone Setups

JAGJIT SINGH DHATTERWAL,[1] KULDEEP SINGH KASWAN,[2] and
AMIT PANDEY[3]

[1]*Department of Artificial Intelligence & Data Science,
Koneru Lakshmaiah Education Foundation, Vaddeswaram, AP, India,
E-mail: jagjits247@gmail.com*

[2]*School of Computing Science and Engineering, Galgotias University,
Greater Noida, Uttar Pradesh, India, E-mail: kaswankuldeep@gmail.com*

[3]*Computer Science Department, College of Informatics, Bule Hora
University, Bule Hora, Ethiopia, E-mail: amit.pandey@live.com*

ABSTRACT

The relationship with drones and wireless networks poses two key questions; first, how wireless networks help drone personal or technical use. Secondly, how drones can enhance performance of the network, i.e., increase on-demand capabilities, increase coverage, increase capacity, and improve agility as an aerial node. From a networking point of view, the very first category of drones is listed as mobile phones (MEDs) and broadband infrastructure drones as wireless ones (WIDs). This chapter discusses both the MED and WID cases under practical limits of 5G at the start of 5G Phase I completion (Rel-15). We also address possible solutions to transparent challenges which have been highlighted, either by implementing existing guidelines or presenting recommendations for further changes. While drones can easily be inserted into cell phone networks, tend to 4G LTE-A or 5G Rel-15 substantial achievements constructing main structures. Nevertheless, it is important to

The Internet of Drones: AI Applications for Smart Solutions. Arun Solanki, PhD, Sandhya Tarar, PhD,
Simar Preet Singh, PhD & Akash Tayal, PhD (Eds.)

fine-tune potential launches by looking at current approaches from the MED and WID aspect and to fill the holes with new technologies.

3.1 INTRODUCTION

Many commercial applications of drones in all sectors have led to a growing appetite for wireless communication for drones, as well as their penetration of the consumer electronics industry [6]. The best way to provide safe, stable broad-based connectivity to drones is with mobile networks [1]. In the meantime, focus has been paid to airborne contact by the need for omnibus networking for a wide spectrum of consumer equipment (EUs) including driverless vehicles, sensors, and eMBB computers. Different types of aerial nodes have proved innovative in unparalleled ways in enhancing 5G efficiency, agility, and flexibility [2, 9].

From a networking point of view, this paragraph characterizes mobile networks improved drones as mobile devices and aeries serving cellular entries as mobile devices (WIDs). The chosen category of drones for both types is identical to medium sized systems with modest capacities, e.g., squares or unmanned resolved aircraft (HWP) [2, 7] except where otherwise defined, e.g., HWPs. Be mindful that WIDs are a sub-set of networked moving mechanisms that also include NTNs (TR 38.811).

Drones have varied features, from nanorobot to wider spindly legs than Boeing [2]. Drones are available at various speeds. This variation represents the vocabulary in which a variety of words are formed, for example unmanned air cars or systems (UAV or UAS, respectively) [3]. "drone" standardization efforts in 3 GPP are continuing to provide wireless communication to personal drones and commercial drones via mobile networks, i.e., in contrast 'drone' is unclear to any of the time frame; a controlled remotely device can operate on any channel (air, water or land). But the most usual and portable word is always "drone." In this report, two goals are explored about the interconnectivity aspects of the MEDs and WIDs. Firstly, the structure on drone penetration in current, and future, mobile networks illustrate substantial information on the standards of 3GPP [4]. Secondly, to respond to unresolved concerns through introducing current methods or by suggesting more changes. We will assist concerned scientists with the necessary sources in this study to access the standardization papers. One of the aims of this chapter is to provide an exhaustive review of the progress of 5G normalization behavior by creating a bridge between academia and

industrial researchers. This is, according to our understanding, the first literature to deal with the use of drones primarily from the RAN and fundamental
architectural structures (SA) [5].

3.2 MOBILE NETWORKS-BASED DRONES HANDLING TECHNIQUE

Drones are able to handle different sizes from nanorobots to aircraft with
wingspans bigger than a Boeing [2]. This variation represents the vocabulary
in which a variety of words are formed, for example unmanned air cars or
systems (UAV or UAS, respectively). Compared to other words, "drone"
means a vague system that can be run in remotely controlled (land, air,
or water), but it is also the most prominent portable terms in this chapter
[6]. 3GPP continuous standardization activities to offer improved Wireless
communication to personal and commercial drones through mobile networks
are supported in this chapter. In contrast, "drone" is also the most common
term.

3.2.1 MOBILE-ENABLED DRONES

In this group, drones are UEs, i.e., enterprise users from broadband networks
that also have personal drone networks. Drones are currently used in comprehensive applications for inspections and inspections, transportation, and
logistics, and tracking. For rural and metropolitan scenarios, the TR36.777
specifies a 300 m elevated limit and 160 km/h high speed. The "commercial
air-to-ground" of TR38.913 is not meant to confound these usage cases, as
the word implies to offer on-board commercial aircraft mobile communication services to citizens and machinery [8].

Drones have historically been using a non-licensed connection to access
a GCS or drone pilot station [10]. However, unlicensed contacts, which are
mainly forbidden, are only restricted in their durability and range as the
Federal Aviation Administration (FAA) in operations outside the visual-loos
(BVLOS) transmission impairment of remote monitoring signals is primarily
prohibited. In comparison, thanks to its long wavelengths and consistency,
mobile networks can allow BVLOS operations. In addition, safe messaging
(TS 33.501), lawfully permitted intercept power (TS 33.107), position
authentication, and trustworthy identity (TR 33.899) are additional benefits
that can be achieved by existing and future mobile network standardizations
[11].

MEDs create two forms of connections to GCS: firstly, the remote piloting and telemetry, identification, and navigation command structure links (TR 36.777). Although remote piloting can involve a video transmission to provide the pilot with almost a sensation, tele command and telemetry links fall under one single non-payload contact umbrella. As many regulators do not allow fully self-contained Drones and robotics may operate only semi automatically underneath the control of drone operators (due to operational consequences), there are vital ties to command and control [10]. An application connection providing information, for example, sensor data, video, audio, and images is the second type of link formed with a GCS. Notice that mainly payload communication capabilities are required for the application. Application info, on the other hand, is less important in certain situations than command and control [13].

3.2.2   WIRELESS INFRASTRUCTURE DRONES

WIDs serve to maximize network availability, for example by raising reach or capacity, as opposed to MEDs. WIDs can be categorized according to their characteristics and demands:

1. **Aerial based Drone-BSs:** Drone-BSs are used to establish any of all BS capabilities [12]. For drone-BSs, it could be appropriate both wirelesses backhaul and fronthaul [19], or either of these can be supplied via tethering. Registered or unauthorized may be the downlink and/or uplink radio access [14]. Drone-BSs have many shifting patterns including, for example, spinning, spinning, stationary or at the required positions following particular routes (e.g., top of buildings). The patterns vary depending on the circumstances, machines, and specifications for contact.
2. **Aerial-Relays (AR) Communication:** Users or operators can deploy ARs. In the previous example, a non-licensed spectrum should Usage for user-AR communications, and AR is a UE cellular network communication [15]. The ARs of the operator should be much more total because properly incorporated. They will serve either as intermediary hops for IAB (sec. III-B3) or as an analog replay with up/down converter topologies [16].
3. **ATND:** Air transport network drones type NTNs (ATN). There has been a rapid rise in interest in ATNs [17]. The hybrid solutions appear most effective whilst both licensed and non-licensed solutions are

available. As an integral part of 5G NTN, TR 38.811 recognizes satellites [18]. They are not, though, taken into account here, since the height approaches the limits within the limited cell-like procedures.

3.3   5G STANDARDIZATION ADVANCEMENT AND WHICH IS MORE RELEVANT IN COMPARISON TO THE DRONES PROCESS

5G network capability is addressed in this chapter in conjunction with the establishment of the Rel-15 Functional Classes of Run (RAN1, RAN2, RAN3) and SA (SA2, SA5). This segment has the following organization: First, we explain the slice principle. 5G Run topics are then discussed, which are mostly relevant. Themes involve standardized MEDs operations, architectural functions for MEDs and WIDs, built-in connectivity and backhaul tests, and approved and unauthorized drone choices. Finally, 5G central and 5G network optimization experiments will be evaluated in terms of promising options for drone integration [20].

3.3.1   SIMPLICITY AND VERSATILITY WIRELESS NETWORKS

The slice framework allows for the simplicity and versatility of the service-oriented set-up of wireless networks [11]. The network is also arranged to assist multiple drone service providers, for example a slice for the use of MED applications or a slice to separate WID traffic (Figure 3.1).

For SA2 and RAN, the slice is a "logical network with some network functionality and network features"; moreover, it calls the "network slice managed instances (NSIs)" with various components, i.e., managed nets slice subnet instances (NSSIs) for example in relation to the domain (RAN, core) or location (Ottawa, Toronto, etc.) [14, 21].

3.3.2   5G RAN

RAN promotes thorough cutting, for instance, enabling hard splitting and commodity separation among strips, QoS extracting in chunks but knowledge about the identity of the slices. There will be more advances in RAN and its importance for drones [22]:

1. **Standardization for MEDs:** TR 36,777, which addresses concepts, situations, performance, and metrics, interference epidemic, and possible cures, is the most detailed analysis for MEDs. In future versions details and setups will be presented concerning the outcomes of field tests, device level and mobility measurements and rapidly fading models [23].

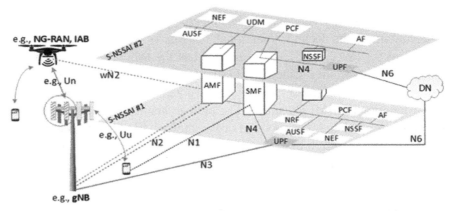

FIGURE 3.1   Structure of wireless network.

Most RAN professional bodies conduct standardization exercises for MEDs, though SA2/5 does not have a clear MED report. Though IoT, URLLC, and eMBB consider MEDs, they may be imprecise [24].

2. **Architectural Roles:**
   5G architecture benefits from the disconnection of network components' functions, capacities, and functionalities to deliver modular technological frameworks that can be applied to individual requirements. Different integration options are also possible for WIDs of different degrees of usability, expense, protection, and complexity [24]. Rel-15 comprises three types of RAN nodes:
   a. **eNB:** E-UTRA termination of protocol for the EU user plan and control plane, linked via S1 interface to the EPC heart (TS 36.300).
   b. **NG-eNB:** The EU termination of E-UTRA user plane and plane, linked NG interface to 5G-core (TS 38.300) [25].

c. gNB: New radio (NR) termination of the EU user plane and control plane protocol linked to the NG gui 5G-core (TS 38.300) [26].

Often, the gNB module is called the NG-RAN node. To ensure a smooth operation, interoperability's between 4G and 5G are necessary before complete 5G is implemented and 5G UEs are made accessible. In either standalone or non-standalone mode, interoperability may be feasible. If it is not standalone, a WID serves as an NR node for the use of large data rates for the EU, while the NB is managed by the terrestrial control plane. Notice that only eNB and service-gateway or user-plane feature are connected to the NR node. WIDs with a cellular backhaul may be NGeNB or gNBs while WID Both cells can be penetrated by LTE/LTE-A and 5G compatibility [12].

BSs can be introduced as master and secondary nodes of hierarchy in the case of multi-RAT dual networking (DC) (TS 36.340). The other Memory Controller could be Ng-eNB or GNB, known as NGEN-DC or NE-DC (TS 37.340). EN-DC indicates that an eNB is the most significant NG-RAN network main controller. While all DC solutions can be practicable for WIDs, in theory, it is more effective for WIDs, because of wireless front/backhaul, to be a secondary node to reduce their complexities [27].

WIDs can therefore be air-conditioned DUs with a limited Functionality of the central unit (CU). The CU/DU division is planned for achieve cloud-RAN technologies with multiple split (TR 38.801) options that allow for centralization versus control delivery and capability to be arranged based on each wireless networks' condition, e.g., support of large quantities of UEs, bandwidth excess or lack, tolerance to delay or sensitivity, and extending the available bandwidth [28].

Option-2 and option-3 are primarily discussed among 8 split options (TR 38.801 in). Option 2, including the DU radio contacts and the PDCP protocol for CU radio source data, the service data adaptation protocol (SDAP) for data radio transmitting and the CU radio source monitoring protocols transmissions are available too. Radio connection control (RLC) is available with DU (TS 38.401). When option-2.1 (3 C-like split) is used, there is no split for signaling radio bearers; whereby DUs must have any layer and the requisite control plane capability. Choice 3 separates both the signaling radios, and the data radios, into RLCs with a CU-based high-RLC and DU-based low-RLCs. Therefore, Option3 makes lighter DUs than Option 2. However, as seen in TS 38.401 procedures Option-3 spends fronthaul bandwidth on the monitoring and maintenance of radio capital standards [29].

In TR 38.816, RAN1 studies the possibilities of lower levels-split. Driving downlink speeds of 4.1 to 18.2 Gbps and 37.6 to 454.6 Gbps are expected for Option-6 and Option-7.1 based on measurements with sets of parameters includes uplink/downlink, propagation system, antennas ports number. Despite the great variation in required frontal speeds, the LOS opportunities and broad range of NR make it possible to reach high rates from aerial DUs to CUs [30].

When separate alternatives are regarded for WIDs, there are several compromises. Second, lower layer split increases the requirement of bandwidth and reduces latency resistance of the frontal link in contrast to upper layer splits. It raises the difficulty of transmitting the border-wide signal (especially for PHY layer). The bottom layer split, however, decreases the device difficulty of DU considerably. If front end connections with high-SINR and broad bandwidth are created, air time can increase. Finally, the narrower split layer increases centralization and the number of EU countries the DU will serve [31].

3. **Integrated Access:** Recently, IAB studies things for promoting wireless networking without scaling transport networks is accepted by RAN Working Groups. IAB's value for WIDs is twofold: a satellite backhaul that is natively enabled by 5G networks can be used as a first air relay. Second, a WID will reduce the number of hops as an Intermediate IAB node and allow LOS and mobility to provide topological stability [32, 33].

TR 38.874 eliminates visible light communications when considering the frequency of the carriers up to 100 GHz. The WID will have effective solutions [13] by soothing the versatility of frequent switching and multi connectivity requirements. WID's simplicity facilitates a coherent topology architecture and alleviates the problem of mobile coverage deficiencies. There may be sufficient slight adjustments in the PHY layer as LOS is probable. Co-channel interference is nevertheless the limiting factor, particularly for band backhauling [9]. In the Drone IAB background, practical solutions can be sought through the revisit and investigation of the L2/L3 relay architecture, for instance using modification layer routing advice relays on RLC, drones co-orders and details relevant to the QoS [13].

4. **Licensed/Unlicensed Options for MEDs and WIDs:** Provided the approved access (LAA) requirements for the unlicensed range, the issue arises: will LAA support drones? Interested question arises. Determined by the maximum transmission power requirements it is possible to manipulate those structures by having small cells, integration thorough IPsec tunnel actually reacts. Changes depending on the planned drone application range and altitude compared with the Reportage for certain LAA cells not exceeding 150 m. In Rel-15 an NR-based accessibility to the unlicensed spectrum has been accepted with the goal of taking the improvements of NRs to spectrum resources either below or below 7 GHz, such as versatile numerology (subcarrier distance), miniature slot, frame structure or wideband (TS 38.889). The use of greater distance between subcontractors and/or mini slots increases performance, but decreases energy per symbol for same transmitting power and, thus, reduces coverage. Access connections for some drone applications are also harder to support.

Although Wave and NR frequency hopping techniques will boost received SINR in drone by narrow beams and elimination of interference, the received signal strength (RSS) in a non-licensed band cannot be improved because a power reverse transmission will be needed for full EIRP requirements to be complied with. This basically means 5G NR-U air link for radio links cannot be configured NR-U MEDs so they must stay similar to protocol stack differences between the two types of nodes.

Micro/pico cells to be served. It can, by comparison, be sufficient for the radio access connections of non-standalone WID NR-U UEs. In the photo. A drone secondary gNB (drone-SgNB) serving device can maintain near proximity to the DC-UEs, whereas a wide macro master-eNB (MeNB) protection or master-gNB reportage (MgNB) would make its user/control-plane connections over the licensed band.

The UEs feature in required EN-DC options-3/3A should be provided if the routing protocol data is transmitted by wireless Xx-C to the MeNB. First, if an option-3 is used between the MeNB and DroneSgNB, A User Plane data transfer Wireless Xx-U link to the DroneSgNB NR RLC Layer via the LTE PDP layer. Second, via PDCP Drone-SgNB NR layer, the S2U wireless link to the EPC provides the user-plane data directly, while EPC is transferred to a secondary (SCG) carrier bearer (EPC Choice 3A). These two cases are especially important as the early stages of conversion from LTE to 5G are expected to be close. Likewise, in the UEs operations. In case of wireless Xn-C link to the makers-gNB carrying the control-plane data, 2b) is supported with the stated full-5G options 4/4A. If the Master-GNB is divided into the Drone-SgNB When (option4), the NR SDAP, required to map bearer packets into QoS groups before the PDCP layer, can be carried via the Xn-U wireless link. Whereas a (shot) SCG carrier (option 4A) is supplied directly via the drone SgNB's NR SDAP sublayer with a wireless N3 link to its 5G core.

Cell network connectivity to the unlicensed band may only have been configured to satisfy the regulatory criteria by LTE LAA or 5G NR-U. However, it does not guarantee equal co-existence with incumbent RATs such as Wi-Fi. Technological innovations defined before LAA for unlicensed band by Carrier Wi-Fi are still trust-worthy innovations and are also required to take advantage of 5G networks as seen because of their intrinsic equivalence with Wi-Fi. The inclusion of PDCP is a choice of NR and WLAN (NWA). In any event, a drone-WIFI termination can only be applied without being collocated. This is the easiest and most inexpensive WID conver-gence scenario for unauthorized entry, clearly.

The second approach is NR and WLAN (NWIP) IP-level inte-gration where UEs run in multi-home mode and accommodate two separate IP flows through all air interfaces. The IP flow unloaded to drone AP is, however, reasonably unencrypted, and hence an IPsec tunnel is formed by integrating the NR PDU using a NWIPEP sublayer between the Master-gNB and the UE via a drone-AP in Figure 3.2.

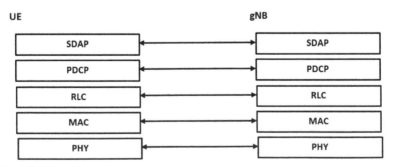

FIGURE 3.2   Control plane protocol.

As defined, NR-Unlicensed, and the NR-WLAN aggregation/integration options provide a broad range of options for WID integration when using the wide open, free unlicensed continuum despite its limitations for MEDs especially exceeding 7 GHz. In addition, the feasibility of such options could be studied across several scenarios from complex IoT and URLLC deployments.

3.3.3   5G CORE

5G Core Networks develop a service-oriented infrastructure to further leverage the advantages of NFV and SDN to promote automation and maximize its reliability. 5G-core architecture concepts permit an NF to speak directly to other NFs, promote multi-provider aggregation, and allow multiple slices with specific configurations.

Fifth-generation (5G networks improve existing features by dissecting them to new features. For example, the EPC Movability Agency is categorized into AMF and SMF (AMF is a non-access stratum termination which encompasses functionality like mobility control, verification of access and legitimate interception (TS 23.501). Likewise, SMF has MME roaming and Operation and packet-gateway control plan capabilities, like the allocation and maintenance of the UE IP address. Most NFs on the control-planes are improved to slice consciousness, for example, when you connect UEs, you pick network slice by AMF (TS 38.801). Single slice network collection help knowledge (S-NSSAI) defines a slice for the network. The S-NSSAI is made up of a slice/service sort (SST) which optionally differentiates a slice within the same SST, referring to NS key features, including eMBB, uRLLC, and a slice differentiator (SD).

Enhanced virtual machine and core slicing support facilitates the introduction of new features, especially MEDs and WIDs supportive [2]. There are still several challenges with applying this flexibility. Since MED traffic varies from TL to greedy bandwidth multimedia content, MEDs can be connected to multiple parts. There can be eight incidents connected simultaneous to a UE and depends on the availability and the nature of the slices whether or not they are appropriate for a MED. Recently, TS 23.401 provides high-level aerial-UE assistance for subscriptions and the way they transmit during transfers to the EPC. Sliding elements, such as a particular SST or SD value, are not included in MED assistance. Network exposure feature (NEF) enables control plane NF functionality to be exposed to external actors in controlled manner, such as an untrusted device function, edge computing and other vendor controller NFs. NEF will improve the cost-effectiveness and speed of deploying edge computing to support MED operations.

As for WIDs, traffic without disrupting current infrastructure is an important concern. One approach is to build a WLAN slice that could have several NFs and non-shared user plane functions in common controller planes. It does however not suffice to build just one slice, as UEs with multiple services may exist. There are then several tradeoffs in this situation: because operations of drone-BS are costly, the goal is to use drone-BS for the greatest number of facilities. Therefore, for each WID, several 5G core slices can be required. This is expensive and difficult to administer and raises wireless upgrade pressures for NSSAIs, such as N2 and N1. When current diagrams are shared, it is difficult to provide appropriate insulation; the traffic monitoring for certain facilities requires high granularity. For incorporation of drone fleets, the complexity grows exponentially, requiring successful monitoring systems.

3.3.4 5G NETWORK MANAGEMENT

FCAPS, i.e., errors, billing, authorization, efficiency, and security monitoring are typically the responsibility of network management. In recent years, demanding management criteria of versatile networks have led to current management capabilities and operationally oriented administration and instrumentation framework. NSI, NSSI, and NF Control Functions are other utilities, such as positive decision (NSMF, NSSMF, and NFMF) (PM). A service provider and his customer may have different relationships in this architecture, as seen in.

Depending on the position of the aircraft, drones may then be operated by different management agencies. The text image give an example of a slice distribution procedure where drones, NSSI, NF, and infrastructure suppliers are shown as slice customers. TS 28.530 actually follows the various network slice as-a-service (NSaaS) information and management exposures. However, existing guidelines do not directly support NF and infrastructure vendors.

In case of different business options and a network of layers of components at various levels, upgrade TSs can be needed for integration of WIDs. New model network services entities, for example, may be helpful, comparable to the TS 28.541 additional templates for CU/DU. In addition, there is no comprehensive review of FCAPS WID specifications. How to set up a WID for calculating and reporting load information, for example? Does WID or MEDs have to lift external warnings, such as remaining gasoline, Failure?? Furthermore, preparations are being made to develop network monitoring systems by collecting and evaluating later part data [14]. As an intelligence provider to settle on, for example, the role of WIDs, network management is vital for incorporation of Drones. However, the frameworks for automation have yet to be implemented and there is no consensus on the allocation of roles, for example between SA2 and SA5. Indeed, only consistent definitions of WID functions in the 3GPP networks and collaboration between standardization agencies and working groups will help to identify solutions to these problems in reality. Issues which require more than teamwork, but significant analysis, will be addressed next.

3.4 STANDARDIZATION PROSPECT AND DIGITIZATION

In percentage Implementation of Rel-15 study on the facets of the far is highlighted.

A new slice or a current one update can be generated to respond to an NSI request. Requests for provision from the corresponding administration service providers have been decomposed into their constituents in Figure 3.3. Drones may have differing functions as slice consumers, slice providers, and slice constituent in mobile networks, for example.

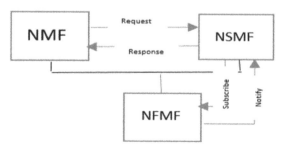

FIGURE 3.3 Drone network management overview. [*Abbreviations:* NMF: Network management function; NSMF: Network subnet management function; NFMF: Network function management function].

While the Rel-15, along with those from the Rel16 recommendations for more normalization, is a pretty early step of standardization to concentrate on practical strategies for drone operations, the Rel-15 unfinished studies are. Included among standardization and testing subjects may be changes of Basic protocols for MEDs, new terminal forms (e.g., Aerial European Registry), network resource separation studies of the safety and efficiency aspects of drone. Next, chosen subjects will be comprehensive.

3.4.1 NETWORK CONFIGURATION AND SLICE DESIGN FOR WIDS

In the process of communication network and slice design, current services, such as utilities, transportation network, would typically be considered. This implies the predetermined set of Run nodes and NFs. However, WIDs' mobility adds flexibility and calls for innovative design methods and programming techniques:

1. **RAN NSS Design:** The most suitable architectural function for WIDs must be chosen. For example, CU/DU cannot be optional when the terrestrial nodes are eNBs; AR can be preferred. WIDs that can run any control plane and create stable Xn connections to eNBs are therefore necessary. WID may be implemented as a DU if the gNB and the 5G UEs are plurality. The split choice could in this case be decided on the basis of the fronthaul communication capabilities, technological WID requirements and network needs (e.g., congestion relief or expanded coverage). Properly built Nodes of contributors, stable modulation schemes including distribution methods for services (e.g., hard slice, soft slice, upgrading RRMs).

2. **CN NSS Design:** By employing scalable networking strategies and fulfilling connectivity needs including protection, isolation, and latency, the effect of WIDs integration on 5GC must be minimized. Modifications to current CN NSS may be made, for example, to initiate new NFs or to expand the ability of existing tools. TN capacity should be analyzed and, if necessary, extra capacity assigned. If a newly generated slice of the WID integration segment is used for the re-configuration of AMF, NSSF, etc., and new UPFs. For latency reduction, SMF can be pre-configured with UPF collection. In addition, MEC functions can be strategically implemented to provide extra WID/MED computing.

    Automated network requirement assurance architecture is the secret to agile networking. Besides the technological challenges of architecture, the next subject is the market position of WIDs.

3.4.2   SUPPORT FOR NEW BUSINESS MODELS

MEDs are related business models to UEs. In case of congestion, special services can also be allocated for a MED to help avoid outage, for example, the control connection may be a high priority link. In the final review, MED cases should be assisted by incremental optimization, implementation, and methods of operation.

In comparison it can be difficult and diverse, business models of WIDs:

1. **Aerial-IaaS:** This is true when WID functions like an NF node. WIDs may be used as entry points or as VNFs, i.e., as VNF drones. For example, a drone may be used with MME functionality if MME needs error, congestion, or extra reliability. You should get a drone used to close the application feature via drone-as-MEC for development to reduce. In the situations. Three and WID is a fitting architectural function as an entry point.
2. **Aerial-SaaS:** Aerial network slice subnet (A-NSSA) may be created by multiple WIDs, for Example of a present RAN-NSS or 5GC-NSS overlay (e.g., UPFs, MMEs, and MEC applications).

If a service provider (SS) supplies a WID to an operator, the operators are negotiated between them so that information is not revealed and SPs that do not assign management capacities. If, for example, an SP operates an A-NSS, the operator needs the SP for fine-tuning management, e.g., PM, to reveal details, such as service types/needs, user contexts. Although different market models make WID networks scalable and minimize burden on operators, information exposure and management skills are crucial problems.

3.4.3   NEW QOS AND KPI PARAMETERS

New KPI and QoS parameters for MEDs and WIDs would probably be needed. When WIDs are deployed on demand, for example, their services can produce appropriate revenues. Even if A nominal KPI parameter may be the number of users served, it cannot be sufficient to determine the profitability of WID operations due to more complex charge schemes. Therefore, a new WID KPI, which involves charging policies, operating costs and the number of users served in the formulation, can also be considered. Including the 5GC influence (for instance additional load and reporting), the RAN impact (for example, allocation of resources, interference), output deterioration of UEs in donor node cells (for example, due to interference, or resource short aging), topology quality, we suggest incorporation efficacy as a KPI-Design (e.g., number of hops, reliability). Drone KPIs would have to be combined with new KPIs [15].

3.5   CONCLUSIONS

5G offers a full range of cloud-RAN connection design elements. None is their clear solution about how to use drones. There is no answer. Evaluation of implementation strategies based on network knowledge, and the demand/ need characteristics. Although standardization activities for aerial EUs have already begun in RAN, DC, IAB, and NR-U studies will accelerate the incor- poration Enhanced power generation, flight times and smooth integration of WIDs. Kernel channels as network applications are not removed from the WIDs several problems from slice selection to scalability can be identified. Help for network slicing and modularity, however, is a way of overcoming these problems.

After milestones, for each part of the network many problems persist. Current approaches are to be tested on the angle of WIDs, e.g., split alternatives, IAB processes, maintaining separation and flexible network architecture, data collection and analysis from end-to-end network, setting new KPIs and QoS criteria, and endorsing new business models. One more important factor is the assessment of current drone architecture and the design of modern drones as WIDs. However, the horizon for drones is a wired prospect.

KEYWORDS

- **5G core**
- **5G network management**
- **aerial relays**
- **exposure function**
- **mobile network**
- **network**

REFERENCES

1. *Mobile-Enabled Unmanned Aircraft-How Mobile Networks Can Support Unmanned Aircraft Operations.* GSMA, Tech. Rep.: 2018–04–02. [Online]. Available: https://www. gsma.com/iot/ mobile-enabled-unmanned-aircraft/ (accessed on 19 November 2021).
2. Bor-Yaliniz, I., & Yanikomeroglu, H., (2016). The new frontier in RAN heterogeneity: Multitier drone-cells. *IEEE Commun. Mag., vol. 54*(11), 48–55.
3. Guvenc, I., Koohifar, F., Singh, S., Sichitiu, M. L., & Matolak, D., (2018). Detection, tracking, and interdiction for amateur drones. *IEEE Commun. Mag., 56*(4), 75–81.
4. Bor-Yaliniz, I., El-Keyi, A., & Yanikomeroglu, H. (2019). Spatial configuration of agile wireless networks with drone-BSs and user-in-the-loop. *Trans. Wireless Commun.* To Appear.
5. Mozaffari, M., Saad, W., Bennis, M., & Debbah, M., (2017). Mobile unmanned aerial vehicles (UAVs) for energy-efficient internet of things communications. *IEEE Trans. Wireless Commun., 16*(11), 7574–7589.
6. Lyu, J., Zeng, Y., & Zhang, R., (2018). UAV-aided offloading for cellular hotspot. *IEEE Trans. Wireless Commun.* Early access.
7. Tomic, T., Schmid, K., Lutz, P., Domel, A., Kassecker, M., Mair, E., Grixa, I. L., Ruess, F., Suppa, M., & Burschka, D., (2012). Toward a fully autonomous UAV: Research platform for indoor and outdoor urban search and rescue. *IEEE Robot. Autom. Mag., 19*(3), 46–56.
8. Zhu, Y., Zheng, G., & Fitch, M., (2018). Secrecy rate analysis of UAV-enabled mmWave networks using matérn hardcore point processes. *IEEE J. Sel. Areas Commun.* Early Access.
9. Kalantari, E., Bor-Yaliniz, I., Yongacoglu, A., & Yanikomeroglu, H., (2017). User association and bandwidth allocation for terrestrial and aerial base stations with backhaul considerations. In: *IEEE 28th Annual International Symposium on Personal, Indoor, and Mobile Radio Communications (PIMRC)* (pp. 1–6).
10. Hayat, S., Yanmaz, E., & Muzaffar, R., (2016). Survey on unmanned aerial vehicle networks for civil applications: A communications viewpoint. *IEEE Commun. Surveys Tuts., 18*, 2624–2661.
11. Zhang, H., Vrzic, S., Senarath, G., Dào, N. D., Farmanbar, H., Rao, J., Peng, C., & Zhuang, H., (2015). 5G wireless network: MyNET and SONAC. *IEEE Network, 29*(4), 14–23.
12. Lien, S. Y., Shieh, S. L., Huang, Y., Su, B., Hsu, Y. L., & Wei, H. Y., (2017). 5G New radio: Waveform, frame structure, multiple access, and initial access. *IEEE Commun. Mag., 55*(6), 64–71.
13. "Overview on IAB. 3GPP RAN WG-3, Tech. Rep. R3–181998, April, 2018. [Online]. Available: https://portal.3gpp.org/ngppapp/TdocList.aspx?meetingId=18782 (accessed on 19 November 2021).
14. ETSI Zero Touch Network and Service Management, (2017). *Zero-Touch Network and Service Management-Operators' View on the Necessity of Automation in End-to-End Network and Service Management, and Operation.* Accessed [Online]. Available: https://portal.etsi.org/TBSiteMap/ZSM/OperatorWhitePaper (accessed on 19 November 2021).
15. "Documentation of requirements and KPIs and definition of suitable evaluation criteria". *5G Mobile Network Architecture,* Tech. Rep. Deliverable D6.1, Sept. 2017. [Online].

Available: https://5g-monarch.eu/wp-content/uploads/2017/10/5G-MoNArch_761445_ D6.1_Documentation_of_Requirements_and_KPIs_and_Definition_of_Suitable_ Evaluation_Criteria_v1.0.pdf (accessed on 19 November 2021).

16. 5G PPP Architecture Working Group, (2016). *"View on 5G Architecture"* (PDF).

17. De Looper, C., (2020). *What is 5G? The Next-Generation Network Explained.* Digital Trends.

18. Forest Interactive (2020). *Positive 5G Outlook Post COVID-19: What Does It Mean for Avid Gamers?* Retrieved November 13, 2020.

19. Hoffman, C., (2019). *What is 5G, and How Fast Will it Be".* How-To geek website. How-To Geek LLC.

20. Horwitz, J., (2019). *The Definitive Guide to 5G Low, Mid, and High band Speeds.* VentureBeat online magazine.

21. Davies, D., (2020). *Small Cells-Big in 5G.* Nokia.

22. Shatrughan, S., (2019). *Eight Reasons Why 5G Is Better Than 4G.* Altran.

23. Forum, C. L. X., (2019). *1 Million IoT Devices per Square Km-Are We Ready for the 5G Transformation?* Medium.

24. Segan, S., (2019). *What is 5G?* PC Magazine online. Ziff-Davis.

25. Rappaport, T. S., Sun, S., Mayzus, R., Zhao, H., Azar, Y., Wang, K., Wong, G. N., et al., (2013). Millimeter wave mobile communications for 5G cellular: It will work!. *IEEE Access, 1,* 335–349. doi: 10.1109/ACCESS.2013.2260813. ISSN 2169-3536.

26. Nordrum, A., & Clark, K., (2017). *Everything you Need to Know About 5G.* IEEE Spectrum magazine. Institute of Electrical and Electronic Engineers.

27. I Am Crazy About Massive MIMO, (2020). *Kitihara of Softbank Ordering 1,000's of Massive MIMO Bases.* wirelessone.news.

28. Kevin Shatzkamer (2019). *Not Just Another G: What Exactly is 5G? blog on Dell Technologies.*

29. Arm Ltd. (2020). Managing the Future of Cellular: What 5G Means for the Radio Access Network (RAN) available on https://www.arm.com/solutions/5g (accessed 06 December 2021).

30. Yu, H., Lee, H., & Jeon, H., (2017). What is 5G? Emerging 5G mobile services and Network Requirements. *Sustainability, 9*(10), 1848. doi: 10.3390/su9101848.

31. Aicha Evans (2018). *Intel Accelerates the Future with World's First Global 5G Modem.* Intel Newsroom.

32. Don Butler (2019). *"Ford: Self-Driving Cars "will be Fully Capable of Operating Without C-V2X".* Wireless one. news.

33. Gemalto EN (2018). What is the Difference Between 4G and 5G? available on https://justaskthales.com/en/difference-4g-5g/ (accessed 06 December 2021).

PART II

DRONE AUTOMATION SOLUTIONS FOR SECURITY AND SURVEILLANCE

CHAPTER 4

Security Issues in the Internet of Drones (IoDs)

NAMISHA BHASIN,[1] SANDHYA TARAR,[1] and KORHAN CENGIZ[2]

[1]*Gautam Buddha University, Greater Noida, Uttar Pradesh, India*

[2]*Trakya University, Edirne – 22030, Turkey*

ABSTRACT

Drones are small flying vehicle which have changed life in many ways by providing information transmission, surveillance of a place where it was difficult to travel earlier by humans, knowing about hazard and forwarding that information instantly to relevant stakeholders. These days people want to connect with each other without wasting any time which is challenging but has become a necessity. The medium used to connect with each other is through a network medium which can be easily hacked and exploited hence security is a major concern in this network. Except for a physical attack the attackers by exploiting kits can hack the system. To make the system secure various approaches are followed and many more are required.

4.1 INTRODUCTION

A small lightweight flying vehicle as shown in Figure 4.1, which can be used for any task either simple or complex which can be indoor packet delivery, military operations at dangerous locations, hazard detection, setting an ad-hoc network, monitoring of a location where otherwise tasks were not feasible or costly, operations at undermine or under sea and many more

The Internet of Drones: AI Applications for Smart Solutions. Arun Solanki, PhD, Sandhya Tarar, PhD, Simar Preet Singh, PhD & Akash Tayal, PhD (Eds.)

[1]. Sometimes, when the target is simple the operations are performed by a single drone and sometimes when target is complex a group of drones also known as internet of drones (IoDs) is used to achieve the required objective.

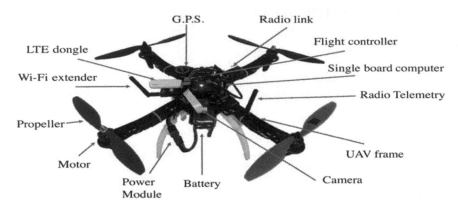

FIGURE 4.1   Different parts of a UAV [76].

To achieve the target these drones are well connected either through Blue-tooth, Wi-Fi, Zigbee, control area network (CAN), radio waves, automatic dependent surveillance-broadcast (ADS-B) or by cloud-based architecture to manage the drones through the Internet by using micro air vehicle link (MAV Link) protocol as shown in Figure 4.2 [2, 3, 5].

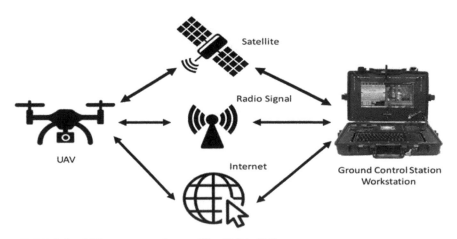

FIGURE 4.2   Different ways of controlling UAVs [76].

IoDs should be well coordinated to achieve the target in optimized manner. IoDs are connected to the control room for further processing of the information and are able to perform the required task with the help of cameras, sensors, actuators, communication module, processing, and storage capabilities. Various types of drones exist as in Table 4.1.

TABLE 4.1 Types of Drones

Drone Type	Pros	Cons	Application
Fixed-wing	Performed well for long distance operations	Cannot be used for still operation like aerial photography	Surveying an area, structural inspection
Fixed wing hybrid VTOL	Work in automated mode	Vertical lift is used to lift the drones	Amazon is using drones for prime delivery service
Single rotor helicopter	They have higher flying times	Need special training to fly them on air to avoid any accident	Single rotor drones are much efficient than multi-rotor drones
Multi-rotor drones	Manufacturing is easy and they are cheapest	Not suitable for long distance aerial mappings	Aerial photography, aerial video surveillance

4.1.1 ARCHITECTURE

Layer protocols model for IoDs is shown in Figure 4.3, where the first layer airspace layer provides the facility of: (i) map-used to know about other nodes and intersections; (ii) airspace broadcast and track used to broadcast vehicle location and future path to avoid any collision; (iii) plan trajectory used to have a path to be followed so that proper coordination among IoDs is maintained; (iv) airspace precise control used to provide an environment for emergency which can be landing, instant path change or holding the position for some time; (v) collision avoidance used to avoid collision among drones and with bird; (vi) as weather condition also has an important role specially in case of IoDs therefore it cannot be ignored for future path planning;

Second layer is Node-to-Node layer which provides the facility of: (i) zone graph-provides an overall information like drone's current position to other drone's position and current and future path as the airspace is divided into zones and further into sectors. Zones can be managed by one or multiple service provider as shown in Figure 4.4; (ii) N2N broadcast and track-drones broadcast information about themselves including fuel time left; (iii) plan pathway and contingency-a partial or full path transection is broadcasted by drones; (iv) refuel-path to be followed for fueling; (v) N2N precise control-for holding, landing or change of path for drones; (vi) as congestion notification plays an important part for any application, the whole system is enabled with predefined routes which make it conflict-free, but in Emergency situations drones need to avoid congested zones hence this notification plays an important role; Third layer End-to-End layer deals with: (i) interzone graph-as architecture is divided into various zones hence information that coordinating drone is in which zone is vital; (ii) routing-plays a significant role to select the safe in addition to the shortest path; (iii) handoff-as drones move from one zone to another hence, how drones will be allotted that space area is handled in this layer;, and forth layer service layer deals with: (iv) zone broadcast-used in case if any information need to be broadcast to drones; fifth and last layer is Application layer-kept for future applications so that new applications can become a part easily [1, 2].

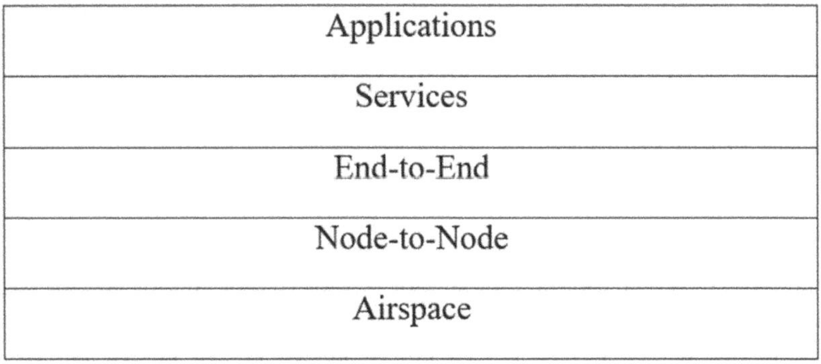

Applications
Services
End-to-End
Node-to-Node
Airspace

FIGURE 4.3   Layer protocol model for IoD [1].

Source: Reprinted from Ref. [1]. Copyright © 2016, IEEE. Open access.

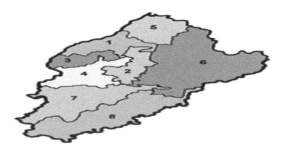

FIGURE 4.4   Zone-wise architecture.

Source: Adapted from Ref. [1].

Management of drones using Internet and Cloud have several benefits like: (1) virtualize access to UAs resources where virtual UAs are mapped to actual UAs because of abstract interface provided by cloud infrastructure; (2) offload computations to the cloud from the UAVs as they have small processing and storage capabilities as shown in Figure 4.5 [2].

FIGURE 4.5   Cloud-based architecture.

IoDs can play various roles but for Home infrastructure it can be used for security and entertainment. Here IoDs are connected to the environment through Gateways. They also play an important role for community by being utilized in factory and smart meter. IoDs are also helpful in transport by taking care of parking and emergency services. In utility and infrastructure also IoDs have an important role as shown in Figure 4.6.

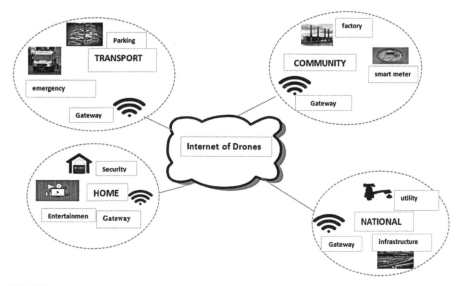

FIGURE 4.6   Architecture of generic IoD network.

Several difficulties are faced for Video analysis in real time by small autonomous drones like wireless bandwidth which is lossy in nature; small processing capacity hence for complex computation need help from other processing units; energy consumption as drones is battery operated and small in size hence need refueling very frequently; result accuracy, and timeliness of results. But, storing data at cloud for analysis is time consuming which needs improvement and with the help of edge computing target can be achieved [3, 4]. Here data is collected and processed locally as data is received from different scenarios; according to the requirements it can use the facilities made available from both private and public Cloud platforms respectively. The aim for all the cases is to guarantee a high flexibility-hence, whatever facility is available can be used; robustness and adaptability-can be available for any platform; level of service. While in hierarchical based architecture drones form a cluster and supervised by a head drone known as cluster head as shown in Figure 4.7.

FIGURE 4.7   Hierarchical-based architecture.

For UAV surveillance system as shown in Figure 4.8, a drone is used to catch a scene of a place where it is difficult to visit for humans. If operated manually the user is responsible for the movement of drone, after that it collects the signals from that location and then call closest ground control station (GCS) through Internet to surveillance center (SC) using the private network. Finally, data is utilized by GC to analyze the data so that proper action can be taken [4].

FIGURE 4.8   UAV connecting to GCS [4].

Source: Adapted from Ref. [4].

4.1.2   PROBLEM DESCRIPTION

As can be seen from the above discussion that IoDs can play a significant role in coming future and no area will remain untouched by having a drone as integral part of the architecture [3]. While looking at the architectures proposed so for had discussed how drones navigate and communicate in optimized manner but nothing is proposed to make the system secure. As IoDs are connected to each other or to the system through lossy links hence, there is always a necessity of identification of the drone where two similar drones are there; localization when current location of the drone plays an important role, securing either from physical or cyber-attack, and providing IP addresses to the growing number of devices [8]. The main purpose of the system should be able to facilitate the user with the services that must be reliable (confidentiality), authenticate, integrated, and accepted by everyone (non-repudiation). An attacker can disturb the services by denial of service (DoS), replay, Stolen/Breach, guessing the password and identity off-line, user, and node impersonate, user traceability, attack by privileged insider and stolen-verifier, Log-in by many users with the same login-id, password change by unauthorized ways. Applying conventional security techniques may enhance the security but may restrict availability of the network as the overheads caused by such mechanisms on these devices may cause a delay in communication and computation time. Single-channel communication may be unpredictable in terms of reliability specially in multi-hop usage scenarios. Similarly, the source privacy of location is one of the privacy challenges as attacker can come to know about the location of victim by observing the pattern of communication despite user being efficient with encryption tool. As in Multicast communication a single packet is sent instead of sending one instruction message for each device in the group where an attacker can be a part and can compromised the whole system. As one of major issue in IoD is the security for many devices. Recent cyber-attacks have highlighted the shortcomings of many IoD devices. Many of these device manufacturers simply wanted to be the first in a niche market, ignoring the importance of security. Proper security implementation in IoD has only been done by a minority of designers and manufacturers. Numerous security techniques are discussed and shown to properly protect the data that will pass through many of these devices. The overall goal is to have an overall security solution that overcomes the current shortfalls of IoD devices, lessening the concern for IoD's future use in our everyday lives. Challenges based on their impact to IoD security are: 1. Standardization 2. Trust and Authentication 3. Privacy 4.

Information Security 5. Network Attacks 6. Latency 7. Wireless Communications 8. Version Control and Updates 9. Physical Attacks 10. Size, Weight, and Power.

4.2 CONFIDENTIALITY, INTEGRITY, AUTHENTICATION, AND AVAILABILITY ATTACK

As usage of Drones has increased so they can be seen anywhere and anytime doing some useful work. At the same time chances of cyber-attack has also increased. Drones use frequency of 2.5 or 5 GHz for communication [69]. Drones are composed of Flight controller, GCS, Data Links-used for communication between UAV and GCS having either by direct radio waves or through satellite [70]. Drones' communication can be like drone-to-drone (D2D) communication which are vulnerable to sybil, jamming, and DDOS attacks, drone to network communication which uses any network for communication which make it prone to any attack, drone to satellite communication which uses Global positioning system (GPS) which is very secure or drone to ground station which uses Bluetooth or Wi-Fi connection for communication which make it vulnerable to eavesdropping or Man-in-the-middle attack [71–74].

Crashes among UAV can be avoided with the help of radio frequency identification (RFID) and RF. To avoid collision among UAVs collision avoidance algorithms have been developed which can be applicable both in 2D and 3D environment [75, 76]. In 3D environment three constraints are followed: temporal, physical, and geometric. Obstacle collision algorithms are used by UAVs to identify objects in the vicinity [77]. Drones can be a victim of malware and backdoor attacks. Drones security concern is different from mobile ad-hoc network (MANET), where a network is setup with the help of mobile devices and there is no requirement of center agency, and WSNs, where a set of sensors are used to form a network, as UAVs cover boarder area and topology changes speedily. Hence, a network which is delay tolerant and capable of storing and forwarding is required. Drones can be used to access Wi-Fi or Bluetooth enabled devices.

The system must be trust-worthy and being efficient with encryption tool which is the basic requirement, but still the system suffers from various attacks as shown in Figure 4.9 [37]. Data sent from sender instead of reaching to the proper person is received by someone else which is an attack on the availability and message is read by the third party. It is an attack on confidentiality then when third party managed to change the message. When attack is on the integrity the attacker successfully sends data to the destination by exploiting the identity of others. Fabrication then there is attack on authenticity. Most of the time IoDs work on real-time based applications like fire hazard detection, military operations, etc. [3]. Hence, there is always a great need of a secure and efficient system which follows drone authentication approach in which an authorized drone in the IoD environment only can approach another authorized drone. Hence, any new deployed drones must authenticate themselves to access the system. Sometimes, drones are inaccessible because of many reasons like weather, fuel, etc., but when they become approachable, they must be in a trusted state and system should be enough self-protected that adversary should not be able to modify the primary code. Machine learning (ML) plays a significant role to make the system secure. By using this technique system is able to detect the unusual behavior of the system with better reliability and with less efforts that too in less amount of time.

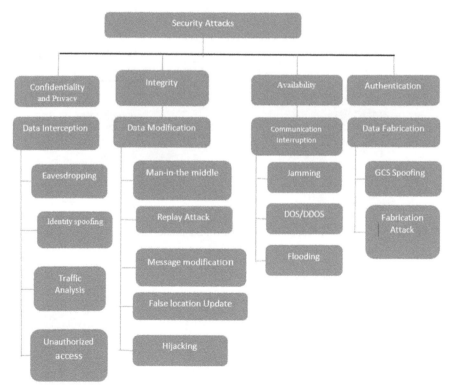

FIGURE 4.9 Security threats and attacks.

Source: Modified from Ref. [103]. https://creativecommons.org/licenses/by/4.0/

4.2.1   DATA INTERCEPTION

As system operates using various transmission technologies like ZigBee, CAN, Bluetooth, and Z-Waves, etc., which make the system lossy hence, attackers can easily exploit the system as shown in Figures 4.10 and 4.11. In that direction frequency at which drone is functioning plays an important role. Similarly, other threats are jamming and replay of message. The various solutions provided are: (i) directional radio beam; (ii) use range of frequencies instead of one frequency; (iii) use of light weight encryption algorithms; (iv) use of one-time password; (v) recognizing jammers as they use strong signals hence can be traced easily [10]. To decode the signals GNU Radio tool is used and following parameters are considered as shown below in Table 4.2.

TABLE 4.2   Features for Data Decoding Using Rf Frequency Caught by GNU Radio Tool

Frequency
Sampling Rate
File Name
Waterfall graph

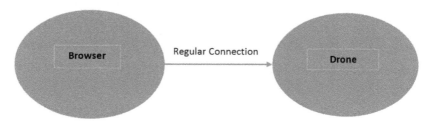

FIGURE 4.10   Regular connection to connect with the drone.

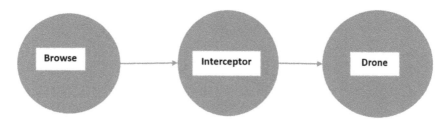

FIGURE 4.11   Interceptor listen to the data communicating between browser and drone.

In that direction the next step used for secure transmission is by either using secret keys or certificates issued from the trusted third-party authority for transmission with its neighbor drones lightweight authentication scheme for IoD deployment [11, 45, 46], was proposed using fuzzy extractor method, hash functions and bitwise XOR operations for the verification. Message Validation was accomplished by the use of message authentication and cyclic redundancy code (CRCs) where predefined functions are used and message value is divided by these code value if remainder is zero then message received is same as was send, or encryption methods such as AES Galois also known as Counter Mode where a block of 128 bits are used for validation. Device diagnosis can be performed with event log which contain all the information of the events which can be about failed or successful boot-up transaction, how much data transmission occurs, and other important information's can also be checked [18].

4.2.2 MALICIOUS DATA INJECTION

To induce wrong information an attacker can compromise a node by many ways like through its physical interfaces, by tampering the node hardware or by manipulating the sensed environment [19]. All this kind of attacks are malicious data injections with an aim to exploit the system. Sometimes, bad data detector (BDD) is unable to find malicious data as shown in Figures 4.12 and 4.13. By applying supervised algorithm, i.e., support vector machine (SVM) was used to categorize label data between malicious and no-malicious and also for unsupervised technique as shown in Figure 4.13, used Gaussian Probability density function with a threshold value after applying principal component analysis (PCA) which is used for reducing dimension [41].

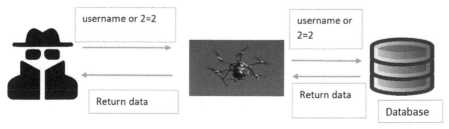

FIGURE 4.12 Malicious data injection.

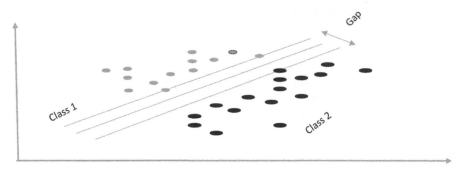

FIGURE 4.13   Support vector machine classification for malicious and non-malicious data.

4.2.3   DIGITAL SIGNATURES

Digital signature as shown in Figure 4.14, used to verify the authority of a user, and composed of two phases: (i) the signature generation; and (ii) verification algorithms [23, 27]. It performs singing using private key and verification by public key. Any illegal modifications can be checked with the help of hash values. They can be validated either by: (i) users only and no third party is involved; or by (ii) third party which is also known as arbiter to validate the signatures [33].

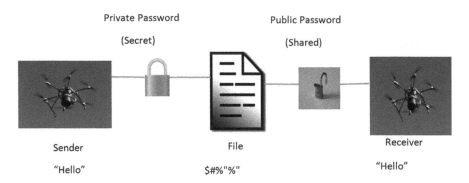

FIGURE 4.14   Digital signature.

4.2.4   DIGITAL SIGNATURE ALGORITHM (DSA)

Digital signature algorithm (DSA) composed of following steps: (i) domain parameter generation where two prime numbers are selected; (ii) key pair

Generation where sender public and private key is calculated; (iii) signatures are generated to sign a message using secure hash algorithm with the help of random number; (iv) signatures are verified [24]. But the solution is complex one and consumes a lot of memory which is replaced with "*is*DSA" algorithm [24, 33]. It removed the complex modular inverse operation and change the way calculation is performed in signature and verification. One more light-weight solution proposed for smart devices was shortened complex digital signature algorithms (SCDSA) where for signature and verification process it used small complex numbers instead of real numbers in addition to secure hash functions [33].

4.2.5 CRYPTOGRAPHIC

It is a two-step process followed for secure transmission of a message: (i) encryption where normal text is converted to obfuscate form by using a key; (ii) decryption where obfuscate text is converted to normal text in the readable [30] form only by authorized user [26–29]. Two categories of cryptographic are: (a) symmetric-key cryptographic where sender and receiver both use single key/secret key to encode and decode a message; (b) asymmetric-key cryptographic where message is encrypted with public key and for decryption private key is used. Following are the ways of encryption and decryption:

1. **Electronic Code Book (ECB):** In this from the whole data blocks of 64 bits are created which are encrypted independently and separately one at a time [31]. Hence, there is no influence of one on other during encryption process as shown in Eqn. (1).

$$C_i = F(B_i, K_i) \tag{1}$$

   ➢ **State-of-the-Art:** The disadvantage of the algorithm is it not being able to diffuse the data properly as can be seen in Figures 4.15 and 4.16 that similar data changes into similar encrypted form which make it easy to extract the data.

2. **CBC (Cipher Block Chaining):** In this each block of 64 bits is XORed with the next plaintext block, which makes all the blocks dependent on all the previous blocks [35]. For the first block no previous ciphertext exists hence, XORed with Initialization Vector, or IV for short shown in Eqn. (2).

$$\{C_0 = F(B_0 \oplus IV, K_0) \; C_i + 1 = F(B_i + 1 \oplus C_i, K_i)\} \tag{2}$$

FIGURE 4.15   Original image data [32].

Source: Reprinted with permission from Ref. [32]. lewing@isc.tamu.edu and The GIMP.

FIGURE 4.16   Image data after encryption [32].

Source: Reprinted with permission from Ref. [32]. lewing@isc.tamu.edu and The GIMP.

➢ **State-of-the-Art:** The data can be retrieved after an attack as
this algorithm also suffered from diffusion problem [33]. All
this review represents the need of lightweight and more secure
cryptographic algorithms. The symmetry key algorithm named
secure IoT (SIT) uses hybrid approach which is based on Feistel
and Substitution-Permutation networks [34]. In this to make the
message confuse and diffuse 64-bits key is used with mathematical function which is taken from the user which also act as input
for the key expansion block to generate unique keys-k1, k2, k3,
k4, and where k5 is generated by XOR of previous keys. These
keys are used for encryption/decryption process. Confusion and
diffusion themselves are combination of logical operations like

left shifting where bits are shifted to left; swapping used to swap the values and substitution where are values are substitute with other values. A lightweight identity-based cryptography method where message is encrypted by appending readings from the sensors with day and time of doctors visit at the hospital with a limitation of secret key is generated by a third party [33].

4.2.6 MEMORY PROTECTION UNIT

As various attacks exist like file less malware where virus can remain at RAM without downloading any file hence protection of memory is an essential part which can be achieved with the help of access permission [41]. Because of simple object access protocol (SOAP) sometimes these types of attacks are successful. Hence, different memory units follow different rules for accessibility and can be helpful in mitigating buffer overflow attack.

4.2.7 MAN-IN-THE MIDDLE ATTACKS/EAVESDROPPING/REPLAY ATTACKS

The attack is possible because of some flaws in the protocols [37–40]. As when new drone comes into system then it tries to connect with the system using hello message but at that time this message is answered by malicious drone which represents itself as cluster head and to cluster head as the new joining node. In this way the malicious drone able to secretly listen the communication between the nodes. When type of set-up is ad-hoc it worsens the situation and malicious node can easily connect to the system. In this attack the attackers can secretly relays, alters, or delay the message while users believe that they are having a direct communication with each other but in realty this is not happening [48]. Instead, data transmission was maliciously or fraudulently repeated-previous packet is resented instead of new one; or delayed-so that information cannot reach at right time. In the process of data communications, although data is encrypted, there is still a possibility that it can be known to outsider. This technique is called man-in-the-middle-attack or eavesdropping or replay attack [49]. Till now various explanations have been proposed. One of them was Interlock protocol. In this protocol interlock method is used where users exchange their public keys which is used by the user A to encrypt the message using the key received from the other user B and send a part of the message to the user B. The user B also follows the same steps with key received from the user A. In next step, user A sends another part of the message encrypted with the key then user B combines both the parts. A new dynamic IoD security system which is enabled with sensor for collecting data and actuator used for taking action, exchange the data via a secured MQTT protocol in a fog network then encrypts the MQTT payloads with ECC and adds a timestamp to the payloads to avoid data tampering and eavesdropping and uses wake-up patterns to make Replay attacks inactive [42]. Also, depending upon the residual energy capacity key-strength of the used ECC keep changing dynamically and also, wake-up pattern is helpful in reducing the received replicated packets. In one approach RC5 algorithm is used which uses only XOR and shift operation to secure the link [43]. Two more ciphers used in lightweight cryptographic operations are: PRESENT and CLEFIA. PRESENT uses a key length of either 80 or 128 bits to encrypt 64-bit data. CLEFIA is a 128-bit Feistel structure encryption algorithm. It provides more security against attacks by employing a diffusion procedure to lessen the number of rounds which is either by using differential or linear method. Alike to AES it can have a key of length either 128, 192, or 256 bits. It is a good example of balance between speed, area, and security. ARP can be used for cracking the encryption keys.

➢ **ARP Cache Poison:** used to change the network traffic and for these a malicious script "Scapy" is executed to disconnect the drone from the network [94].
➢ **Drone Detection:** there are various ways to detect the drone as follows:
  • by the sound produced by motor of drone;
  • by taking their pictures;
  • through their speed;
  • by the gasses they produce;
  • detection by radar;
  • by RF signals they transmit while communicating with the Ground station.

4.2.8 SPOOFING

By using ML techniques to find spoofing by collecting samples from mobile radio channel then pass these samples through low pass filters and results improved by 8% of the previous techniques [20]. A supervised ML method which uses five features namely: (i) satellite number; (ii) carrier phase; (iii) pseudo range (PR); (iv) doppler shifts (DO); and (v) SNR and algorithm artificial neural networks (ANN) to diagnose GPS signals and a decision is made regarding presence or absence of the attack with an accuracy of 98% [21].

4.2.9 TRAFFIC ANALYSIS/DRONE DETECTION

Analyzing the traffic to identify the device is performed by various ways, the device identification is done within a network by analyzing and classifying network traffic data but there is no standard for identification of brands or types of devices [35]. The following features are used for the identification of the device as shown in Table 4.3.

TABLE 4.3 Features Used in Traffic Analysis for Traffic Identification

Source IP
Destination IP
Source port number
Destination port number

Various factors like drone detection, Drone state identification, Detection delay, Packet Loss. For (1) drone detection in heterogenous environment following features were used as shown in Table 4.4; (2) drone state which can be flying, not flying, i.e., lying on the ground is checked by eavesdropping the Wi-Fi spectrum and collected data packets; (3) Detection delay: it was found that number of samples and the detection delay are linear proportion to each other and detection rate is 99.68% at a distance of 200 m; (4) Packet Loss: as distance increases packet loss increases, i.e., at a distance of only 200 m packet loss percentage is 73.8%. The following features were considered for the analysis [36] as shown in Table 4.4.

TABLE 4.4 Features Used for Traffic Analysis

Interarrival time
Packet size
Mean interarrival time
Standard deviation interarrival time
Mean packet size
Standard deviation packet size

4.2.10 UNAUTHORIZED ACCESS

to the drone data: can be detected using inverse weight clustering and C4.5 decision tree [35]. Inverse weight clustering: It is an improved form of k-means clustering algorithm which executes till get a centroid which make the cluster stable. C4.5 decision tree used to classify the data which uses training data of already classified data and then testing is performed on the remaining data. When pen test was conducted on parrot bebop UAV then it was found that they are vulnerable to address resolution protocol, which is used to know IP address of a device, and cache poisoning attacks [91]. The real-time data is done using recursive least squares method (RLSM) that is able to detect cyber-attacks. For fault detection and isolation neural network is performed on real-time data. Decision tree classifies the given data features through various branches. The split can be for categorical or continuous values [69].

4.2.11 CHANNEL JAMMING

Jamming as shown in Figure 4.17, is a type of noise which surpasses the frequency used by genuine user so that the users face difficulty in accessing the system [53]. There are several purposes except malicious activities for jamming the target's radio communications [54]. To find the Jammer location various techniques are proposed for calculating the position of a radio receiver. One solution proposed is by calculating radio signals propagation strength because the strength of noise used by attacker is high. In another method two factors: (i) the value of received signal strength (RSS) received by the receiver; and (ii) use of trilateration technique to infer the distance to the transmitter of the jammer. For this radio propagation model and RSS indicator is used [55]. In another method [56] time-of-arrival (ToA) when packet was received; angle-of-arrival (AoA) used to know the direction of received signal; and the RSS-the power strength of signal are considered to determine the distances between the jammer and the nodes and position of the jammer in the IoD. Different techniques and equipment can be used to prevent the drone from communicating with the remote controller. In various experiments it was found that if antenna used was not omnidirectional then its signals can be capture by attacker easily. Hence, by using focused antenna this problem can be removed [56]. For flying ad-hoc network (FANET) to get rid from jamming multipath routing protocol known as Jarm Rout, was proposed. The Jarm Rout uses three schemes: (i) link quality scheme: where link qualities used by a node are different from its neighbor nodes; (ii) for calculation of traffic load current traffic is required which need channel contention information-where data can be send after sensing the channel; and the number of packets stored in the buffer; (iii) The spatial distance scheme used to compute the spatial distance of multiple paths to calculate the optimized path between source and destination nodes. In another approach a tool JAM-ME used to fulfill its mission even if anti-drone jamming protection system is there but with an average overhead of 70% [57, 58].

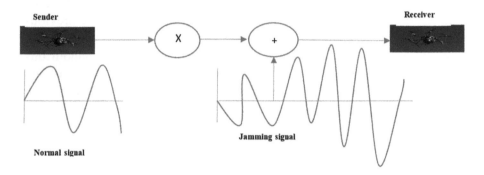

FIGURE 4.17   Jammer used for the drone.

Source: Adapted from Ref. [77].

4.2.12   MAVLINK

MAVLink is a header and use message-marshaling library as communication protocol. In order to make the system secure for sending information encryption is applied at MAVLink protocol. MAVLink suffered from DoS, eavesdropping, message forgery and hijacking of UAV issues. They are discussed as in further sections.

4.2.13   DENIAL OF SERVICE (DOS) ATTACKS/DISTRIBUTED DENIAL OF SERVICE (DDOS)

Adversaries main aim as shown in Figure 4.18, is blocking the path of communication so that a genuine user has no access to the services which can last for minutes or even days [9]. When a malicious node blocks the bandwidth by sending the packets which is more than capacity of the channel then congestion occurs. Similarly, when more than one malicious node is involved in overusing the resources called distributed denial of services (DDOS) and in order to remove it packets of data are dropped from the system which leads to decrease in throughput and causes damage to the victim. In this attack most of the time same and simple query is send again and again to flood the system.

DDOS attacking power has increased enormously with IoDs [8]. As drones are connected to Internet which make them vulnerable to DOS/DDOS attack. These attacks are difficult to prevent because attackers replay encrypted data to gain access of the victim's system. Hence, system needs to be sanitized from these faults and if attacker tries to encrypt the previous message it should not produce the same ciphertext. For this problem the solution proposed uses XORing the encrypted plaintext with random bits or number of messages [3]. This number is a part of MAVLink header and sender and receiver have a count of the packets used in communication. Most of the time the data packet received is classified as malicious or not by classification algorithms. For this various approach can be followed like: (i) statistical; (ii) machine learning; and (iii) neural network. Most important thing is number and type of features. As in the approach the features considered to find DDOS attack [10, 11] as shown in Table 4.5 are used.

TABLE 4.5   Features Used to Find DDOS Attack

Source IP
Destination IP
Transport layer protocol
Source Port
Destination Port

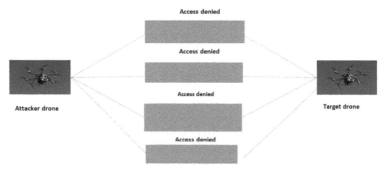

FIGURE 4.18   DoS attack.

Various ways are used to find an attack:

1. **Static Approach of Classification:** In this approach authors use ports numbers which were used earlier in attacks as their feature to classify that packet received is malicious or not. Hence, Port-based classification is performed to classify the data [12]. If their signatures are identical then they are classified as malicious. But this approach is not durable as many applications use dynamic port numbers.

2. **Device Identity Approach:** Then authors use device addresses as a feature to find either received packet is malicious or not. The devices which were used earlier in attack is taken as a criterion to judge either the packet is malicious or not. The techniques utilize to determine the identity of a device to classify either the device is good or bad but this technique is very complicated and also take a lot of classification time [13].

3. **Payload based Inspection:** After this the classification of a packet is malicious or not performed on the basis of payload [12]. In this payload inspection techniques where used to check for malicious content. But obfuscation and encryption techniques are used which make it difficult to classify the packets as malicious or not.

4. **Behavior based Analysis:** In this technique data packet is kept under observation and its behavior is monitored to classify either the packet received is malicious or not.

5. **Rule-based Intrusion Detection:** For finding false data injection attacks rule-based concept is used to provide secure communication between drone and ground station [81]. Specification based detection technique is used to detect random, reckless, and opportunistic attacks with minimum error containing false positive, and false negative cases. An adaptive behavior-rule specification-based IDS, BRUIDS, was used to detect malicious drone provides higher detection results [82]. Rules need to be created to handle the attacks but they are unable to provide any solution for unknown attack.

4.2.14 ANOMALY DETECTION

In this a system with normal behavior is build and any deviation from that pattern comes under anomaly [14]. Hence, able to remove limitation of rule-based intrusion detection mechanism which is not able to detect unknown attacks. It used to prevent UAVs form jamming and DDoS attack. To prevent drone motor from overheating reinforcement learning method is applied to temperature which is calculated using DS18B20 sensors and processed by raspberry-pi CPU [83, 84]. They behave differently than normal traffic and also helpful in detecting attacks which were previously unknown. Despite this, sometimes they have misclassified non-malicious traffic (false positives). An anomaly can be categorized as: (i) point anomaly: when a datapoint deviates from its normal pattern; (ii) contextual anomaly: when datapoint behaves differently in a particular context; (iii) collective anomaly: when a collection of dataset has different pattern with respect to entire dataset. Researchers use different datasets to perform their research. Some perform real-time experiments, simulation-based experiments or in-build datasets as used by in research work Dataset available at the center for applied internet data analysis (CAIDA) or DDoS Attack 2007 dataset and also at ISCX 2012 DDoS dataset used for their research [15, 16]. The researcher used following dataset to check either received packet is by DDOS attack or not. These features describe about: (i) the quantity of information sent in one direction; and (ii) the duration of a connection. These properties were used to illustrate the standard behavior of a host and deviations from this can be an indication of inconsistent behavior, such as a DDoS attack as shown in Table 4.6 [14].

TABLE 4.6   Features Used for DDOS Attack

Feature	Purpose
totalSourceBytes	Bytes send by source
log(totalSourceBytes)	Base 10 logarithm value of totalSourceBytes
totalDestinationBytes	Bytes sent by destination
totalSourcePackets	Number of packets send by source
FlowDuration	Time of bidirectional communication in seconds
log(FlowDuration)	Base 10 logarithm value of flow duration
SourceBytes-per-Packet	Number of bytes divide by number of packets send by source
DestinationBytes-per-Packet	Number of bytes divide by number of packets send by destination

Source: Adapted from Ref. [14].

4.2.15   MESSAGE FORGERY

As data is transmitted among drones and they are enabled with small storage and processing capabilities so they can suffer from message forgery attack. The real purpose of sending a forge message is to exploit the drone as message enclose virus. Various tools like AUTOFORGE are used to exploit the system and these tools are cheap and easily available which can automatically forge security vulnerabilities such as brute-forcing-doing forcefully; leaked username and password probing-while probing get the login details because of flaws in the system; and access token hijacking-where attacker steal the token of the session to steal the important information [52]. And results proved that vulnerability because of password brute forcing attacks was 86% of servers, leaked username and password probing attacks was 100%, and vulnerability to Facebook access token hijacking attack was 12%. To keep message integrity SNAIL stack, deceive was used ECC algorithm for message encryption and for hashing MD5 after encapsulation of the data [53].

4.2.16 PRIVACY

As UAVs can be used by criminals for tracing which can be noticing the location of the target or taking pictures which can be used for blackmailing. The threat can be of type: (i) physical where picture of video of target is taken; (ii) location where location of the target is noticed; (iii) behavior change because target is under surveillance; (iv) identity privacy which requires true identity of drone is kept secret but in case of any dispute it can be authenticate by the authorities [78, 79]. By the use of pseudonym certificates secure message authentication environment is provided but with high computation cost which cannot be possible in case of drone's network as they have limited resources [5]. Adversaries can locate a device with the help of few monitoring devices and they just required low power range of listening and simple traffic analysis techniques which can leads to severe location privacy breach [6]. Even architecture proposed leads to location privacy breach because drones have to broadcast their geographical address to ZSPs in order to avoid congestion [7]. There is always a requirement to check which type of data is coming from outside to inside but there is equally requirement to check the type of data is flowing from inside to outside as adversary can also be a part of the internet.

4.2.17 SOFTWARE DEFINED NETWORK (SDN)

One of the current methods of network management here programming is involved with network management layer which makes the system dynamic and also helpful in determining the nature of network data. These types of attacks are very easy and cheap as they are having special kits for the attack [18, 19].

4.2.18   IDENTITY-BASED ATTACK

In the identity spoofing, the malicious drone get access of IoD network by using spoofing Id. Now, attacker gain all the rights to do listen what is going on in the network; can participate in voting and many more. Lack of encryption makes the communication vulnerable to interception and spoofing attack. Here signals can be spoofed which lead to false information regarding position of UAVs. When spoofer is nearby then it is called *proximity spoofing attack otherwise known as distant spoofing attack.*

4.2.19   FALSE INFORMATION DISSEMINATION/FABRICATION ATTACK

Where a malicious drone spread false information, which can create problem for genuine drones [85]. For Physical layer security interactive algorithm based upon block coordinate decent and convex optimization methods is performed [86].

4.2.20   SPOOFING ATTACKS/SYBIL ATTACK

The attacker drone as shown in Figure 4.19, can imitate its identity to launch any attack like denial-of-service (DoS) attacks or send wrong data. To compromise the IoD they can even broadcast services falsely to other nodes in vicinity or announce false routing and control information. GPS plays an important role in several operations of UAV [17]. But GPS system is not secure as unencrypted signals are used in communication which make it vulnerable to several attacks. In GPS spoofing attack attackers transmit bad signals with high power to the victim and also can store the satellite transmitted signal at it is of any autonomous system [80]. Features used in the classification are shown in Table 4.7 and various solutions proposed as shown in Table 4.8.

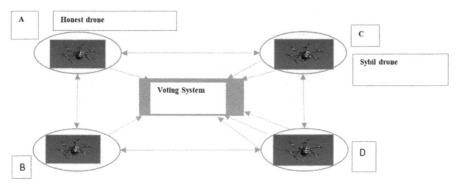

FIGURE 4.19   Sybil attack.

Source: Adapted from Ref. [24].

TABLE 4.7   Features Used for Spoofing Attack

Satellite vehicle number (SVN)	Satellite unique number
Signal-to-noise ratio (SNR)	Determines strength of satellite signals.
Pseudo range (PR)	How much far away is satellite from the receiver
Doppler shift (DO)	It is (the received GPS signal frequency-the reference signal)
Carrier phase shift (CP)	Used to synchronize the transmitter and receiver clocks.

Source: Adapted from Ref. [21].

Generative adversarial network (GAN) as shown in Figure 4.20, is used to differentiate between spoofing attack in a simulation environment and success rate is 76.2% [22]. Using GAN system is trained to fool the attacker defense mechanism such as RF finger printing and from the defender point of view, defense mechanism is trained against potential spoofing attacks when an attacker pair of transmitter and receiver cooperates [43].

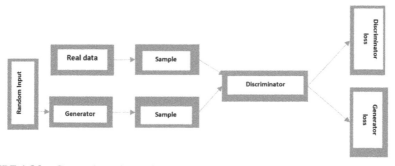

FIGURE 4.20   Generative adversarial network (GAN).

Source: Adapted from Google, Generative Adversarial Networks.

TABLE 4.8 Various Solutions Proposed for Sybil Attack

Types of Sybil Attacks	Defense Methods
Routing	Graph-based detection methods-use the information gathered from social network to find the attacker drone
Distributed storage	Machine learning (ML) techniques-to find the sybil drone uses supervised, unsupervised, and semi-supervised algorithms
Data aggregation	ML techniques-to find the sybil drone uses supervised, unsupervised, and semi-supervised algorithms
Voting/reputation systems	Graph-based techniques-use the information gathered from social network to find the attacker drone
Resource allocation	Prevention schemes and graph-based detection methods-use of Captcha or one-time password
Misbehavior detection	Graph-based detection and manual verification

Source: Adapted from Ref. [24].

4.2.21 THREE-WAY HANDSHAKE ATTACK

the attack can be used to crack the password while handshake occur between new device and access point. Hence, attackers can jam or de-authenticate the communication [92].

4.2.22 WI-FI AIR CRACK ATTACK

Wi-Fi attack enable the attacker to hijack it. The attacker gains full control over the device. The attack take place in three steps: (i) sniffing tool such as "airodump" is used to find wireless network; (ii) injection tool named "airplay" used to increase the traffic; (iii) "aircrack" used to disturb the network [93].

4.2.23 FORENSIC-BY-DESIGN

As currently available digital forensic (DF) tools and techniques are unable to comply with the diversification and distributed quality of the IoDs infrastructures hence, digital forensic investigations (DFIs) has not been able to fully adapt to DF techniques due to this fact [44]. As a result, gathering, examining, and analyzing probable evidence from IoDs environments that may be used as admissible evidence poses a challenge to DF investigators. The process can be: (i) proactive which includes planning and preparing before potential security IoT incidents can happen. So, for this to occur source identification, plan for incident identification, digital evidence collection and storage are performed; (ii) storage to network to device forensic is consider; (iii) reactive which occurs after the incident hence acquisition to investigation steps are performed. The technique is usually used in IoDs. Where a framework is used for the examination of network traffic and anomalies are detected in the traffic. Digital investigation process uses multi-tier hierarchy-based approach. First tier is used for data collection, analysis, and present the findings. Second tier is an object-based sub phase. In event-based digital forensic investigation evidences are collected by going through different phases like: (i) the readiness and deployment; (ii) physical crime scene investigation; (iii) digital crime scene investigation; and (iv) presentation. During preparation phase where conventional and digital evidences are combined for investigation used to identify the device. In second phase data retrieval is performed by extracting data cards. In third phase analysis of data is performed and data is presented in human readable form. Waterfall model goes through different phases like: (i) preparation and identification; (ii) weight measurement and customization check; (iii) fingerprints; (iv) memory card; (v) geo-location; and (vi) Wi-Fi and Bluetooth [87–90].

4.2.24 SECURE DATA AGGREGATION

As low-cost drones have inadequate capabilities of sensing, computation, and communication hence it is significant that the data used in communication can be reduced to minimize bandwidth utilization. Data aggregation is the process of briefing and combining data gathered by drones. This is required so that the amount of data broadcast in the network can be minimized which can be a reason of security breach also. Drones have many limitations like battery, small processing power, low bandwidth. So, without data aggregation technique number of packets traveled are large as compare to an architecture with it. For example, if 3 nodes are connected to a head and as such there are 3 more groups then at first level 9 packets will be pass on to the head and at the next level 9+3=12 packets will travel. At next level 12+3=15 packets will travel. Hence, a total of 9+12+15=36 packets travel without aggregation technique. But if data aggregation technique is followed then 9+3+1=13 packets travel as shown in Figure 4.21. There are various data aggregation techniques:

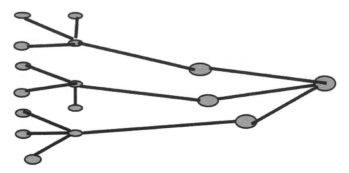

FIGURE 4.21   Data aggregation architecture.

4.2.25   TREE-BASED AGGREGATION TECHNIQUE

Greedy incremental tree (GIT) which is a data-centric routing protocol uses the concept of Directed Diffusion for data aggregation [45, 46]. Shortest path tree routing considers number of nodes visited while moving from source to destination hence, if drone priority to reach at minimum time, then it follows the same procedure [47]. This application also considers the limitation of battery thereby allowing an adaptable sleep plan for drones by monitoring applications. Information flows from parent to children drone and parents keep their children's data to avoid any loss of data. Data aggregation takes place in 2 phases. For the first step, known as distribution phase, queries are dispersed to the drones by the base station (BS) and for the next phase known as collection phase data is collected. And during the distribution phase, a message is disseminated by the BS requiring drone to establish a routing tree so that the BS or cluster head of IoDs can transmit its queries. There is a field in each message that can store the information of level or distance from the root of the child drone. If this message is received by a drone that does not fit to any level it sets its own level by incrementing the current level in the message by one and assigns the sender drone as its parent. This continues until all drone nodes in the network join the tree and have a parent. In order to update the tree structure this is updated on a regular basis. Once the tree is completed the BS questions the network via the aggregation tree. In Ref. [48], drone nodes use the cryptographic algorithms when there is a doubt of cheating activity. For monitoring data aggregators, a secure aggregation tree (SAT) is built. In SAT, any child drone can listen to the incoming data of its parent drone. But where there is doubt on the way of data aggregation, a weighted voting scheme is employed to know if it has cheated or behaving properly. If the data aggregator is found guilty, then tree is rebuilt locally so that the malicious drone data aggregator can be omitted from the system.

4.2.26   Cluster-Based Data Aggregation Protocols

This is a procedure in IoDs where drone nodes form clusters and in each cluster a cluster head is chosen for execution of many operations locally [49]. Cluster heads can communicate directly with the sink through long range radio transmission channels using a technique called HEED, for cluster head selection, which uses the concept of multiple power levels of drones [50]. It requires on an average minimum power within the cluster to reach the level of cluster head. This is called average minimum reachability power (AMRP). AMPR is used to find the cost of communication in each cluster. In order to choose cluster heads, each drone calculates its probability of becoming the cluster head as in Eqn. (3):

$$P(ch) = CX \frac{\text{Eresidual}}{\text{Emax}} \tag{3}$$

where; C and $E_{residual}$ and E_{max} denote the initial percentage of cluster heads, the current residual, and initial energy of the drone respectively. Each drone broadcasts a cluster head message, and drone with the lowest AMRP is selected as cluster head. This process recursively continues until every node is assigned to a cluster head. In one more way periodic per hop data aggregation is used which is called Cougar and it is suitable for applications where drones continuously generate interrelated data [51].

4.2.27   DYNAMIC TOPOLOGIES AND ADAPTIVE ROUTING MECHANISMS

Various drones using Bluetooth, CAN or ZigBee connect with each other because of many reasons to form a network called flying ad-hoc networks (FANETs) [59]. As UAVs are highly mobile so they change their topology frequent also their space of movement is 3D hence, routing is a challenging task in FANETs. Their applications are mostly happening because of multi-hop environment, where it was found that average speed of a UAV was almost 30–460 km/h in a three-dimensional environment [60]. As the topology changes quickly, so network faces link variation problem [61]. Various routing protocols have been proposed as follows:

4.2.27.1   DESTINATION-SEQUENCED DISTANCE VECTOR (DSDV)

Based on the Bellman-Ford-Moore algorithm provides loop free path; it is a proactive routing protocol-where table is updated periodically and each node is having information for all the other nodes in the network. Little changes in the algorithm also make it more suitable for FANETs. Each UAV in DSDV must know absolutely everything about all of the other UAVs connected to the network as they exchange their information with each other hence, provide real time routing information [62]. It is a time-consuming protocol so not recommendable for the situations when transmission time matters. Sybil or spoofing attack can take place very easily.

4.2.27.2   OPTIMIZED LINK STATE ROUTING (OLSR)

It is a proactive routing protocol hence decision of path is available before sending the packet and it is updated periodically or when needed [63]. In this protocol man-in-the middle attack can take place easily.

4.2.27.3   DYNAMIC SOURCE ROUTING (DSR)

It is a reactive routing protocol hence decision of path is taken when there is a need for transmission of the data. This routing protocol is a characteristic of reactive routing where a network can be self-configured, self-organized, and without infrastructure [64]. But as route changes very frequently because of drones' movement so has to take care of this. They suffer from Interruption and spoofing attacks.

4.2.27.4   AD-HOC ON-DEMAND DISTANCE VECTOR (AODV)

It is a reactive routing protocol. AODV does not retain the routes to destinations which are not active during the communication process hence, they update their tables when required [65].

4.2.27.5 TIME-SLOTTED ON-DEMAND ROUTING (TSODR)

It is a reactive protocol. This routing algorithm is basically a time-slotted form of AODV [66].

4.2.27.6 ZONE ROUTING PROTOCOL (ZRP)

It is hybrid routing protocol. This routing protocol is best suited for the dissimilar mobility patterns of UAVs and is basically based on the concept of "zones" [67]. They suffer from black-hole attack.

4.2.27.7 TEMPORARILY ORDERED ROUTING ALGORITHM (TORA)

It is hybrid routing protocol. It is highly adaptive and the most suitable on-demand routing protocol for multi-hop networks. Here, each UAV can only update routing tables about the neighboring UAVs [68].

4.2.28 NETWORK TOPOLOGY ATTACK

With the help of virtual network functions an abstract layer is provided which is capable of reducing the dependency of the devices for functioning of network [95]. With the capability of providing dynamic services the system is boon but also make it eligible for different attacks which can happen while virtual router communicates with virtual UAV without firewall. Attackers can also obtain information about different-sites of infrastructure [96]. To make the system trustworthy platforms are verified remotely with the help of tool open cloud integrity tool named as openCIT [101]. In other solutions release of patches, disable ports and services which are not required, usage of strong password is recommended.

4.2.29 DNS AMPLIFICATION ATTACK

While in DoS attack when all the resources are overused there is a need of additional virtual domain name server. Hence, while deploying it attacker can spoof IP addresses and by using it can launch nasty DNS queries which will again overload the system. To manage the system administrator will again launch the one more additional DNS server. Thereafter victim will get multiple reply from multiple domain name servers [97].

4.2.30   ATTACK ON SECURITY LOG

With the increase in network traffic during attack amount of security logs make buffer overflow which may create problem for the system. System will start removing the entries which multiply the problem as after that it became difficult for the system to understand the nature of traffic. There is always a possibility of losing sensitive information [98]. The solution suggested was by back-tracing the packets which can help to know the root cause of bugs by recreating the series of events [99]. For getting packet level information FleXam tool is used [100]. To generate normal traffic Ostinato tool is used. For collection and labeling the data Security Onion is used and Argus tool to have data in MySQL and ML techniques are applied and system was able to recognize the attack in 99.9% of cases [102].

4.2.31   MALICIOUS INSIDER

When attacker is no other than but from administrator only which can have user's information like SSH Keys, IDs, password either from the records of memory dump or hard-drive volume [98].

4.3   CONCLUSION AND FUTURE SCOPE

In this chapter various threats are discussed while considering IoD. As drones have a promising future because they are small, lightweight, inexpensive, suitable for various applications and can reach any place. Now no field is untouched with the usage of drones but as they are not having great power of computation or storage hence, they are linked with a medium to either other drones forming IoDs or FANET or sensors for collecting data or Internet act as a medium between the source and Internet for fast transmission of data. As way of communication is can be anything from Bluetooth to ZigBee, MVALink, etc., hence way of communication is not safe and system can be hacked by the hacker easily. In this chapter various approaches have been discussed and many more is required to make the system more secure. As system is having lossy connections so approach should be best fitted in that direction and also storage capacity is small hence lightweight techniques which can make the system secure are required. In future, more sophisticated techniques are required so that any flaw of the system can be recovered in a decent time period. As fake signals can be generated using deep learning (DL) techniques which are very much similar to the original signal pattern. Hence, it is going to be a challenging task to differentiate between them, so more advance techniques are required to make the system safe and secure.

KEYWORDS

- cybersecurity
- ground control station
- internet of drones
- micro air vehicle
- unmanned aerial vehicle
- vulnerabilities

REFERENCES

1. Gharibi, M., Boutaba, R., & Waslander, S. L., (2016). Internet of drones. *IEEE Access, 4*, 1148–1162.
2. Qureshi, B., Koubâa, A., Sriti, M. F., Javed, Y., & Alajlan, M., (2016). Dronemap-a cloud-based architecture for the internet-of-drones. In: *International Conference on Embedded Wireless Systems and Networks*.
3. Devasia, S., & Lee, A., (2016). Scalable low-cost unmanned-aerial-vehicle traffic network. *Journal of Air Transportation, 24*(3), 74–83.
4. Lin, C., He, D., Kumar, N., Choo, K. K. R., Vinel, A., & Huang, X., (2018). *Security and Privacy for the Internet of Drones: Challenges and Solutions, 56*(1), 64–69. IEEE Communications Magazine.
5. Förster, D., Kargl, F., & Löhr, H., (2014). PUCA: A pseudonym scheme with user-controlled anonymity for vehicular ad-hoc networks (VANET). In: *2014 IEEE Vehicular Networking Conference (VNC)* (pp. 25–32). IEEE.
6. Mahmoud, M. E., & Shen, X., (2012). A novel traffic-analysis back tracing attack for locating source nodes in wireless sensor networks. In: *2012 IEEE International Conference on Communications (ICC)* (pp. 939–943). IEEE.
7. Mukherjee, A., Dey, N., & De, D., (2020). EdgeDrone: QoS aware MQTT middleware for mobile edge computing in opportunistic Internet of Drone Things. *Computer Communications, 152*, 93–108.
8. Bakker, J., (2017). *Intelligent Traffic Classification for Detecting DDoS Attacks using SDN/OpenFlow*. Victoria University of Wellington.

9. Tan, Z., Jamdagni, A., He, X., Nanda, P., & Liu, R. P., (2013). A system for denial-of-service attack detection based on multivariate correlation analysis. *IEEE Transactions on Parallel and Distributed Systems, 25*(2), 447–456.

10. Fulkerson, B. (1995). Machine learning, neural and statistical classification. *Technometrics, 37,* 459.

11. Mirkovic, J., & Reiher, P., (2004). A taxonomy of DDoS attack and DDoS defense mechanisms. *ACM SIGCOMM Computer Communication Review, 34*(2), 39–53.

12. Nguyen, T. T., & Armitage, G., (2008). A survey of techniques for internet traffic classification using machine learning. *IEEE Communications Surveys & Tutorials, 10*(4), 56–76.

13. Hayes, M., (2014). *Traffic Classification in Enterprise Networks with the Era of IoT.* COMP489 Report, Victoria University of Wellington.

14. Dilli, R., (2019). Anomaly detection based on machine learning techniques. *International Journal of Recent Technology and Engineering* (IJRTE) ISSN: 2277-3878, *8*(4), November 2019.

15. *"The CAIDA "DDoS Attack 2007" Dataset.* (2007). https://www.caida.org/catalog/datasets/ddos-20070804_dataset/#H2875 (accessed on 1 December 2021).

16. Shiravi, A., Shiravi, H., Tavallaee, M., & Ghorbani, A. A., (2012). Toward developing a systematic approach to generate benchmark datasets for intrusion detection. *Computers & Security, 31*(3), 357–374.

17. Renfro, B. A., Stein, M., Boeker, N., & Terry, A., (2018). *An Analysis of Global Positioning System (GPS) Standard Positioning Service (SPS) Performance for 2017.* https://www.gps.gov/systems/gps/performance/2014-GPS-SPS-performance-analysis.pdf (accessed on 19 November 2021).

18. Humphreys, T. E., Ledvina, B. M., Psiaki, M. L., O'Hanlon, B. W., & Kintner, P. M., (2008). Assessing the spoofing threat: Development of a portable GPS civilian spoofer. In: *Radionavigation Laboratory Conference Proceedings.*

19. Tippenhauer, N. O., Pöpper, C., Rasmussen, K. B., & Capkun, S., (2011). On the requirements for successful GPS spoofing attacks. In: *Proceedings of the 18*th *ACM Conference on Computer and Communications Security* (pp. 75–86).

20. De Lima, P. E. M., Lachowski, R., Pellenz, M. E., Penna, M. C., & Souza, R. D., (2018). A machine learning approach for detecting spoofing attacks in wireless sensor networks. In: *2018 IEEE 32*nd *International Conference on Advanced Information Networking and Applications (AINA)* (pp. 752–758). IEEE.

21. Manesh, M. R., Kenney, J., Hu, W. C., Devabhaktuni, V. K., & Kaabouch, N., (2019). Detection of GPS spoofing attacks on unmanned aerial systems. In: *2019 16*th *IEEE Annual Consumer Communications & Networking Conference (CCNC)* (pp. 1–6). IEEE.

22. Shi, Y., Davaslioglu, K., & Sagduyu, Y. E., (2019). Generative adversarial network for wireless signal spoofing. In: *Proceedings of the ACM Workshop on Wireless Security and Machine Learning* (pp. 55–60).

23. Tan, C. C., Wang, H., Zhong, S., & Li, Q., (2009). IBE-Lite: A lightweight identity-based cryptography for body sensor networks. *IEEE Transactions on Information Technology in Biomedicine, 13*(6), 926–932.

24. Alharbi, A., Zohdy, M., Debnath, D., Olawoyin, R., & Corser, G., (2018). Sybil attacks and defenses in internet of things and mobile social networks. *International Journal of Computer Science Issues (IJCSI), 15*(6), 36–41.

25. Illiano, V. P., & Lupu, E. C., (2015). Detecting malicious data injections in wireless sensor networks: A survey. *ACM Computing Surveys (CSUR), 48*(2), 1–33.
26. Esmalifalak, M., Liu, L., Nguyen, N., Zheng, R., & Han, Z., (2014). Detecting stealthy false data injection using machine learning in smart grid. *IEEE Systems Journal, 11*(3), 1644–1652.
27. Alrehily, A. D., Alotaibi, A. F., Almutairy, S. B., Alqhtani, M. S., & Kar, J., (2015). Conventional and improved digital signature scheme: A comparative study. *Journal of information Security, 6*(01), 59.
28. Zhang, H., Li, R., Li, L., & Dong, Y., (2013). Improved speed Digital Signature Algorithm based on modular inverse. In: *Proceedings of 2013 2nd International Conference on Measurement, Information and Control* (Vol. 1, pp. 706–710). IEEE.
29. Mughal, M. A., Luo, X., Ullah, A., Ullah, S., & Mahmood, Z., (2018). A lightweight digital signature based security scheme for human-centered internet of things. *IEEE Access, 6*, 31630–31643.
30. Thakur, J., & Kumar, N., (2011). DES, AES and Blowfish: Symmetric key cryptography algorithms simulation based performance analysis. *International Journal of Emerging Technology and Advanced Engineering, 1*(2), 6–12.
31. National Research Council, (1996). *Cryptography's Role in Securing the Information Society*. National Academies Press.
32. En.wikipedia.org. (2020). *Block Cipher Mode of Operation*. [online] Available at: https://en.wikipedia.org/wiki/Block_cipher_modes_of_operation (accessed on 19 November 2021).
33. Usman, M., Ahmed, I., Aslam, M. I., Khan, S., & Shah, U. A., (2017). *SIT: A Lightweight Encryption Algorithm for Secure Internet of Things*. arXiv preprint arXiv:1704.08688.
34. Palisse, A., Le Bouder, H., Lanet, J. L., Le Guernic, C., & Legay, A., (2016). Ransomware and the legacy crypto API. In: *International Conference on Risks and Security of Internet and Systems* (pp. 11–28). Springer, Cham.
35. Meidan, Y., Bohadana, M., Shabtai, A., Guarnizo, J. D., Ochoa, M., Tippenhauer, N. O., & Elovici, Y., (2017). ProfilIoT: A machine learning approach for IoT device identification based on network traffic analysis. In: *Proceedings of the Symposium on Applied Computing* (pp. 506–509).
36. Sciancalepore, S., Ibrahim, O. A., Oligeri, G., & Di Pietro, R., (2020). PiNcH: An effective, efficient, and robust solution to drone detection via network traffic analysis. *Computer Networks, 168*, 107044.
37. Jeong, S., Bito, J., & Tentzeris, M. M., (2017). Design of a novel wireless power system using machine learning techniques for drone applications. In: *2017 IEEE Wireless Power Transfer Conference (WPTC)* (pp. 1–4). IEEE.
38. Park, J., Kim, Y., & Seok, J., (2016). Prediction of information propagation in a drone network by using machine learning. In: *2016 International Conference on Information and Communication Technology Convergence (ICTC)* (pp. 147–149). IEEE.
39. Shahid, M. R., Blanc, G., Zhang, Z., & Debar, H., (2018). IoT devices recognition through network traffic analysis. In: *2018 IEEE International Conference on Big Data (Big Data)* (pp. 5187–5192). IEEE.
40. Wazid, M., Das, A. K., & Lee, J. H., (2018). Authentication protocols for the internet of drones: Taxonomy, analysis and future directions. *Journal of Ambient Intelligence and Humanized Computing*, 1–10.

41. Wazid, M., Das, A. K., Kumar, N., Vasilakos, A. V., & Rodrigues, J. J., (2018). Design and analysis of secure lightweight remote user authentication and key agreement scheme in Internet of drones deployment. *IEEE Internet of Things Journal, 6*(2), 3572–3584.

42. Rahim, R., (2017). Man-in-the-middle-attack prevention using interlock protocol method. *ARPN J. Eng. Appl. Sci., 12*(22), 6483–6487.

43. De Rango, F., Potrino, G., Tropea, M., & Fazio, P., (2020). Energy-aware dynamic internet of things security system based on elliptic curve cryptography and message queue telemetry Transport protocol for mitigating replay attacks. *Pervasive and Mobile Computing, 61*, 101105.

44. Kebande, V. R., & Ray, I., (2016). A generic digital forensic investigation framework for internet of things (IoT). In: *2016 IEEE 4th International Conference on Future Internet of Things and Cloud (FiCloud)* (pp. 356–362). IEEE.

45. Intanagonwiwat, C., Estrin, D., Govindan, R., & Heidemann, J., (2002). Impact of network density on data aggregation in wireless sensor networks. In: *Proceedings 22nd international conference on distributed computing systems* (pp. 457–458). IEEE.

46. Intanagonwiwat, C., Govindan, R., Estrin, D., Heidemann, J., & Silva, F., (2003). Directed diffusion for wireless sensor networking. *IEEE/ACM Transactions on Networking, 11*(1), 2–16.

47. Madden, S., Franklin, M. J., Hellerstein, J. M., & Hong, W., (2002). TAG: A tiny aggregation service for ad-hoc sensor networks. *ACM SIGOPS Operating Systems Review, 36*(SI), 131–146.

48. Wu, K., Dreef, D., Sun, B., & Xiao, Y., (2007). Secure data aggregation without persistent cryptographic operations in wireless sensor networks. *Ad Hoc Networks, 5*(1), 100–111.

49. Ozdemir, S., & Xiao, Y., (2009). Secure data aggregation in wireless sensor networks: A comprehensive overview. *Computer Networks, 53*(12), 2022–2037.

50. Younis, O., & Fahmy, S., (2004). HEED: A hybrid, energy-efficient, distributed clustering approach for ad hoc sensor networks. *IEEE Transactions on Mobile Computing, 3*(4), 366–379.

51. Yao, Y., & Gehrke, J., (2002). The cougar approach to in-network query processing in sensor networks. *ACM Sigmod. Record, 31*(3), 9–18.

52. Zuo, C., Wang, W., Lin, Z., & Wang, R., (2016). *Automatic Forgery of Cryptographically Consistent Messages to Identify Security Vulnerabilities in Mobile Services*. In NDSS.

53. Jung, W., Hong, S., Ha, M., Kim, Y. J., & Kim, D., (2009). SSL-based lightweight security of IP-based wireless sensor networks. In: *2009 International Conference on Advanced Information Networking and Applications Workshops* (pp. 1112–1117). IEEE.

54. Matyszczyk, C., (2013). *Truck Driver has GPS Jammer, Accidentally Jams Newark Airport* (Vol. 11). CNET News.

55. Barsocchi, P., Lenzi, S., Chessa, S., & Giunta, G., (2009). A novel approach to indoor RSSI localization by automatic calibration of the wireless propagation model. In: *VTC Spring 2009-IEEE 69th Vehicular Technology Conference* (pp. 1–5). IEEE.

56. Patwari, N., Ash, J. N., Kyperountas, S., Hero, A. O., Moses, R. L., & Correal, N. S., (2005). *Locating the Nodes: Cooperative Localization in Wireless Sensor Networks, 22*(4), 54–69. IEEE Signal processing magazine.

57. Pu, C., (2018). Jamming-resilient multipath routing protocol for flying ad hoc networks. *IEEE Access, 6*, 68472–68486.

58. Tedeschi, P., Oligeri, G., & Di Pietro, R., (2019). Leveraging jamming to help drones complete their mission. *IEEE Access, 8*, 5049–5064.

59. Khan, M. A., Khan, I. U., Safi, A., & Quershi, I. M., (2018). Dynamic routing in flying ad-hoc networks using topology-based routing protocols. *Drones, 2*(3), 27.
60. Gankhuyag, G., Shrestha, A. P., & Yoo, S. J., (2017). Robust and reliable predictive routing strategy for flying ad-hoc networks. *IEEE Access, 5,* 643–654.
61. Frew, E. W., & Brown, T. X., (2009). Networking issues for small unmanned aircraft systems. *Journal of Intelligent and Robotic Systems, 54*(1–3), 21–37.
62. Yassein, M. B., & Alhuda, N., (2016). Flying ad-hoc networks: Routing protocols, mobility models, issues. *International Journal of Advanced Computer Science and Applications (IJACSA), 7*(6).
63. Singh, J., & Mahajan, R., (2013). Performance analysis of AODV and OLSR using OPNET. *Int. J. Comput. Trends Technol., 5*(3), 114–117.
64. Johnson, D. B., & Maltz, D. A., (1996). Dynamic source routing in ad hoc wireless networks. In: *Mobile Computing* (pp. 153–181). Springer, Boston, MA.
65. Shobana, M., & Karthik, S., (2013). A performance analysis and comparison of various routing protocols in MANET. In: *2013 International Conference on Pattern Recognition, Informatics and Mobile Engineering* (pp. 391–393). IEEE.
66. Perkins, C. E., & Royer, E. M., (1999). Ad-hoc on-demand distance vector routing. In: *Proceedings WMCSA'99. Second IEEE Workshop on Mobile Computing Systems and Applications* (pp. 90–100). IEEE.
67. Haas, Z. J., & Pearlman, M. R., (2001). Zone routing protocol (ZRP) a hybrid framework for routing in ad hoc networks. *Ad hoc Networking* (2nd edn., Vol. 1). Chapter 7.
68. Thakrar, P. M., Singh, V., & Kotecha, K., (2020). Improved route selection algorithm based on TORA over mobile ad hoc network. *Journal of Discrete Mathematical Sciences and Cryptography, 23*(2), 617–629.
69. Yaacoub, J. P., & Salman, O., (2020). Security analysis of drones systems: Attacks, limitations, and recommendations. *Internet of Things*, 100218.
70. Altawy, R., & Youssef, A. M., (2016). Security, privacy, and safety aspects of civilian drones: A survey. *ACM Transactions on Cyber-Physical Systems, 1*(2), 1–25.
71. Dinger, J., & Hartenstein, H., (2006). *Availability, Reliability, and Security*. ARES-2006, The First International Conference.
72. Kleinberg, R., (2007). Geographic routing using hyperbolic space, *IEEE INFOCOM 2007 – 26th IEEE International Conference on Computer Communications,* 1902–1909, doi: 10.1109/INFCOM.2007.221.
73. Lee, W. C., Xu, J., Li, J., & Silvestri, F., (2009). Scalable information systems. *Future Generation Computer Systems, 25*(1), 51, 52.
74. Chen, M., Challita, U., Saad, W., Yin, C., & Debbah, M., (2017). *Machine Learning for Wireless Networks with Artificial Intelligence: A Tutorial on Neural Networks*. arXiv preprint arXiv:1710.02913.
75. Zeitlin, A., & McLaughlin, M., (2006). *Modeling for UAS Collision Avoidance*. AUVSI Unmanned Systems North America, Orlando.
76. Primatesta, S., Guglieri, G., & Rizzo, A., (2019). A risk-aware path planning strategy for UAVs in urban environments. *Journal of Intelligent & Robotic Systems, 95*(2), 629–643.
77. Awrejcewicz, J., (2011). *Numerical Analysis: Theory and Application*. BoD–Books on Demand.
78. Gutwirth, S., Leenes, R., De Hert, P., & Poullet, Y., (2012). *European Data Protection: Coming of Age*. Springer Science & Business Media.

79. Clarke, R., (2014). The regulation of civilian drones' impacts on behavioral privacy. *Computer Law & Security Review, 30*(3), 286–305.

80. Lubin, A., (2018). In: Minárik, T., Jakschis, R., & Lindström, L., (eds.), *10*th *International Conference on Cyber Conflict CyCon X: Maximizing Effects.*

81. 81. Zamboni, D., (2008). 2008 Proceeding Editor: Diego Zamboni Publisher: Springer-Verlag, Berlin, Heidelberg Conference: Paris France July 10–11, 2008 ISBN: 978-3-540-70541-3.

82. Mitchell, R., & Chen, R., (2013). Adaptive intrusion detection of malicious unmanned air vehicles using behavior rule specifications. *IEEE Transactions on Systems, Man, and Cybernetics: Systems, 44*(5), 593–604.

83. Rani, C., Modares, H., Sriram, R., Mikulski, D., & Lewis, F. L., (2016). Security of unmanned aerial vehicle systems against cyber-physical attacks. *The Journal of Defense Modeling and Simulation, 13*(3), 331–342.

84. Lu, H., Li, Y., Mu, S., Wang, D., Kim, H., & Serikawa, S., (2017). Motor anomaly detection for unmanned aerial vehicles using reinforcement learning. *IEEE Internet of Things Journal, 5*(4), 2315–2322.

85. Sedjelmaci, H., Senouci, S. M., & Ansari, N., (2017). A hierarchical detection and response system to enhance security against lethal cyber-attacks in UAV networks. *IEEE Transactions on Systems, Man, and Cybernetics: Systems, 48*(9), 1594–1606.

86. Zhang, G., Wu, Q., Cui, M., & Zhang, R., (2019). Securing UAV communications via joint trajectory and power control. *IEEE Transactions on Wireless Communications, 18*(2), 1376–1389.

87. Beebe, N. L., & Clark, J. G., (2005). A hierarchical, objectives-based framework for the digital investigations process. *Digital Investigation, 2*(2), 147–167.

88. Jain, U., Rogers, M., & Matson, E., (2017). Drone forensic framework: Sensor and data identification and verification. *2017 IEEE Sensors Applications Symposium (SAS)*, 1–6.

89. Carrier, B., & Spafford, E. H., (2004). Digital forensic research workshop. *An Event-Based Digital Forensic Investigation Framework, 11–13.*

90. Roder, A., Choo, K. K. R., & Le-Khac, N. A., (2018). *Unmanned Aerial Vehicle Forensic Investigation Process: Dji Phantom 3 Drone as a Case Study.* arXiv preprint arXiv:1804.08649.

91. Vaarandi, R., Markus, K., & Mauno, P., (2016). Event log analysis with the log cluster tool. *Milcom 2016–2016 IEEE Military Communications Conference.*

92. Abbaspour, A., Yen, K. K., Forouzannezhad, P., & Sargolzaei, A., (2018). A neural adaptive approach for active fault-tolerant control design in UAV. *IEEE Transactions on Systems, Man, and Cybernetics: Systems.*

93. He, D., Chan, S., & Guizani, M., (2017). *Drone-Assisted Public Safety Networks: THE Security Aspect, 55*(8), 218–223. IEEE Communications Magazine.

94. Booker, M., (2018). *Effects of Hacking an Unmanned Aerial Vehicle Connected to the Cloud* (Doctoral dissertation, The Ohio State University).

95. Kumari, A., Gupta, R., Tanwar, S., & Kumar, N., (2020). A taxonomy of blockchain-enabled softwarization for secure UAV network. *Computer Communications, 161,* 304–323.

96. Jain, K., (2004). Security based on network topology against the wiretapping attack. *IEEE Wireless Communications, 11*(1), 68–71.

97. Han, M., Canh, T. N., Noh, S. C., Yi, J., & Park, M., (2019). "DAAD: DNS Amplification Attack Defender in SDN," *2019 International Conference on Information and*

Communication Technology Convergence (ICTC), 2019, pp. 372–374, doi: 10.1109/ICTC46691.2019.8939897.

98. Kumari, A., Tanwar, S., Tyagi, S., Kumar, N., Obaidat, M. S., & Rodrigues, J. J., (2019). Fog computing for smart grid systems in the 5G environment: Challenges and solutions. *IEEE Wireless Communications, 26*(3), 47–53.

99. Handigol, N., Heller, B., Jeyakumar, V., Maziéres, D., & McKeown, N., (2012). Where is the debugger for my software-defined network?. In: *Proceedings of the First Workshop on Hot Topics in Software Defined Networks* (pp. 55–60).

100. Shirali-Shahreza, S., & Ganjali, Y., (2013). FleXam: Flexible sampling extension for monitoring and security applications in open flow. In: *Proceedings of the Second ACM SIGCOMM Workshop on Hot Topics in Software Defined Networking* (pp. 167–168).

101. Briscoe, B., (2014). *Network Functions Virtualization (NFV)-NFV Security: Problem Statement.* White Paper, ETSI NFV ISG.

102. Moustafa, N., & Jolfaei, A., (2020). Autonomous detection of malicious events using machine learning models in drone networks. In: *Proceedings of the 2nd ACM MobiCom Workshop on Drone Assisted Wireless Communications for 5G and Beyond* (pp. 61–66).

103. Koubâa, A., Allouch, A., Alajlan, M., Javed, Y., Belghith, A., and M. Khalgui, M. Khalgui, Micro Air Vehicle Link (MAVlink) in a Nutshell: A Survey, in IEEE Access, vol. 7, pp. 87658-87680, 2019, doi: 10.1109/ACCESS.2019.2924410.

CHAPTER 5

Real-Time Monitoring and Analysis of Troposphere Pollutants Using a Multipurpose Surveillance Drone

RAMAKANTA CHOUDHURY,[1] NAVNEET YADAV,[1] JAIDEEP KALA,[1] SONALIKA BHANDARI,[1] CHANDRAKANTA SAMAL,[2] and NOOR ZAMAN JHANJHI[3]

[1]*Department of Electronics and Communication Engineering, Maharaja Argrasen Institute of Technology, New Delhi, India, E-mail: rkchoudhury1@gmail.com (R. Choudhury)*

[2]*Department of Computer Science, Acharya Narendradev College, Delhi University, India*

[3]*Department of Computing and IT, Taylor's University, Malaysia*

ABSTRACT

This chapter presents a multi-tasking quadcopter drone that is capable of carrying scientific payloads such as an air quality index (AQI) determining IoT system that can be mounted over the quadcopter to measure tropospheric pollutants such as PM2.5, PM10, CO, SO_2 along with temperature and humidity with a high level of accuracy and a flight time of over 20 minutes. This quadcopter is also capable of carrying 1 kg of additional payload with a restricted flight time to perform other operations such as carrying a thermal imaging camera, advanced sensor equipment or other non-scientific payloads. This platform enables real-time monitoring of air quality at varying altitudes continuously as data from sensors is directly transmitted to a server through

The Internet of Drones: AI Applications for Smart Solutions. Arun Solanki, PhD, Sandhya Tarar, PhD, Simar Preet Singh, PhD & Akash Tayal, PhD (Eds.)

internet connectivity on the raspberry pi0, which acts as the brain of this IoT system. Due to this on-board processing and storing sensor data eliminates the need to land the drone for data collection. This UAV system can give sensor output over a wide range (0–1,024 ppm) and over-temperature ranges from–25°C to +60°C making it suitable for operations in extreme conditions which are only limited by the battery performance and flight time delivered. This quadcopter platform is equipped with PM2.5 sensor, DHT22 sensor and MQ gas sensors and which have been fabricated on a single plate that can be directly mounted on any quadcopter platform hence providing flexibility in operation. We have endeavored to measure the air pollutants through this platform at two sites in New Delhi. The data has been gathered by flying the drone at specified times in the morning, noon, evening, and night in a different month. The data collected from the drone has been compared with the pollution data of government agencies from the same location to ensure the accuracy of the developed system. This Platform also enables measurement of gas leakage and levels (such as ethane, methane, propane) in industrial areas or over gas pipelines hence eliminating the need for human involvement in dangerous and inaccessible sites. The results garnered from the various developmental tests proves the reliability and accuracy of this system.

5.1 INTRODUCTION

Drones which traditionally have been called unmanned aerial vehicles (UAVs), are gaining more popularity around the world today because of their large number of applications in distinct fields. The establishment of drones can be traced back to the period of World War I when the countries in the west were looking for better military transportation and surveillance solutions. The major boost in the technology was observed throughout World War II and into the Cold War as well. In 1982 Israel became the first country to use drones in its war with Syria [1]. Apart from the military usage drones are being used in many other applications such as; law enforcement and surveillance, search, and rescue operations, disaster management, storm tracking and forecasting hurricanes and tornadoes, terrain mapping and planning, buildings' inspections, express shipping and delivery, precision crop monitoring, and aerial photography for journalism and films.

A lot of research work is being carried out in the above-mentioned areas where the use of drones can make the operations more manageable, flexible, economical, and reliable. Several new types of drones are being made to work for different kinds of applications. With these advances in the field of drones, they can now have higher payloads of sensors and can remain airborne for more than 24 hours. The usage of drones or UAVs in the 5G and higher generations of wireless networks has been discussed in Ref. [2]. The drones have the advantage of being deployed anywhere at a much lesser cost as compared to the cost of setting up fixed communication infrastructure. The upgrades in communication technology [3–5] can help establish robust connection links in remote areas using the advantage of controlled mobility of drones.

Due to the mobile nature and payload carrying capacity of drones, many researchers have made use of drones/UAVs in the area of air pollution measurement [6–8]. In a recent study by the World Health Organization, it was found that air pollution is one of the world's leading factors for death, and it is accountable for 5 million deaths each year. Hence the growing air pollution in cosmopolitans attributes 9% of deaths globally and has become significant risk factors for disease burden. Air quality index (AQI) is a measure used by the government to quantify the quality of air in an area. AQI is represented as a number calculated using the concentration level of air pollutants and particulate matter (e.g., PM2.5, PM10, CO, SO_2 and so on). AQI value over 100 is an indication of air quality being on the way of becoming "heavily" or "seriously" polluted, thus leaving a large population/residents of an area exposed to harmful health effects [9]. Sensors accomplish AQI monitoring in real-time at governmental static observation stations, which is then used for generating AQI maps for a location [10]. However, the major disadvantage of these observatories that it carries out limited sampling and updates data after every two hours and it includes a high maintenance cost. Considering the fact that the distance between two nearby stations is significantly large, also the fact that the AQI is being monitored every 2 hours, such a scenario does not cope with the rising concerns. Hence a real-time monitoring system like a drone with its sensor payload can help ensure a more flexible, accurate, and fine monitoring system with a commendably low maintenance cost. In the presented work we have exploited the mobile, nature of drones to address a global concern of rising air pollution near residential areas due to the establishment of industries which is causing enormous environmental and public health concerns. We have endeavored to measure the air pollutants through appropriate sensors in the drones at two sites in Delhi. The data has been gathered by flying the drone at specified times in the morning, noon, evening, and night in a different month. The data collected from the drone has been compared with the pollution data of government agencies from the same location to ensure the accuracy of the developed system.

5.2   STRUCTURE OF A BASIC DRONE

A multi-rotor drone generates it is thrust through a propulsion system consisting of a brushless DC motor, Electronic speed controller and propeller. All major components involved in a drone are shown in Figure 5.1. To produce lift the downward force generated by the propellers must be greater than the all-up weight (AUW) of the multi-rotor. Practically a multi-rotor is designed in such a way that the maximum thrust generated by it more than twice the AUW of the drone, this enables the drone to counter the wind during flight and have a high level of maneuverability [11]. The propulsion system of the drone is defined by the thrust produced by the motors at different RPM with a fixed voltage level and propeller size. Table 5.1 depicts the relation between Throttle percentage, RPM, current drawn and thrust produced by a single motor for a fixed voltage of 11.1 V and propeller size of 10×4.7 inches (Counterclockwise). The motor under consideration is an EMAX GT2215 1,170 kv DC brushless motor which has been used in this

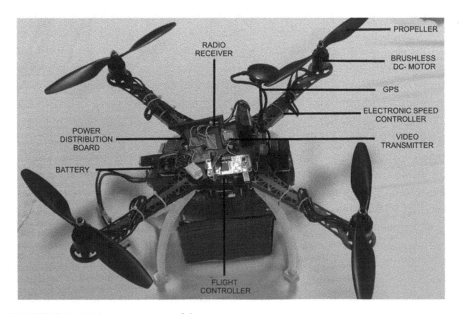

FIGURE 5.1   Main components of drone.

multi-purpose quadcopter. In multi-rotors, a 3-phase brushless DC motor is used which consists of neodymium magnets and can be both in runner (the rotor is surrounded by the stator) and outrunner (the stator is surrounded by the rotor). A bench test apparatus consists of a motor mount and load cells to measure the thrust produced by the motor with varying input throttle. Table 5.1 shows result of bench test of EMAX GT2215 1,170 kV.

The drone in this configuration can lift at about 50% throttle when the total thrust produced by four motors in a quadcopter configuration exceeds the AUW of the drone of about 1.6 kg (without any payload).

TABLE 5.1   Experimental Results of Bench Test of EMAX GT2215 1,170 kv DC Brushless Motor

Throttle	RPM	Current (A)	Thrust (Gms)
25%	2,167	3.7	156
50%	3,950	9.5	405
75%	6,400	14	736
100%	7,435	26	1,250

The processing of the input signal from the radio controller is done by the Flight controller of the UAV, which is a closed-loop control system. A flight controller is a PID controller [12]. PID (proportional, integral, derivative) is part of the flight controller firmware that reads the data from sensors and determines how fast the motors should spin to maintain the desired rotation speed of the quadcopter. The goal of the PID controller is to correct the error, the difference between a measured value (gyro/accelerometer sensor measurements), and the desired setpoint (rotational speed). This error can be minimized by adjusting the control inputs in every loop, which is the speed of the motors [13].

Mission planner software is used for PID tuning of the drone, configuring radio, uploading, and updating firmware of the flight controller and for performing autonomous flights. Figure 5.2 shows the PID tuning window from the mission planner software.

Figure 5.3 represents working of the PID controller which uses feedback mechanism to prevent overshoots and undershoot motor response, enabling smoother flight of an UAV.

5.2.1   MAIN COMPONENTS

1. **Flight Controller:** Flight controller is the main component of the drone. It is responsible for maintaining horizontal and vertical

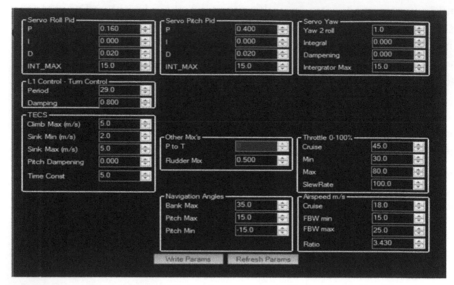

FIGURE 5.2   Tuning of PID parameters in open-source mission planner® software.

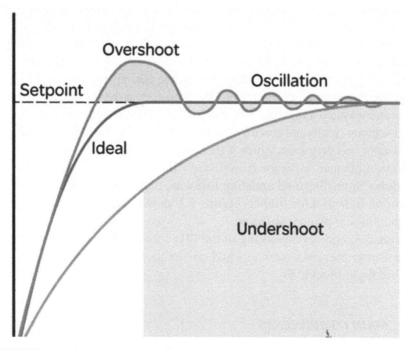

FIGURE 5.3   Working of PID controller.

stability by controlling the rotation of motors. In this quad-copter, Pixhawk 2.4.8 flight-controller is used [14]. The flight controller is also responsible for communication with the external peripherals.

2. **Brushless DC Motor:** EMAX GT2215KV Brushless DC motors are used producing a max thrust of 1,250 gm while drawing 26 A of continuous current at 11.1 V.

3. **Li-Poly/Li-Ion Battery Pack:** The drone is compatible with both 3 cell 5,200 mAh Li-poly battery or 9,000 mAh Li-ion battery pack. These batteries are capable of up to 30 C discharge rate, i.e., over 100 A of continuous current discharge [15].

4. **Electronic Speed Controller:** An ESC receives the signals from the Flight controller to control motor output. Esc's can be for individual motors or 4 in 1 for a quad-copter.

5. **Power Distribution Board:** It is responsible for delivering power to all ESC's, Flight controllers and other peripherals and is directly connected to the battery.

6. **Radio Receiver:** It receives PPM or PWM signal from the radio transmitter and is sent to the flight controller.

7. **Radio Transmitter:** It is controlled by the drone Pilot and provides stick inputs for controlling of drone's flight by providing pitch, roll, yaw, and throttle. It operates at 2.4 Ghz band and has several servo outputs for controlling of camera gimbals and servo motors.

8. **Video Transmitter:** It transmits the video signals from the drone's onboard camera to the video receiver at a 5.8 GHZ band. It consists of a high gain antenna and operates at around 100 mw–800 mw power range.

9. **Propellers:** These comes in various shapes and sizes and this quad-copter twin-blade 10 inches length and 4.7-inch pitch ABS propellers are used. Figure 5.4 depicts the basic circuit diagram of the components involved in the drone [16].

5.3 PROPOSED MULTIPURPOSE SURVEILLANCE DRONE

Drones are mainly being used for camera-based surveillance, and delivery of products, the drone design proposed in this chapter distinguishes from the predominant existing UAV's in two aspects viz. the proposed drone is designed to carry a scientific payload to measure the Air Quality parameters in remote areas, and it then transmits the data in real-time over the internet, along with this the drone is equipped with a sensor array to measure any gas leakage, and in case of high PPM value of CNG, LPG, PNG gas the drone raises an alert in real-time to avoid significant losses [17]. The objective is to perform completely autonomous missions that makes drone suitable for working without human intervention and make them reliable for mobile sensing of air quality and level of harmful gasses in large industrial areas where static government observation centers are not present.

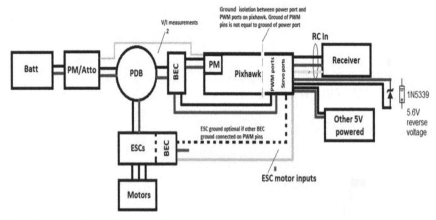

FIGURE 5.4   Circuit diagram of basic components of the drone.

5.3.1   SENSORS USED IN DRONE

The proposed design has been incorporated into a single, coherent device consisting of a microcomputer Raspberry-pi zero, five sensors and a PCB for power distribution that also houses the components for noise cancelation in the circuit. Sensors used for studying the proposed design are discussed in subsections.

5.3.1.1   GP2Y10 PARTICULATE MATTER SENSOR

The GP2Y10 particulate matter sensor series utilizes the phenomenon of scattering of light [18]. The dust density measured includes the concentration of 1 μm, 2.5 μm and 10 μm particles. It is an analog sensor with two primary components (photodiode and LED) and a driver circuit. The LED emits light which gets scattered by the dust particles present inside the circular air chamber. This scattered light is then detected, by the photodiode placed obliquely to the LED inside the sensor. The driver circuit has a resistance and a capacitor which is used to generate the pulse width for the LED to turn ON every 10 ms.

To take correct reading the dust sensor needs calibration which can be done from the electro-optical characteristics of the sensor as shown in Figure 5.5; the output voltage of the sensor corresponding to zero dust density does not equal zero. This is because of the scattering of light due to the stray dust particles inside the sensor. This output value of the voltage is observed to be around 0.6 V and hence is used as an offset and subtracted from the observed voltage reading of the sensor to obtain the correct reading as mentioned in equation 1 where ΔV is the required output, Vo is the output observed and Voc is the offset voltage. The output voltage (Vo) is proportional to the amount of light received by the photodiode.

$$\Delta V = Vo - Voc \tag{1}$$

$$\text{Dust density } (\mu g/m^3) = (\Delta V \times 100) / K \tag{2}$$

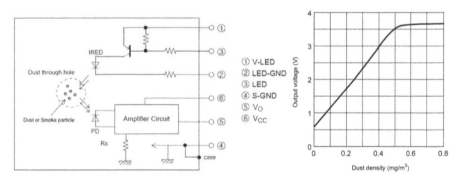

FIGURE 5.5   Internal structure and electro-optical characteristics of the sensor.

5.3.1.2 DHT-22

To measure temperature and humidity in the atmosphere, we have selected DHT-22 sensor due to its high operating range and reliability, as mentioned in Table 5.2. DHT 22 is a digital output sensor comprising of two sensing elements a capacitive element for measuring humidity and a thermistor for measuring temperature. The capacitive sensing element has a moisture absorbing substrate sandwiched between an upper and lower electrode as shown in Figure 5.6, the amount of water vapor absorbed produces ions which leads to change in resistance that is proportional to the relative humidity [19]. While the temperature is sensed, by the second component, i.e., negative thermistor coefficient (NTC), it is a variable resistor [20], whose resistance decreases with increase in temperature. The changes in humidity and temperature are processed by the ADC chip inside the sensor, which then outputs a single 8-bit data to the MCU. The sampling rate of DHT-22 sensor is 0.5 Hz; hence it outputs new data after every 2 seconds. The communication between DHT22 and MCU takes place on a single line data bus where MCU at first initiates read pulse followed by DHT22 sending an acknowledge pulse and then it sends the sensor reading as 40 bits divided into 5 bytes.

TABLE 5.2   Technical Specification of DHT 22 Sensor

Model	DHT 22
Input power	3.3–6 Volts DC
Communication bus	Digital signal via single-bus
Operating range	Humidity 0–100% RH
	Temperature –40°C~80°C
Accuracy %	Humidity +–2% RH (Max +–5% RH)
	Temperature <+–0.5°C
Resolution	Humidity 0.1% RH
	Temperature 0.1°C

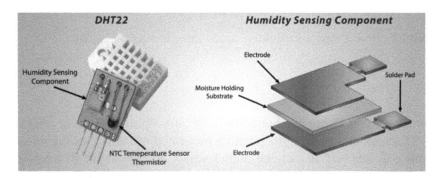

FIGURE 5.6   Diagram of DHT 22 and humidity sensing component.

5.3.1.3 MQ-SENSORS

Large industrial settings contain many types of dangerous chemicals and flammable gasses. Accidental gas leakage can pose serious hazards to a factory, its workers, and the communities around it. On the other hand, many biochemical industries like fermenting industry release their untreated gas waste like CO and CO_2 into the atmosphere, which is a great reason for concern. Around 64% of oil field incidents are caused by combustibles and toxic gasses. To track these industrial areas in real-time mobile gas detection and alarm raising system incorporating various MQ sensors has been suggested using a drone [21].

All MQ sensors have a similar structure consisting of a sensing material laid between the layer of metal mesh. The sensing material is mostly a Metal Oxide Semiconductor whose resistance changes when exposed to the reacting gas. The metal mesh around the sensor acts as protection against unwanted particles which can moderate the sensor reading, hence to ensure proper working of the sensor only gaseous elements can cross the mesh.

5.3.1.4 MQ-2 SENSOR

MQ-2 sensor detects combustible gasses, namely Liquified Petroleum Gas, C_3H_8 and H_2 because of the reducing property of these gasses. The reactive material inside the MQ-2 [22] sensor is a ceramic of Al_2O_3 with an external layer of tin oxide (SnO_2). In case of low concentration region of detectable gasses, the heated element of sensor absorbs oxygen which reduces the amount of current through the sensor due to lack of free ions. While when the concentration of combustible gasses is higher a reduction reaction takes place that increases the sensor's conductivity. This change in conductivity corresponds to the change in voltage signal and the sensitivity characteristics help determine the concentration of gas ranging from 200–10,000 ppm. The typical characteristic curve of the MQ-2 is shown in Figure 5.7, the ordinate axis shows the resistance ratio of the sensor (Rs/Ro), the abscissa depicts the concentration of gasses, where; Rs is the observed sensor resistance; Ro is the sensor resistance at 1,000 ppm H_2.

Figure 5.7 shows the internal structure of an MQ sensor that consists of six leads connected with the central sensing element with Platinum wires. The pair of H lead is used for heating the sensor element while the other two pairs of A and B leads are responsible for carrying the output signal to the microcontroller.

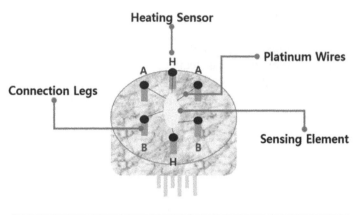

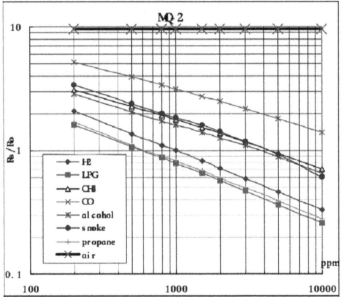

FIGURE 5.7   Sensitivity characteristics and internal structure of MQ-2 sensor.

5.3.1.5   MQ-4 SENSOR

MQ-4 is highly sensitive to natural gas, CH_4 and LNG; it can measure gas concentrations from 200–10,000 ppm. It has an operating voltage of 5V that preheats the sensor element for accurate functioning. The structure and design of the sensor is identical to MQ-2, and it is comprising of Tin Dioxide (SnO_2) as sensing layer. Figure 5.8 shows the sensitivity characteristics of the MQ-4. The detector is recommended to be calibrated for 5,000 ppm of

CH$_4$ concentration in the air with the value of load resistance (R$_L$) equals 20 KΩ. The connection schematic for the sensor is as shown in Figure 5.9.
    Sensor resistance can be calculated as:

$$Rs = (Vc / V_{RL} - 1) \times R_L \qquad\qquad (3)$$

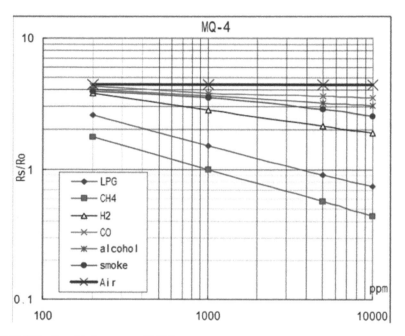

FIGURE 5.8   Sensitivity curve of MQ4.

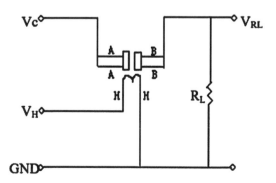

FIGURE 5.9   Schematic for MQ4 connections.

5.3.1.6   MQ-7 SENSOR

MQ-7 sensor is primarily used for detecting the concentration of Carbon Monoxide (CO) varying from 0 to 10,000 ppm. It uses SnO_2 as sensor's reactive material as it exhibits low conductivity in clean air. The sensor works on the cyclic method of detection by high and low temperature, maintained by heating at changing the voltage of 1.5 V and 5 v respectively. It detects CO at low temperature and a rise in conductivity is observed with the rise in CO gas concentration [23]. The higher temperature of the sensing element is used for the removal of other gases adsorbed at low temperature. The change in conductivity causes voltage change proportional to the gas concentration which is read by the MCU. The relationship from the sensitivity curve shown in Figure 5.10 should be used to calibrate the sensor by taking the initial value of RL equals to 10 kΩ.

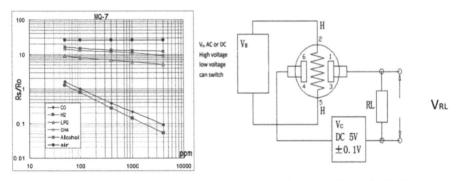

FIGURE 5.10   Sensitivity characteristics and schematic for connections of MQ-7 sensor.

5.3.2   INTERFACING OF SENSORS WITH RASPBERRY PI

All of the above sensors used as part of the Air quality monitoring system shown in Figure 5.11 output analog signals except the DHT22, i.e., Temperature, and Humidity sensors which gives Digital signal output. Hence before the signals are processed in Raspberry Pi [24], they are supplied to a 12-bit analog to digital converter (ADC) which provides a Digital output through its SDA (serial data) and SCL (serial clock) pins to the Raspberry pi. These sensors are connected to the ADC to its 4 Analog input channels. The readings from these sensors are processed by a python script which runs as soon as the Raspberry pi boots hence the system starts taking the readings when the drone is powered on. The readings from each sensor are restricted to a range of 0–1,024 ppm value for all sensors except PM2.5 sensors [25, 26].

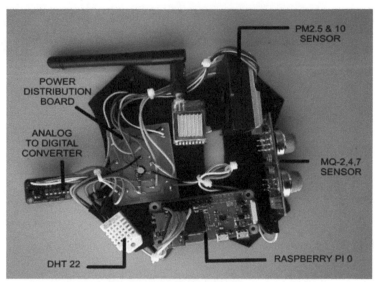

FIGURE 5.11 Mountable AQI measurement system.

The python script stores all readings in a CSV file along with the date, time, and GPM coordinates of the location from where readings are being taken. The GPS used is an Ublox M8N GPS which gives precise location with up-to 2 meters of accuracy. The CSV files can be retrieved from the internal memory of raspberry pi through remote access and the data is also posted on a remote server through an API post request generated by the python script [27]. An internet dongle is connected to the raspberry pi though its USB port which transmits the data to the remote server. Hence data can be accessed during the operation enabling real time monitoring of AQI. This gives immense advantage when measuring gas leakage in a factory area or over gas pipelines. Figure 5.12 shows the flowchart of the operations carried by the sensing unit.

This entire AQI monitoring system can be detached from the top plate of the drone and act as add on for any drone system. Entire set-up is powered by 5v DC input to the raspberry pi and such regulated 5 V DC sources are abundant in drone system. Table 5.3 depicts the CSV file generated for the sensor data by Raspberry pi [28].

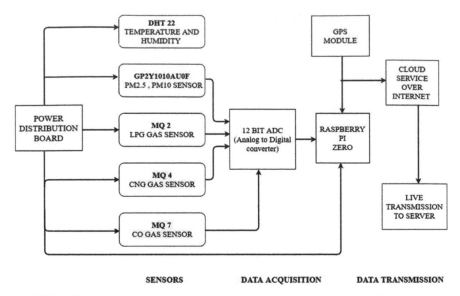

FIGURE 5.12   Interconnection of various sensors with raspberry pi and ADC.

TABLE 5.3   Sample Data from the CSV File Generated

Date & Time	PM2.5	Temperature	Humidity	Methane Level	LPG Level	Carbon Monoxide Level
2/9/2020; 2:30:20	210	22	39	1.404899536	2.751241547	5
2/9/2020; 2:30:23	210	22	38	1.40469824	2.751326808	5
2/9/2020; 2:30:26	208	22	38	1.404682973	2.751579738	4

5.3.3   STRUCTURAL FEATURES OF THE PRESENTED DRONE

The Quad-copter has a Fibreglass center plate with the arms made of ABS material with Carbon fiber tube through it to provide additional strength. The empty weight of the quadcopter weighs about 400 gm's. The AUW of the quadcopter with a 550 gm's LI-ion battery pack is 1,600 gm's including a 120-gm payload box. Figure 5.13 shows the structure of the quadcopter with all the components [29].

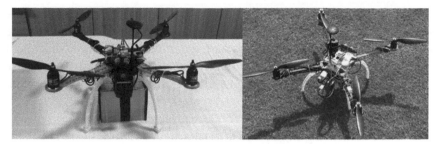

FIGURE 5.13   Mobile platform for AQI monitoring system.

This Multi-purpose quadcopter system is designed to carry out multiple tasks at once, Along with carrying the Scientific payload of the Air quality monitoring system which weighs about 150 gm's the quadcopter has a 3-d printed ABS payload box attached to its bottom plate between the landing gears designed [30] to carry an additional payload of about 1 Kg. The various flight time offered by the quadcopter system along with its All up weight with a 9,000 mAh Li-ion battery is shown in Table 5.4 [31].

TABLE 5.4   Flight Time and Payload Characteristics of Drone

Payload Type	All Up Weight of Quad-Copter	Payload Weight	Flight Time (min)
Without payload	1,600 gm	0 gm	22
AQI system	1,750 gm	150 gm	19
Secondary payload	2,600 gm	1,000 gm	10

5.4   TROPOSPHERE PARAMETERS MEASUREMENT

The national capital, Delhi has always been among the world's top polluted cities, with a total of 24 government approved industrial areas in the capital. Monitoring air quality in these areas must be a prime concern for the healthy lives of residents [32] in the surroundings. Our system caters to this need, and it can measure the parameters, namely total particulate matter, i.e., PM2.5 and PM10, Carbon Monoxide (CO) levels, temperature, and humidity in the tropospheric region of the atmosphere. Alongside it can monitor gas leakages for Natural gas and liquefied petroleum gas. Vehicular and factory exhausts have high levels of Carbon Monoxide, which is poisonous to humans [33]. This gas cannot be detected by humans due to the absence of any taste or smell hence it must be monitored using sensors in all areas.

The World Health Organization suggests a potential health risk to people exposed to more than nine ppm CO for a large period of their lives. The parameters being measured by the system can give potential insights about the air quality of an area where static government monitoring systems are not present and help authorities to generate localized air quality distribution maps for the city.

The two chosen locations for the mobile analysis of tropospheric air quality were Sri Aurobindo Marg which is an arterial road in South Delhi and Shaheed Sukhdev College of Business Studies, which is close to an industrial area in North West Delhi. The idea behind this experiment was to draw a contrast between the air quality of a residential area near an industrial hub and a residential area in near an affluent location. Both the locations have government monitoring centers where data updates after every hour; this will help in checking the accuracy of the readings of the proposed scientific payload. The details of the experimental model are shown in Table 5.5, and the drone was flown at the two locations on different days at five different times during the day, the average of the sample data collected was compared with the average of two consecutive hours of government data. Tables 5.6 and 5.7 depicts the sample data from location 1 and location 2 collected by the system.

TABLE 5.5   Details of Experimental Model

Location Details		Time Slot				
Location 1	Sri Aurobindo Marg	8:30 AM	11:30 AM	2:30 PM	5:30 PM	8:30 PM
Location 2	Shaheed Sukhdev College of Business Studies	8:30 AM	11:30 AM	2:30 PM	5:30 PM	8:30 PM

Sample Reading [Location 2 Readings (Shaheed Sukhdev College of Business Studies)—Date: 10 November 2019, 8:30 PM]

Figure 5.14 shows the sample reading of PM2.5 while Figure 5.15 shows the reading of temperature, humidity, and CO recorded by the AQI monitoring system.

5.5   COMPARISON OF DRONE DATA AND WEATHER DEPARTMENT DATA

In this section, a detailed quantitative analysis of the experimental setup is discussed along with a comparison of data from the government AQI monitoring systems installed near both the test locations. It was observed during work that the quality of air has immensely deteriorated for location 2, which is closer to an industrial area, thus exposing the people living in the surroundings to relatively higher health risks.

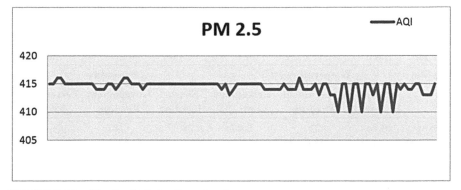

FIGURE 5.14   Plot for PM2.5 at Location 2.

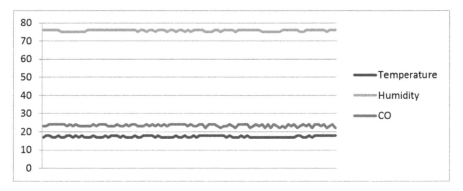

FIGURE 5.15   Plot for temperature, humidity, and CO at Location 2.

Tables 5.6 and 5.7 lists the observations recorded and transmitted for location 1 and location 2, respectively. The listed readings in the table are an average of over 270 values recorded for five different time slots with each flight spanning about 15 minutes.

TABLE 5.6   Readings at Location-1 (Sri Aurobindo Marg)

Time Slot	PM 2.5 (PPM)			Co Level (PPM)		
	Sensor Reading	Government Reading	Percentage Error	Sensor Reading	Government Reading	Percentage Error
8:30 AM	143.25	144	0.52	6.23	6	3.83
11:30 AM	142.47	142	0.33	3.10	3	3.33
2:30 PM	117.43	117	0.36	4.14	4	3.5
5:30 PM	109.89	109	0.81	1.94	2	3.0
8:30 PM	128.6	129	0.31	2.06	2	3.0
Time Slot	Temperature (°C)			Humidity		
	Sensor Reading	Government Reading	Percentage Error (%)	Sensor Reading	Government Reading	Percentage Error
8:30 AM	20.86	21	0.67	78.37	78	0.47
11:30 AM	27.31	27	1.14	52.45	52	0.86
2:30 PM	30.23	30	0.76	37.77	38	0.60
5:30 PM	27.75	28	0.89	58.64	59	0.61
8:30 PM	25.29	25	1.16	74.22	74	0.29

TABLE 5.7   Readings at Location-2 (Shaheed Sukhdev College of Business Studies)

Time Slot	PM 2.5			Co Level		
	Sensor Reading	Government Reading	Percentage Error	Sensor Reading	Government Reading	Percentage Error
8:30 AM	295.29	296	0.23	10.2	10	2
11:30 AM	206.35	207	0.31	7.8	8	2.5
2:30 PM	210.51	210	0.24	4.14	4	3.5
5:30 PM	167.23	168	0.48	3.87	4	3.25
8:30 PM	204.24	203	0.61	17.67	18	1.83
Time Slot	Temperature			Humidity		
	Sensor Reading	Government Reading	Percentage Error	Sensor Reading	Government Reading	Percentage Error
8:30 AM	11.12	11	1.09	81.37	81	0.45
11:30 AM	20.81	21	0.9	63.49	64	0.79
2:30 PM	22.75	23	1.08	37.77	38	0.60
5:30 PM	21.10.5	21	0.65	41.32	41	0.78
8:30 PM	14.87	15	0.86	64.24	64	0.37

Data from the table describes the hazardous concentration of particles like PM-2.5 and carbon monoxide (CO) in parts per million presents in the atmosphere. Also, it shows that the ambient air parameters like temperature and humidity are also affected by the growth of industrial regions. To validate the data of the experiment, the error percentage is calculated for the average reading of the scientific payload onboard and actual government data. The result shows that the error for all measured quantities is within the permissible error limit of 4%. This makes the recorded data usable for checking and generating localized air quality maps for inaccessible industrial areas around the country.

The other sensor readings of both MQ-2 and MQ-4 responsible for detecting LPG and CH_4 / LNG gas respectively remains unaffected with location and time due to inherently low concentrations of these gasses in the atmosphere. Hence these sensors are of use only in case of raising the alarm for any gas leakage at a plant or mishap in the surroundings.

The collected data also highlights the trend in AQI throughout the day as shown in Figure 5.16, where it can be observed that the morning and late evening hours see relatively higher pollution levels as compared to mid-day hours. The prime factor responsible for such a trend is the varying vehicular activities [34, 35] throughout the day. The pattern of pollution is inevitably distinct for two densely populated locations in the national capital this must raise concern towards reallocating and penalizing the responsible factory setups in the area.

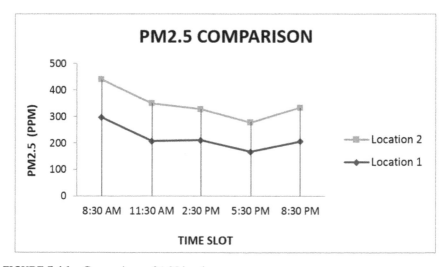

FIGURE 5.16   Comparison of AQI level.

5.6   CONCLUSION

The AQI measuring payload of the quadcopter is not only useful for measuring pollutants like particulate matter but also for detecting gas leakages in an industrial complex such as methane and propane. This system can particularly be useful in places where human access is limited, such as river banks where factory waste is released into rivers. The concentration and nature of pollutants in the factory waste can be regularly checked using the quadcopter to determine the type of gasses and its concentration (in ppm) generated by the waste [36]. The data can be directly transmitted to the control center, and a wide variety of analog or digital sensors can be used for detecting the different types of gasses giving a wide variety of application to the system. Since the error in readings were minimal (<4%) when data were compared with data provided by the metrological department hence the accuracy of the system is validated for commercial applications. This quadcopter is capable of carrying payload up to 1 kg hence more sophisticated sensors can be used in different configurations depending on the requirement. Such systems will also eliminate the need of having multiple gasses and AQI measuring stations at a close location as the advantages this mobile platform provides is that of longer flight time, up to 22 minutes depending on the payload although quadcopters these days easily provide a flight time of over 1 hour with lithium-ion batteries hence these systems can be used for recording data at multiple locations and at different altitudes (up to 1 km) in a single flight, increasing the practicality of the UAV and eliminating the need of multiple AQI measuring stations and sensors in factories or chemical plants. With swappable batteries, the requirement of multiple flights can be met without having multiple UAV's [37]. This UAV system can give sensor output over a wide range (0–1,024 ppm) and over-temperature ranges from –25°C to +60C making it suitable for operations in extreme conditions which are only limited by the battery performance and flight time delivered.

Future upgrades and possibilities for such AQI measuring drones may include the addition of sensors that detect harmful industrial gases such as phosgene, chlorine, nitrogen dioxide and other carcinogenic industrial wastes that may be present in large quantities due to leakage in a particular area. Surprise checks can be done over factories that deal with such bi-products. This AQI system can be easily separated from this multi-purpose drone and be used as a separate entity as it just requires a 5 V DC supply for operation and is independent of drone's operation and working hence this system can be widely used in any commercially available UAV with proper mount for the sensors, and its subsystems. Longer range telemetry and data transmission from AQI system can be met by enabling 4G connectivity on Raspberry pi mini-computer to which the sensors are interfacing hence eliminating the need of high power 2.4/5.8 GHz transmitters and receivers for sensor data gathering. The data recorded can be directly updated to any website or cloud space through the raspberry pi thus providing wide access to the data gathered.

KEYWORDS

- **artificial neural network**
- **gas sensors**
- **internet of things**
- **monitoring**
- **PM 2.5**
- **quadcopter**
- **tropospheric pollutants**

REFERENCES

1. Kindervater, K. H., (2016). The emergence of lethal surveillance: Watching and killing in the history of drone technology. Security Dialogue, 47(3), 223–238.
2. Li, B., Fei, Z., & Zhang, Y., (2019). UAV communications for 5G and beyond: Recent advances and future trends. In: *IEEE Internet of Things Journal* (Vol. 6, No. 2, pp. 2241–2263).
3. Zeng, Y., Zhang, R., & Lim, T. J., (2016). Wireless Communications with Unmanned Aerial Vehicles: Opportunities and Challenges, 54(5), 36–42. In: *IEEE Communications Magazine*.
4. Huo, Y., Dong, X., Lu, T., Xu, W., & Yuen, M., (2019). Distributed and multilayer UAV networks for next-generation wireless communication and power transfer: A feasibility study. In: *IEEE Internet of Things Journal* (Vol. 6, No. 4, pp. 7103–7115).
5. Mozaffari, M., Saad, W., Bennis, M., & Debbah, M., (2017). wireless communication using unmanned aerial vehicles (UAVs): Optimal transport theory for hover time optimization. In: *IEEE Transactions on Wireless Communications* (Vol. 16, No. 12, pp. 8052–8066).
6. Bolla, G. M., *et al.*, (2018). ARIA: Air pollutants monitoring using UAVs. In: *2018 5th IEEE International Workshop on Metrology for AeroSpace (MetroAeroSpace)* (pp. 225–229). Rome.

7. Tosato, P., Facinelli, D., Prada, M., Gemma, L., Rossi, M., & Brunelli, D., (2019). An autonomous swarm of drones for industrial gas sensing applications. In: *2019 IEEE 20th International Symposium on a World of Wireless, Mobile and Multimedia Networks (WoWMoM)* (pp. 1–6). Washington, DC, USA.

8. Carrozzo, M., *et al.*, (2018). UAV intelligent chemical multisensor payload for networked and impromptu gas monitoring tasks. In: *2018 5th IEEE International Workshop on Metrology for AeroSpace (MetroAeroSpace)* (pp. 112–116). Rome.

9. Park, S. K., O'Neill, M. S., Stunder, B. J., Vokonas, P. S., Sparrow, D., Koutrakis, P., & Schwartz, J., (2007). Source location of air pollution and cardiac autonomic function: Trajectory cluster analysis for exposure assessment. In: Journal of Exposure Science & Environmental Epidemiology (Vol. 17, No. 5, pp. 488–497).

10. Yang, Y., Zheng, Z., Bian, K., Song, L., & Han, Z., (2018). Real-time profiling of fine-grained air quality index distribution using UAV sensing. In: *IEEE Internet of Things Journal* (Vol. 5, No. 1, pp. 186–198).

11. Shen, C. H., Albert, F. Y. C., Ang, C. K., Teck, D. J., & Chan, K. P., (2017). Theoretical development and study of takeoff constraint thrust equation for a drone. In: *2017 IEEE 15th Student Conference on Research and Development (SCOReD)* (pp. 18–22). Putrajaya.

12. Salih, A. L., Moghavvemi, M., Mohamed, H. A., & Gaeid, K. S., (2010). Flight PID controller design for a UAV quadrotor. In: Scientific Research and Essays (Vol. 5, No. 23, pp. 3660–3667).

13. Ostojić, G., Stankovski, S., Tejić, B., Đukić, N., & Tegeltija, S., (2015). Design, control and application of quadcopter. In: International Journal of Industrial Engineering and Management (Vol. 6, No. 1, pp. 43–48).

14. Ebeid, E., Skriver, M., Terkildsen, K. H., Jensen, K., & Schultz, U. P., (2018). A survey of open-source UAV flight controllers and flight simulators. In: Microprocessors and Microsystems (Vol. 61, pp. 11–20).

15. Khofiyah, N. A., Maret, S., Sutopo, W., & Nugroho, B. D. A., (2018). Goldsmith's commercialization model for feasibility study of technology lithium battery pack drone. In: *2018 5th International Conference on Electric Vehicular Technology (ICEVT)* (pp. 147–151). Surakarta, Indonesia.

16. Berrahal, S., Kim, J. H., Rekhis, S., Boudriga, N., Wilkins, D., & Acevedo, J., (2016). Border surveillance monitoring using quadcopter UAV-aided wireless sensor networks. In: Journal of Communications Software and Systems (Vol. 12, No. 1, pp. 67–82).

17. Rohi, G., & Ofualagba, G., (2020). Autonomous monitoring, analysis, and countering of air pollution using environmental drones. In: Heliyon (Vol. 6, No. 1, p. 03252).

18. Wang, T., Han, W., Zhang, M., Yao, X., Zhang, L., Peng, X., Li, C., & Dan, X., (2020). Unmanned aerial vehicle-borne sensor system for atmosphere-particulate-matter measurements: Design and experiments. In: Sensors (Vol. 20, No. 1, pp. 57–58).

19. Bogdan, M., (2016). How to use the DHT22 Sensor for Measuring Temperature and Humidity with the Arduino Board, 68(1), 22–25. In: ACTA Universitatis Cibiniensis.

20. Feteira, A., (2009). Negative temperature coefficient resistance (NTCR) ceramic thermistors: An industrial perspective. In: Journal of the American Ceramic Society (Vol. 92, No. 5, pp. 967–83).

21. Tombeng, M. T., (2017). Prototype of gas leak detector system using microcontroller and SMS gateway. In: Cogito Smart Journal (Vol. 3, No. 1, pp. 132–138).

22. Trisnawan, I. K. N., Jati, A. N., Istiqomah, N., & Wasisto, I., (2019). Detection of gas leaks using the MQ-2 gas sensor on the autonomous mobile sensor. In: *2019 International Conference on Computer, Control, Informatics and its Applications (IC3INA)* (pp. 177–180). Tangerang, Indonesia.
23. Ibrahim, A. A., (2018). Carbon dioxide and carbon monoxide level detector. In: *2018 21st International Conference of Computer and Information Technology (ICCIT)* (pp. 1–5). Dhaka, Bangladesh.
24. Serrezuela, R. R., Cardozo, M. Á. T., Ardila, D. L., & Perdomo, C. A. C., (2017). Design of a gas sensor based on the concept of digital interconnection IoT for the emergency broadcast system. In: Journal of Engineering and Applied Sciences (Vol. 12, No. 22, pp. 6352–6356).
25. Rathod, M., Gite, R., Pawar, A., Singh, S., & Kelkar, P., (2017). An air pollutant vehicle tracker system using gas sensor and GPS. In: *2017 International Conference of Electronics, Communication and Aerospace Technology (ICECA)* (pp. 494–498). Coimbatore.
26. Maksimović, M., Vujović, V., Davidović, N., Milošević, V., & Perišić, B., (2014). Raspberry Pi as Internet of things hardware: Performances and constraints. In: Design Issues (Vol. 3, No. 8).
27. Russell, S. F., & Paradiso, J. A., (2014). Hypermedia APIs for Sensor Data: A Pragmatic Approach to the Web of Things. In MIT Open Access Articles.
28. Zhao, C. W., Jegatheesan, J., & Loon, S. C., (2015). Exploring IoT application using raspberry pi. In: International Journal of Computer Networks and Applications (Vol. 2, No.1, pp. 27–34).
29. Bertrand, B., Colin, Y., Patron, A., & Pho, V., (2016). Inventors; FatdoorInc, assignee. Quadcopter with a Printable Payload Extension System and Method. United States patent US 9,457,901.
30. Culler, E., Thomas, G., & Lee, C., (2012). A perching landing gear for a quadcopter. In: 53rd AIAA/ASME/ASCE/AHS/ASC Structures, Structural Dynamics and Materials Conference 20th AIAA/ASME/AHS Adaptive Structures Conference 14th AIAA (p. 1722).
31. Yadav, S., Sharma, M., & Borad, A., (2017). Thrust efficiency of drones (quadcopter) with different propellers and there payload capacity. In: International Journal of Aerospace and Mechanical Engineering (Vol. 4, No. 2, pp. 18–23).
32. Wen, X. J., Balluz, L., & Mokdad, A., (2009). Association between media alerts of air quality index and change of outdoor activity among adult asthma in six states, BRFSS, 2005. In: Journal of Community Health (Vol. 34, No. 1, pp. 40–46).
33. Lave, L. B., & Seskin, E., (1973). Air pollution and human health. In: Readings in Biology and Man (pp. 169–294).
34. Kathuria, V., (2002). Vehicular pollution control in Delhi. Transportation Research Part D: Transport and Environment (Vol. 7, No. 5, pp. 373–287).
35. Goyal, S. K., Ghatge, S. V., Nema, P. S., & Tamhane, S. M., (2006). Understanding urban vehicular pollution problem vis-a-vis ambient air quality-case study of a megacity (Delhi, India). In: Environmental Monitoring and Assessment (Vol. 119, pp. 557–569).
36. Seiber, C., Nowlin, D., Landowski, B., & Tolentino, M. E., (2018). Tracking hazardous aerial plumes using IoT-enabled drone swarms. In: *2018 IEEE 4th World Forum on Internet of Things (WF-IoT)* (pp. 377–382). Singapore.

37. Chae, H., Park, J., Song, H., Kim, Y., & Jeong, H., (2015). The IoT based automate landing system of a drone for the round-the-clock surveillance solution. In: *2015 IEEE International Conference on Advanced Intelligent Mechatronics (AIM)* (pp. 1575–1580). Busan.
38. Multi-purpose AQI quad-copter prototype was presented at MHRD organized Smart India Hackathon (2019). Hardware Edition (Ministry of Innovation) Winning 1st position. Link: https://www.sih.gov.in/pdf/past_events/hardware_2019_winner.pdf Entry No.28 (Team falcon 7) (accessed on 19 November 2021).

CHAPTER 6

Advanced Object Detection Methods for Drone Vision

MAHENDRA KUMAR GOURISARIA,[1] G. M. HARSHVARDHAN,[1]
NITIN S. GOJE,[2] SOUBHAGYA SANKAR BARPANDA,[3] and
SACHI NANDAN MOHANTY[4]

[1]School of Computer Engineering, KIIT Deemed to be University,
Bhubaneswar, Odisha, India, E-mail: mkgourisaria2010@gmail.com
(M. K. Gourisaria)

[2]Department of Computer Science, Webster University, Tashkent,
Uzbekistan

[3]School of Computer Science and Engineering, VIT-AP University,
Amravati, Maharashtra, India

[4]Department of Computer Science and Technology, IcfaiTech, ICFAI
Foundation for Higher Education, Hyderabad, Telangana, India

ABSTRACT

The use of unmanned aerial vehicles (UAVs) is increasing by the day, be it
in the public or the private sector. The use of UAVs requires a well-defined
architecture at its core for coordinated and proper functioning; a framework
called the internet of drones (IoD). Most UAVs are equipped with on-board
cameras useful for monitoring of potential obstacles by the controller of the
UAV from a remote location. Moreover, today we have work galore in the
computer vision sector which focus on development of various deep learning
(DL) algorithms for object detection and even image segmentation. Some
researchers have even started speculating the challenges involved of using

The Internet of Drones: AI Applications for Smart Solutions. Arun Solanki, PhD, Sandhya Tarar, PhD,
Simar Preet Singh, PhD & Akash Tayal, PhD (Eds.)

such algorithms in drone vision. This chapter describes the potential of such DL techniques applied on real-time footage gathered by the on-board cameras on the UAVs. These technologies will aid object tracking, self-navigation, and obstacle detection and collision avoidance capabilities. Machine learning (ML) algorithms have found immense applications in almost every area of modern research. A specialized branch of ML is DL which mainly deals with neural networks. These neural networks are designed to mimic the human brain and form the basis of many advanced artificial intelligence (AI) applications, such as smart computer vision. The DL computer vision architectures we study in this chapter rely heavily on the use of neural networks, which we shall describe in detail.

In particular, we discuss three state-of-the-art DL computer vision architectures namely Faster-R convolutional neural networks (faster R-CNN), single shot detection (SSD) and you only look once (YOLO). We discover how advanced computer vision can help UAVs maintain a real-time database of land-based entities with the use of different ways of annotations (2D annotations, 3D annotations, polygon annotation and semantic segmentation). For example, many UAVs operating over the same zone with the same IoD architectures may, with the use of the above technologies of algorithms mentioned, enumerate vehicles on different major roads in an urban area to maintain a real time traffic intensity projection over different sectors of the urban area. In other instances, drones with such capabilities may detect birds and measure their distances and take necessary steps to avoid a collision. These algorithms are very popular in computer vision using DL today and have been used in many applications.

6.1   INTRODUCTION

Governments use UAVs for military surveillance related activities, surveying a large geographical area for possible suitable terrain to build infrastructure, monitoring the population of endangered species in wildlife reserves, rescue operations, etc. [1, 2]. Whereas the private sector uses these UAVs for cinematography, delivering services, checking the ripening of crops in agriculture, etc. In agriculture, UAVs may use technologies such as LiDAR, which is a technique to measure distances through laser to check the height of crops. Moreover, UAVs equipped with thermal sensors can be used to determine the temperature of livestock, or the presence / temperature of water. Big companies like Amazon, Google, Matternet, and DHL have taken up the use of drones for delivering goods to customers and plan to leverage their advantages [10–13]. Gharibi et al. [14] lay out an IoD architecture with an IoD system. The architecture comprises various sets of rules and vocabulary like airways, intersections, zones, inbound, and outbound gates, etc., useful in the traffic control of a high number of drones operating simultaneously. The IoD system, on the other hand, refers to a set of protocols and algorithms working atop the architecture for proper coordinated function of drones. We discuss the navigational structure of the IoD architecture in detail in this chapter. The structure of this chapter is as follows: In Section 6.2, we review the related work where people have applied YOLO, Faster R-CNN, and SSD in different domains [6–9]. Section 6.3 talks about the different kinds of annotations that are possible in computer vision which can potentially be applied to drone vision. Section 6.4 describes the structure of IoD as proposed by Gharibi et al. [14]. Section 6.5 introduces the fundamental concepts of deep learning (DL) by giving brief notes about the artificial neural network (ANN) and convolutional neural network (CNN) which reside at the very core of the advanced computer vision algorithms that we describe in Sections 6.6–6.8 (YOLO [5], Faster R-CNN [3] and SSD [4], respectively). Finally, we state the conclusion and future scopes of this chapter in Section 6.9. The organization of the chapter is illustrated by Figure 6.1.

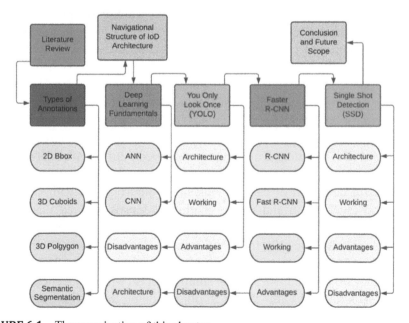

FIGURE 6.1   The organization of this chapter.

6.2   LITERATURE REVIEW

In this section, we will be reviewing how YOLO, Faster R-CNN and SSD have been applied in various object detection applications. To the best of our knowledge, these models have not been applied to specific drone-vision applications, hence, we shall only be discussing generic use-cases. One of these specific applications could be facial recognition for scanning crowds. Jiang et al. [16] used Faster R-CNN on the WIDER dataset [17] with the base CNN as the VGG16 [18] on a training set of 12,880 images comprising 159,424 faces to achieve state-of-the-art accuracy. They report that the region-proposal network (RPN), being end-to-end trainable with its property of sharing of convolutional features between the detector network (in this case, the VGG16), is the reason for its feasibility for face detection. Using the same dataset [17], Wang et al. [19] implemented Faster R-CNN by modifying certain properties such as the loss function and multi-scale training. The approach employs center loss [20] and online hard example mining [21] (OHEM) in addition to multi-scale training to attain state-of-the-art accuracy for the WIDER and FDDB [22] datasets. Zhang et al. [23] use the base network for R-CNN to be ResNet101 [24] which uses skip-connections between alternative layers pre-trained on generic ImageNet images for the detection of faces using multi-scale training technique. They use a light-head Faster R-CNN having a superior inference speed along with a vote-based ensemble prediction technique in the WIDER dataset [17]. In the context of facial recognition, YOLO, and SSD have also been employed by various researchers [25–28].

In the road industry, these algorithms have been applied galore. Studying about their accomplishments in regards to aspects of daily traffic allow us to entrust them for their application in drones for monitoring traffic activity. Xu et al. [29] employed Faster R-CNN with VGG16 [18] as the base network using a UAV (DJI Phantom 2 quadcopter equipped with a gimbal having tri-axial stabilization) to detect moving cars on roadways. They also implemented the successful detection of cars moving in an environment with changing illumination (such as in and out of shadows). It was noticed that Faster R-CNN had a superior performance in comparison to other object detection algorithms such as ViBe [30], frame difference [31], AdaBoost method using Haar-like features [32], and SVM with HOG features (SVM + HOG) [33]. Müller et al. [34] applied the SSD on small objects (which is something the original SSD struggled with)-the small objects being traffic lights using the DriveU Traffic Light Dataset [35]. They could correctly identify traffic lights appearing smaller than 10 pixels on the screen using the base network as inception [36]. In the same use-case, Zuo et al. [37] obtained a mean average precision (mAP) of 0.34 using Faster R-CNN keeping VGG16 [18] as the base network. Lin et al. [38] used YOLO to correctly identify the number of cars crossing through a user-defined checkpoint. This helped solve the problems as a consequence to false detection.

The live detection of objects through drone vision may be done via two approaches, a) on-board and b) remote. The on-board detection of objects is the approach where the drone has an internal CPU that can do the computations and has these algorithms internally implemented, the result of which is finally relayed back to the operator. Whereas remote detection is when the live feed is simply relayed back to the operator who has a CPU installed remotely which can do the computations required for object detection. Either way, sometimes in the case of on-board live detection, the algorithms may exceed normal levels of CPU overhead due to their computationally intensive nature. Due to this, optimized versions of these algorithms have been proposed. Most DL-based applications demand the hardware of graphics processing units (GPUs) for normal pace of functioning. To implement YOLO on mobile devices, Huang et al. [39] introduced YOLO-LITE to achieve state-of-the-art speed of object detection. They argue that batch normalization [40] for shallow networks is not always a boon; it may be detrimental to the speed of inference. This is because the phenomena of covariate shift and vanishing gradient problem are non-existent in small networks such as the YOLO-LITE. Womg et al. [7] proposed a miniaturized SSD architecture for object detection on embedded and mobile devices, namely Tiny SSD. The first component of this network composed of an optimized non-uniform 'Fire' subnet stack which minimized the use of 3×3 convolutional filters along with the input channels to those filters and down sampling the image at later stages. This component feeds a non-uniform subnet of optimized SSD-based auxiliary convolutional feature layers.

6.3   TYPES OF ANNOTATIONS USED IN COMPUTER VISION

In this section we discuss the types of annotations that may be used for real-time object recognition by drones. 2D bounding boxes generate a rectangular point of interest. This is mainly used for satellite-like images which are taken over a landscape. 2D bounding boxes may be used for detection of cars in a parking lot. 3D cuboids are used by advanced computer vision algorithms also used in self-driving cars. They are generated around the objects of interest. Advanced algorithms such as the Mask R-CNN [15] do something similar to this by creating a mask over objects for even more precise object segmentation. Sometimes, objects with different shapes or sizes are to be detected (such as the AC outlets on roofs or vehicles), in this case, polygon annotation works to create a polygon with n edges depending upon the shape of the object. The final form of annotation is image segmentation. Segmenting images semantically helps in classifying and localizing various kinds of objects at once and also especially used for objects that may not fit in a bounding box, such as huge fields or roads that extend on and on.

6.4   NAVIGATIONAL STRUCTURE OF THE IoD ARCHITECTURE

In this section, we introduce a set of vocabulary useful for understanding the operation of drones and how they can navigate under a fixed framework. From the various services that drones may provide, all of them share a common denominator as a requirement-navigation. In this section we only put forward the IoD architecture structure proposed by Gharibi et al. [14]. For this, a vocabulary is formulated and defined further.

1. **Airspace:** The spatial resource used and shared by all the drones. Within an airspace all drone-related services are enclosed and no drone may operate outside the limits of the airspace.
2. **Airways:** These are analogous to roads for vehicles. Just as drivers may not go off-road but follow a network of roads to reach their destination, drones, while navigating, are required to stay on a defined airway.
3. **Intersections:** When (at least) two airways intersect, they form an intersection.
4. **Nodes:** These are defined to be the point-of-interest locations for the drones to, say, deliver their services. The unique quality of nodes is that a drone inside a node is in free-fly mode. This means that there exist no airways in a node and the drone is able to maneuver freely within the limits of the node. Figure 6.2 shows a typical set of airway, intersection, and node.

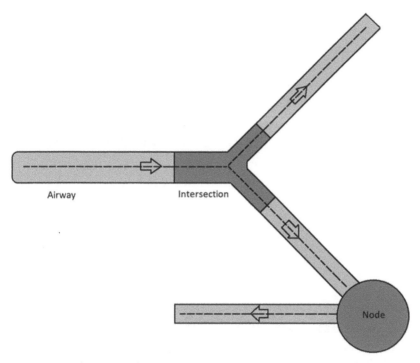

FIGURE 6.2   Demonstration of airways, intersection, and nodes.

5. **Zones:** The partition of an airspace is done through zoning; each zone contains its own set of airways, intersections, and nodes.
6. **Gates:** These allow drones to switch between adjacent zones. They are classified as *inbound* or *outbound* gates. Gates are belonged by both zones between which they lie, and no airway is allowed to cross zonal borders without it being segmented with a gate. A zone graph may be defined as a framework of nodes and intersections as vertices and airways as directed edges. Figure 6.3 demonstrates a zone graph.
7. **Pathway:** Paths in the zone graph are called pathways.
8. **Element:** Any component such as nodes, intersection, and airways. Every element must be reachable from any other element in the graph.
9. **Transits:** When directed airways lead from a source zone to a destination zone, the route is known as a transit.
10. **Interzone Graph:** Substitution of the gates to be intersections between any two zones makes up an interzone graph.

11. **Transit Cost:** The cost of drones moving from zone to zone (transit) through gates. This cost can be anything such as time, distance, energy consumption, etc. Figure 6.4 shows an interzone graph.

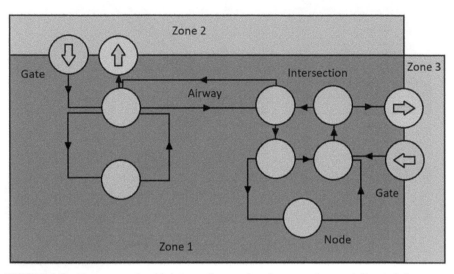

FIGURE 6.3   A zone graph with intersections and nodes as vertices and directed airways as the directed edges. As seen, Zone 1 is adjacent to Zone 2 and 3 and has inbound and outbound gates at the perimeter shared by the corresponding zones.

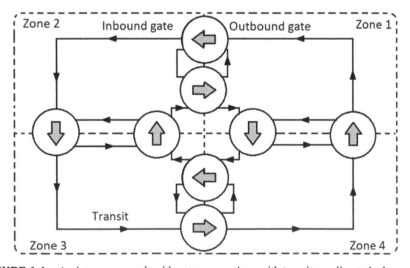

FIGURE 6.4   An interzone graph with gates as vertices with transits as directed edges.

12. **Route:** A path in the interzone graph is called route. This is called so because the paths in the interzone graph lead directly to other zones.
13. **Progress:** Pertains to a metric (such as distance) traveled or increased by a drone in an airway or intersection.
14. **Components and Attributes:** Each vertex and edge in the zone level contains XML information about itself which is unique to the vertex or edge.

6.5 DEEP LEARNING (DL) FUNDAMENTALS

In this section, we describe two fundamental DL models necessary for dealing with YOLO, Faster R-CNN, and SSD later. DL is a sub-section of machine learning (ML) which focuses on learning hidden underlying representations in data more efficiently. It has gained popularity very recently, more specifically, industries globally have started applying DL techniques only after a decade into the 21st century. These algorithms do not require to be told features to look for; as in, these algorithms are able to learn features on their own and decide which ones contribute more to the target and which don't. DL models generally outperform any pure ML model when there is a huge dataset. In almost every DL model there exists a neural network which does the learning. A neural network is an architecture inspired by the human brain and its architecture tries to mimic the actual brain with various neurons interconnected which fire information forward (in a feed-forward multilayer perceptron) in order to predict a target. If there is an error in the prediction, a process known as backpropagation takes place to adjust the weights between the nodal connections in such a way that in the next pass the error is minimized. Based on this fundamental knowledge, we introduce ANN and CNN.

6.5.1 ARTIFICIAL NEURAL NETWORK (ANN)

In this section we give a theoretical overview of ANNs. Most of the commonly used models in DL use the ANN as a full connection layer in some or the other form which makes it the most fundamental neural network. They have found various uses in applications such as speech recognition [41–44], facial recognition [45–48], character recognition [49–53], etc. In CNNs, as we shall see, ANNs are used in combination with some convolutional layers. Sometimes, these are also used with fractionally strided convolutional layers. Figure 6.5 shows the architecture of an ANN which can be divided into 3 major parts: the output, hidden, and the input layers. Directed, weighted connections are made between each layer's nodes, the weights which are adjusted while training. We assume the set of layers $A = \{A_1, A_2, \ldots A_N\}$ which has a total of N layers with $N-2$ hidden layers. We take a set of nodes given by $M = \{M_1, M_2, \ldots M_N\}$ where $M_i = \{M_i^1, M_i^2, \ldots, M_i^{B_i}\}$, $1 \leq i \leq N$ and B_i is the total number of nodes in the A_i layer. We take a set of weights for connections between nodes given by $V = \{V_2, V_3, \ldots V_N\}$ having the starting element as V_2 because the first layer does not have any connections to any previous layer thus those weights are not defined. An element in V is defined as $\{V_i^{1,1}, V_i^{1,2}, \ldots, V_i^{1,B_i}, V_i^{2,B_i}, \ldots, V_i^{B_{i-1},B_i}\}$. As an example, the weight between the 3rd node of A_4 and 2nd node of A_3 would be given by $W_4^{2,3}$. Each neuron receives input from all other nodes in the previous layer through weights, a value which is summed as given in Eqn. (1).

$$S = \sum_{j=1}^{B_{i-1}} W_i^{j,k} M_{i-1}^j, 1 \leq k \leq B_i \tag{1}$$

When the input layer receives inputs, these are propagated through the network in the forward direction with initially set random weights. Each layer is defined to have a particular activation function that is assigned to each of its nodes. This activation function is given by $\Psi(S)$ which decides whether or not to fire (propagate a high or low signal) to the next layers. Many types of activation functions are used, as given:

$$\Psi(x) = \frac{1}{1+e^{-x}} \tag{2}$$

$$\Psi(x) = \frac{1-e^{-2x}}{1+e^{-2x}} \tag{3}$$

$$\Psi(x) = max(0, x) \tag{4}$$

$$\Psi(x) = \begin{cases} 1 \, if \, x \geq 0 \\ 0 \, if \, x < 0 \end{cases} \tag{5}$$

Eqns. (2)–(5) are named as the sigmoid, hyperbolic tangent, rectifier, and threshold function respectively. After a single pass, a set $\hat{y}$ of predicted values at the output layer A_N is formed. A cost function, C is used, defined as:

$$C = \frac{1}{2} \sum_b \sum_a \left(\hat{y}_{a,b} - y_{a,b} \right)^2 \tag{6}$$

In Eqn. (6), b is an index over cases, or input-output pairs, and a is an index over output units. Eqn. (6) denotes one of the most commonly used cost functions in ANNs where y is a set of true values which the model is trying to predict. The main objective of the ANN is to minimize this cost function so that $\hat{y}$ and y are closer to each other. Gradient descent (GD) is the employed technique in this process, where the partial derivative of C with respect to each weight is calculated. This partial derivative is the sum of partial derivatives of each of the b input-output pairs. The partial derivative of C to the output of the output units is given by:

$$\frac{\partial C}{\partial \hat{y}_a} = \hat{y}_a - y_a \tag{7}$$

Applying the chain rule to $\partial C / \partial S_a$ which is the partial derivative of C with respect to the input of the output units representing how a change in the summed input S affects the error:

$$\frac{\partial C}{\partial S_a} = \frac{\partial C}{\partial \hat{y}_a} \frac{\partial \hat{y}_a}{\partial S_a} = \frac{\partial C}{\partial \hat{y}_a} \hat{y}_a \left(1 - \hat{y}_a \right) \tag{8}$$

Now, S is a linear function of weights of the links and states of lower-level nodes. Hence, error variance by a change in weights is given by:

$$\frac{\partial C}{\partial W_i^a} = \frac{\partial C}{\partial S_a} \frac{\partial S_a}{\partial W_i^a} = \frac{\partial C}{\partial S_a} \hat{y}_i \tag{9}$$

The contribution done by the output of the node i to $\partial C / \partial \hat{y}_i$ due to the effect of i on a is given by:

$$\frac{\partial C}{\partial S_a} \frac{\partial S_a}{\partial \hat{y}_a} = \frac{\partial C}{\partial S_a} W_i^a \tag{10}$$

Summing all the contributions of error from all the connections to unit i, we get:

$$\frac{\partial C}{\partial \hat{y}_i} = \sum_a \frac{\partial C}{\partial S_a} W_i^a \tag{11}$$

After sufficient number of feed-forward and backpropagation steps, the ANN is said to have learnt the complex underlying relationships of the input data to become a very powerful model to regress or classify new data inputs.

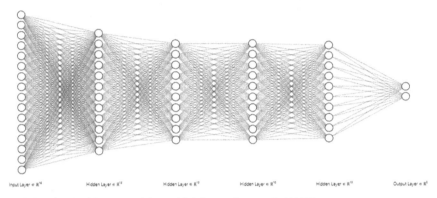

FIGURE 6.5   Architecture of the artificial neural network (ANN).

6.5.2   CONVOLUTIONAL NEURAL NETWORK (CNN)

Apart from ANNs, the most popular DL models are CNNs which are the backbone of computer vision. Proposed by LeCun et al. [54], CNNs compute image data based on spatial, pixel-wise relationships among pixels through the use of convolution of values of the pixels falling in a certain vicinity. They have two layers, a) convolutional layers, and b) fully connected layers (or dense layers). Dense layers are actually nothing but the layers of ANN which give an output. Convolutional layers are used to decrease the dimensionality of the input image and reduce features, only capturing some of the most important features through kernels used for feature detection (which are trained). These kernels are nothing but 2D matrices which slide over the entire image and apply convolution and other operations. Figure 6.6 shows the structure of a typical CNN.

The convolution operation of two functions $f(t)$ and $g(t)$ can be defined as:

$$f(t) * g(t) \triangleq \underbrace{\int_{-\infty}^{\infty} f(\tau) g(t-\tau) d\tau}_{(f*g)(t)} \tag{12}$$

The Conv layers in Figure 6.6 perform convolution operation given by Eqn. (12). Input data is down-sampled by convolutions which is further down-sampled due to the pooling layer. Pooling, usually max pooling, only focuses on the maximum values of the selected convolved features of pixels to retain maximum important information. Pooling helps to provide spatial invariance to CNNs, i.e., images can be classified even if they are flipped, distorted, tilted, etc. Due to the reduction of parameters by more than three quarters, it helps avoid overfitting. FC in Figure 6.6 denotes the full connection which, as we mentioned before, denotes a layer of ANN. The FC layer receives convolved, pooled features post flattening (as in, arrangement of the values into an array suitable to feed an ANN). These values may even by normalized or standardized before feeding the ANN.

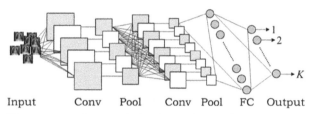

Input   Conv   Pool   Conv   Pool   FC   Output

FIGURE 6.6   The structure of a convolutional neural network.

6.6 YOU ONLY LOOK ONCE (YOLO)

YOLO is, surprisingly, only a deep CNN [5]. Its unified architecture does everything from predicting confidence scores of objects present in the image to actually generating bounding boxes of those objects with high precision. The main advantage of having a unified architecture is that various optimizations can be done to the architecture since there are no complex parts associated which may require optimization post identification of the part that needs it. Redmon et al. formulated the problem of object detection into a regression problem which only looks at the image once to get the segmentations with their class probabilities. Figure 6.7 shows the process that YOLO employs.

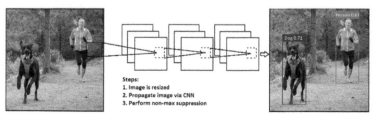

FIGURE 6.7   The images in YOLO are resized to 448×448 pixels on which a CNN is applied and thresholds the detections by the confidence.

Further, we talk about the workings of YOLO before stating its advantages and disadvantages.

6.6.1 ARCHITECTURE

Inspired from Google's LeNet (GoogLeNet [55]), YOLO employs 24 convolutional layers and 2 FC layers. 1×1 and 3×3 reduction layers are used in the convolutional layers. The architecture is shown by Figure 6.8.

FIGURE 6.8   The 24 convolutional layers followed by 2 FC layers in the YOLO architecture.

For the faster version of YOLO, Redmon et al. use 9 convolutional layers with fewer filters. Every layer uses a leaky rectified linear activation function (leaky reLU) on all layers except for the last one, for which the linear activation function is used. Leaky reLU is given by Eqn. (13) as:

$$F(x) = \begin{cases} x, x > 0 \\ 0.1x, x \le 0 \end{cases} \tag{13}$$

One problem faced during training of YOLO was that there were far more grid cells which did not contain an object rather than cells which did. This meant that the contribution of the grid cells having no object to the loss was more than the cells which actually contained objects; leading to divergence of the model early on during training. This problem of model instability is fixed by increasing loss of bounding box coordinate predicting grid cells and decreasing confidence predictions of cells that do not contain objects, denoted by λ_c and λ_n which are set as $\lambda_c = 5$ and $\lambda_n = 0.5$.

Because of the use of sum-squared error, equal weightage was given to deviations of the same scale for both large- and small-scale boxes. This problem was overcome through the use of square root of the bounding box width and height. During training, the multi-part loss function given by Eqn. (14) is optimized.

$$\lambda_c \sum_{i=0}^{G^2} \sum_{j=0}^{B} F_{ij}^{obj} \left[(x_i - \hat{x}_i)^2 + (y_i - \hat{y}_i)^2 \right]$$

$$+\lambda_c \sum_{i=0}^{G^2} \sum_{j=0}^{B} F_{ij}^{obj} \left[(\sqrt{w_i} - \sqrt{\hat{w}_i})^2 + \left(\sqrt{h_i} - \sqrt{\hat{h}_i} \right)^2 \right]$$

$$+ \sum_{i=0}^{G^2} \sum_{j=0}^{B} F_{ij}^{obj} (C_i - \hat{C}_i)^2 + \lambda_n \sum_{i=0}^{S^2} \sum_{j=0}^{B} F_{ij}^{noobj} (C_i - \hat{C}_i)^2$$

$$+ \sum_{i=0}^{S^2} F_i^{obj} \sum_{c \in classes} (p_i(c) - \hat{p}_i(c))^2 \qquad (14)$$

In Eqn. (14), F_i^{obj} is a flag variable that is 1 if the object exists in the ith cell, while F_{ij}^{obj} says that (since there are multiple bounding boxes associated to a single grid cell) the jth bounding box predictor in the ith cell assumes responsibility for the prediction. Figure 6.9 depicts some of the objects detected by YOLO.

FIGURE 6.9   The object detections done through the YOLO object detector.

6.6.2   WORKING

The input image is divided into a $G \times G$ grid. The grid cells in which the centers of objects fall are the cells that will be responsible for generating bounding boxes. In other words, each of the grid cell generates B bounding boxes which have their corresponding confidence scores. Confidence scores point out two things: a) how likely it is for an object to be in a bounding box, and b) how accurate it is in the decision of the object's class. Confidence is defined as:

$$Conf = P(O) \times IoU_{pred}^{truth} \tag{15}$$

In Eqn. (15), $P(O)$ refers to the probability of the occurrence of the object and IoU_{pred}^{truth} is the intersection over union (IoU) of the ground truth and predicted bounding box. $Conf$ is 0 when the grid cell does not contain the center of any object and is equal to IoU_{pred}^{truth} when $P(O)$ is 1, i.e., when the object is present in the cell. Each B bounding boxes contains a 5-tuple $\{x, y, w, h, Conf\}$ where x and y are the coordinates of the center of the bounding box relative to the grid cell (taking the grid cell as origin), w and h are the width and height of the bounding box relative to the full image.

A grid cell contains the conditional class prediction probabilities C over all classes which say which classes are present in the image whose center falls in the grid cell. This entity is given by $P(C_i|O)$. Each grid cell stores a single set of these values over all objects where i lies in the range of 0 to the number of objects in the dataset. During testing, the conditional class probabilities are multiplied with the bounding box confidence predictions as:

$$P(C_i|O) \times P(O) \times IoU_{pred}^{truth} = P(C_i) \times IoU_{pred}^{truth} \tag{16}$$

Eqn. (16) yields the probability scores for each class $P(C_i)$ depending upon the bounding box. Then, after non-max suppression, we get a clearer picture of only the major bounding boxes with actual objects in it with the class probability scores. The image is divided into an $S \times S$ (or $G \times G$) grid where each cell predicts B bounding boxes of varying size and position depending upon whether the center of an object falls in the cell before non-max suppression.

6.6.3 ADVANTAGES OF YOLO

- Many object detection algorithms suffer due to complex moving parts which make the detections slower. Hence, in real-time feeds the algorithms are only able to operate at a sub-par frames per second (FPS). YOLO, however, due to its simple architecture runs at 45 FPS without batch processing on a Titan X GPU. The faster version of YOLO (which is a smaller model-suffers with accuracy, however) operates at a whopping 150 FPS which essentially means there is ≤ 25 ms of delay.

- YOLO achieves more than double the mAP than other real-time detectors. We define precision by:

$$Precision = \frac{TP}{(TP + FP)} \qquad (17)$$

In such a case, true positive (TN) is defined to be the number of detections with IoU ≥ 0.5. For any object detector mechanism, a performance metric namely the IoU is defined to be the ratio of the area of the intersection of the bounding boxes of the ground truth and predicted region and the union of both the boxes. Hence:

$$IoU = \frac{Area\,of\,Overlap}{Area\,of\,Union} \qquad (18)$$

Now, mAP is just the mean of the average precision for each predicted class (including all instances of the bounding boxes predicting a certain class) in the image done by an algorithm.

- There is a global consideration of input images in the case of YOLO. In other words, as opposed to sliding windows and region-proposal based techniques, the whole input image is seen by YOLO which means it encodes contextual information of the classes implicitly. This makes YOLO make a lot lesser errors when background is taken into the context.
- YOLO is highly generalizable in terms of its applications in new domains. This is because the representations learnt by YOLO are highly generalized which make predictions on any variations of natural images not affect its performance.

6.6.4 DISADVANTAGES OF YOLO

- Suffers in accuracy and lags behind some other state-of-the-art models for object detection;

- Incapable of accurately localizing precisely the locations of small objects;
- Struggles to generalize objects in different configurations or unusual aspect ratios;
- Small objects appearing in groups are not clearly detected;
- Using sum-squared error weights error of localization of bounding boxes equally with the classification error may not be ideal to minimize mAP.

6.7 FASTER R-CNN

Faster R-CNN is a product of many iterations of similar models proposed before Ren et al. [3] came up with the algorithm. We shall describe a brief background of the works done by Girshick et al. [56] with R-CNN and then how it was modified to be faster in the Fast R-CNN done by Girshick et al. [57]. Understanding the primitive models helps gain an intuition as to how the Faster R-CNN functions. Later, shall see the advantages and disadvantages of the Faster R-CNN before we get into the details of its workings.

6.7.1 R-CNN

The original region CNN implementation [56] focused on using algorithms such as Selective Search [58] to extract bounding boxes that may potentially contain objects. Using a pre-trained CNN, it extracted features for each bounding box through transfer learning. Once the features were extracted, these high-level data representations were classified using a support vector machine (SVM). Figure 6.10 demonstrates this procedure. Selective Search examines the image at different scales and locations to minimize the number of proposed bounding boxes or region-of-interests (ROIs). It does this through a sliding window that goes over the image to detect objects.

The issue with R-CNN is that it was too slow for real-time object tracking-based applications. The causality being Selective Search which is not very fast. Further approaches were proposed that focused on localizing the objects in the figure through a *neural network* rather than an algorithm like Selective Search that could not be optimized to the needs.

1. Input image   2. Get region proposals (~2k)

FIGURE 6.10 The R-CNN accepts an input image and extracts almost around 2,000 region proposals through selective search to run a pre-trained CNN for computation of features to be fed into an SVM.

6.7.2   FAST R-CNN

Fast R-CNN [32] also used selective search for localization of objects however a *region of interest pooling* mechanism was included as shown in Figure 6.11.

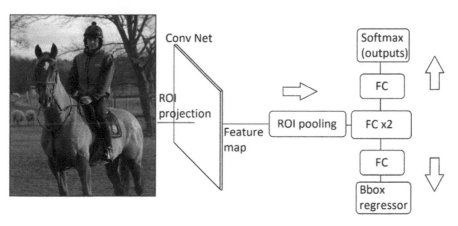

FIGURE 6.11 Faster R-CNN still uses selective search however uses ROI Pooling to extract a fixed-size window from the high-level data representation extracted from the conv feature map before sending into RoI layer to get class predictions and bounding boxes. Here, FC refers to full connection layer.

We will give an extended explanation of ROI pooling when we discuss Faster R-CNN. What ROI pooling does in a nutshell is extract a fixed-size window from the extracted features of the pre-trained CNN to propagate these features to a set of fully-connected layers to get output labels for the ROI. With this, the network was now end-to-end trainable, however, still suffered to get competitive accuracy due to Selective Search region proposal. To do away with this, Faster R-CNN was proposed which incorporated the region proposal into the network.

6.7.3 WORKING OF FASTER R-CNN

Faster R-CNN [3] removes the Selective Search algorithm for generating object proposal regions and instead includes a region proposal network (RPN) as shown in Figure 6.12. Features are extracted from the base network and fed to two components: the RPN, for localization of objects (end-to-end trainable) and the detection network where ROI pooling takes place. Faster R-CNN is just Fast R-CNN with the modification that the object localization is not done through an algorithm such as Selective Search but a fully convolutional network called the RPN. *Anchors* are placed uniformly over the image at varying aspect ratios and scales. This is examined by the RPN to output localization proposals of potential objects.

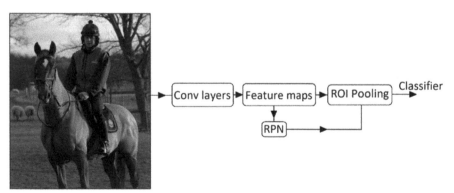

FIGURE 6.12 Structure of faster R-CNN. The object bounding box proposals are now incorporated into the fast R-CNN network itself with the RPN.

It is important to note that the RPN's job is to output an *objectness score* which specifies whether or not an object is present in a region. It does not predict the type of the object. The whole architecture is end-to-end trainable, from region proposal, feature extraction, predicting bounding boxes to finally predicting the class labels for each predicted bounding box.

6.7.4   ADVANTAGES OF FASTER R-CNN

- Object localization technique is trainable now that it is done by a fully convolutional neural network (CNN);
- Higher frame rates for detection (7–10 frames per second) allowing for real-time object detection;
- Sharing of convolutional features between the RPN and detection network leads to reduced computation cost;
- Considers a large number of possible regions (higher than R-CNN).

6.7.5   DISADVANTAGES OF FASTER R-CNN

- Too many complex moving parts involved, making the architecture complicated;
- Although it is end-to-end trainable, fine-tuning Faster R-CNN for different applications may be tedious due to optimization of each component;
- RPN needs to be pre-trained in order to generate bounding boxes;
- Long training time.

Now, we review the architecture of Faster R-CNN component-wise.

6.7.6   ARCHITECTURE

In the base network, the input image is fed to a pre-trained CNN which is done through transfer learning. Ren et al. used VGGNet [18] and ZF [59] as the base networks. The difference in the architecture of the base network CNN from normal CNNs is that the fully connected layers are truncated. Hence, when the input image is fed through the base network, we end up with a feature map. This is shown in Figure 6.13. The convolutional features (feature map) are shared between the ROI pooling module and the RPN. Thus, the base network is a *fully convolutional* component which means that it is fast. Fully connected layers tend to be slower than fully convolutional layers. Another advantage of fully convolutional layers is that they are able to accept images of different resolutions and sizes. This helps to detect small objects, since, if there are fixed image dimensions, during feature reduction through convolutions, tiny objects may appear to be pixelated. Varying sizes of different images helps curb this.

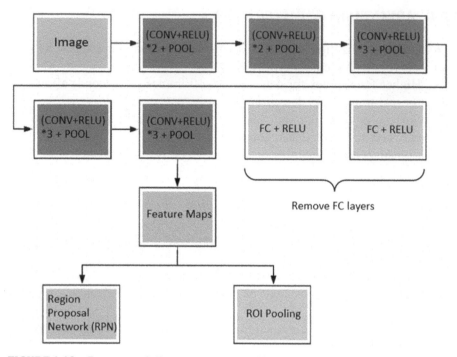

FIGURE 6.13   Base network. Features are extracted from the image through the convolutional layers and then fed into the RPN and ROI pooling, skipping the fully connected layers.

Traditionally, for object detection, the region proposals were either made by a sliding window, image pyramid or a selective search-lie algorithm. Faster R-CNN tackles the problem of predicting the sizes and position of bounding boxes through the use of anchors.

The raw (x, y) coordinates of bounding boxes are not predicted as it is. This would pose obvious problems such as the minimum bound should be less than maximum bound of the coordinates of both axes, and that the network could predict values that may be outside the image boundary. Hence, offset values are predicted that are relative to the position of the reference boxes, such as $\Delta_{x\text{-center}}$, $\Delta_{y\text{-center}}$, Δ_{height} and Δ_{width}. The reference boxes arise from sampling the input image every 17 pixels (typically).

Once the reference boxes are made, each reference box gets an anchor of the combinations of different scales and sizes. Since our image here is of 600×400 pixels, we have resolutions: 64×64, 128×128 and 256×256. Then we have aspect ratios of 1:1, 1:2 and 2:1. This yields a 3×3 combination set. The problem however, is that with this, a lot of anchors are generated.

A pre-trained CNN looping over such a huge amount of anchors would make the network as slow as the original R-CNN. Hence, RPN helps with nullifying the anchors which have low probability of containing an object as we discuss below.

Due to the huge amount of anchors, it is difficult to run a pre-trained CNN over all anchors and still have the network detect objects in real-time. To help with this, the RPN removes the bounding boxes that may not have an object. The RPN takes the convolutional feature map and applies a 3×3 *CONV* to learn 512 filters. The RPN reviews all the bounding boxes and outputs two *objectness scores*-first if the ROI is worth examining (foreground), or if it is not (background). Hence, if we have M anchors fed to the RPN, we get 2×M anchors as the output. Figure 6.14 shows how the RPN outputs objectness scores. The second output is the bounding box regressor which fits around the object, this adjustment is done through 1×1 convolution. The output this time is 4×M since the offsets are $\Delta_{x\text{-center}}, \Delta_{y\text{-center}}, \Delta_{height}$ and Δ_{width}.

FIGURE 6.14   RPN takes the number of anchors as input and predicts the objectness scores of all these anchors.

Further, to tackle overlapping locations of anchors, non-maxima suppression is done before proposal selection which further reduces the number of bounding boxes before feeding the information to ROI pooling module.

What the ROI pooling module does is that it accepts all the object proposals given by the RPN and takes those corresponding feature maps (through array slicing) from the convolutional feature map given by the base network. This process is shown by Figure 6.15.

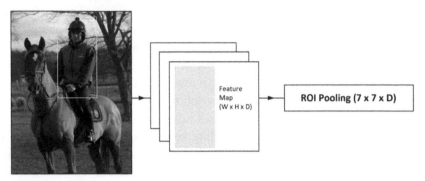

FIGURE 6.15   ROI takes all the proposed localizations and extracts feature maps for those locations and resizes the dimension to $7 \times 7 \times D$ where D is the depth of the feature map.

This extracted feature vector can be fed to the R-CNN detection module to get the correct bounding box coordinates and the class labels.

Once the ROI pooling is done, the extracted, re-sized features are fed to two FC layers to obtain the class labels (with an additional background label) along with the bounding box predictions. In Figure 6.16, we can see that features from ROI pooling are fed into two FC layers each of 4,096-d which output two things, the class labels (N) with the additional background class along with the offset values of the bounding boxes.

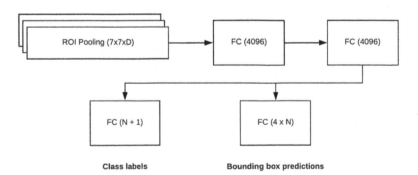

FIGURE 6.16   The detection network with two FC layers being fed the ROI pooling module features which output the class labels (with an additional background label) and the other output being 4 offset bounding box values for each predicted class.

The multi-task loss function that the Faster R-CNN uses to optimize is given as:

$$L\left(\left\{p_i\right\},\left\{t_i\right\}\right) = \frac{1}{N_{cls}}\sum_i L_{cls}\left(p_i,p_i^*\right) + \lambda\frac{1}{N_{reg}}\sum_i p_i^* L_{reg}\left(t_i,t_i^*\right) \tag{19}$$

In Eqn. (19), i denotes the index of the anchor in a mini-batch, p_i is the probability of the ith anchor having an object. p_i^* is a binary variable that takes the value 1 if the ground truth label for the anchor i is that it has an object, and 0 otherwise. t_i represents the four offset parameterized coordinates of the bounding boxes, where t_i^* is the ground-truth. Classification loss is denoted by L_{cls} which is a logarithmic loss over two classes, i.e., whether there is an object or not. Regression loss, denoted by L_{reg} is actually a robust loss function (smooth L_1) This value is multiplied with p_i^* because it only needs to be added when there is an object. The two outputs of the FC layers, namely the *cls* and *reg* layers comprise $\{p_i\}$ and $\{t_i\}$ respectively. λ is a balancing weight and the terms $\{p_i\}$ and $\{t_i\}$ are normalized with N_{cls} and N_{reg}. Figure 6.17 shows results of object detection through Faster R-CNN.

FIGURE 6.17   The bounding boxes imprinted on the input images using Faster R-CNN.

6.8 SINGLE SHOT DETECTION (SSD)

The original R-CNN models [8, 56, 57] were a significant contribution done to the family of object detection algorithms. The problem with these models was that they involved many complexes moving parts and training them was tedious. Even reaching about 7–10 FPS as the detection speed could not be considered to be a fast, real-time object detection. SSD [4] solves these problems by providing a unified framework that does not have any complex moving parts and does both, localization, and detection of objects in a single forward pass of the network. The term single shot means that it does object detection by looking at the image only once (unlike R-CNNs where the model gathers feature maps to feed other components for further processing).

6.8.1   ARCHITECTURE

Single shot detection (SSD) has a base network that is pre-trained just liked Faster R-CNN. Modern CNN architectures such as the VGG16 have proved to be better base networks for the SSD. One point to note is that 80% of the time consumed by SSD is due to the architecture of the base network. In this explanation we use VGG to be the base network. Figure 6.18 shows the architecture of SSD. Up to the Conv6 layer, VGG-16 base network is used to

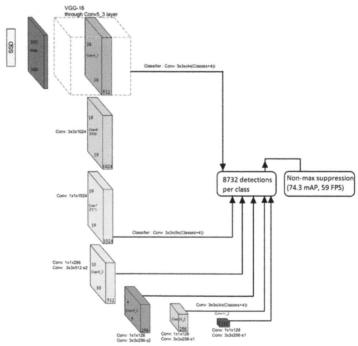

FIGURE 6.18   Architecture of the single shot detection (SSD). VGG-16 is the base network on the far left which is pre-trained. As seen, at every layer there is a connection to the output (FC) layer.

extract features and then we have further convolutional layers, each of which are connected to the output FC layer which give the object detections. Due to the connections made at each layer, SSD can localize images at varying scales, i.e., different dimensions of the selected features of the image.

6.8.2 WORKING

SSD uses the MultiBox algorithm for localizing images [60], however, in a modified manner. MultiBox algorithm uses a series of 1×1 filters and 3×3 to reduce dimensionality and feature rich-learning, respectively. Just as Faster R-CNN used anchors, MultiBox uses *priors* which are bounding boxes of fixed size used before the detection starts. These priors are so placed that the locations of ground-truth bounding boxes and these priors have an IoU at least greater than 0.5 ($\geq$50%). These priors are pre-computed which represent a probability distribution of where objects might appear in an image. This means that the MultiBox algorithm needs to pre-train to compute these priors. This is done through *fixed priors*. The image is split into discrete cells for the generation of feature maps. Each cell has four different bounding boxes defined with different aspect ratios. Feature maps are generated that discretize images into a number of cells which happens as the image is propagated forward through the CNN and the dimensions get smaller and smaller every step of the way.

When the cells are higher in number, smaller objects (such as the cat) can be localized. However, when discretized cells decrease in number, the sizes of these cells increase and thus they are able to detect larger objects (such as the dog) better. Thus, with the various number of convolutional layers, these feature maps have different numbers of cells and for an input image, a variety of sizes of discretizations is considered, making SSD more competent in detecting smaller and bigger objects. Along with the prediction of the localization (bounding box) of the object, the probability of class labels inside the region is also predicted. The bounding box having the largest probability across all classes is kept as the actual bounding box.

SSD uses two loss functions defined by the MultiBox algorithm, which are: a) confidence loss, and b) location loss. For the confidence loss, categorical cross-entropy loss is used which tells us how wrong the model was in predicting the class labels for the bounding box. The location loss is the same (smooth L_1) loss function used in Faster R-CNN. Figure 6.19 shows the SSD in action.

FIGURE 6.19   Bounding boxes generated over different types of vehicles on the road using SSD.

6.8.3   *ADVANTAGES OF SSD*

- Very fast-SSD achieves 59 FPS with a mAP of 74.3% as compared to Faster R-CNN's 7 FPS with a mAP of 73.2% or YOLO's 45 FPS with a mAP of 63.4% on the same dataset.
- Simple network as it eliminates the bounding box proposal networks or any feature resampling stages.
- Achieves high accuracy even with relatively low resolution of inputs which in turn increases speed.
- Instead of anchors (as in Faster R-CNN), SSD uses priors that are initialized to have IoU of over 50% over ground truth bounding boxes.

6.8.4   *DISADVANTAGES OF SSD*

- The main problem with SSD is that it is unable to detect small objects to a high extent. The MultiBox algorithm does not accurately detect objects that do not appear in many feature maps, something that applies to small objects.
- Detection of small objects may be solved by increasing input dimensions of the images but that comes at the cost of reduction of speed of SSD and still is not a very effective solution.

- SSD has problems distinguishing objects of similar categories when the locations of these objects are shared to some extent.

6.9   CONCLUSION AND FUTURE SCOPE

In this chapter, we review potential techniques of object detection and tracking for unmanned aerial vehicle (UAV) related applications. We review an architectural structure of how these drones may navigate proposed by previous researchers and explain some of the segmentation techniques used in computer vision which may also be applied to drone vision. Then, we review the most important DL-based computer vision object tracking and recognition algorithms, namely YOLO (you only look once), faster R-CNN (faster region-convolutional neural network), and SSD. It is noticed that although YOLO may be faster in comparison to Faster R-CNN, it suffers from a general lack of achieving as high accuracy as the Faster R-CNN. Perhaps the best of the three is SSD because it is faster than both Faster R-CNN and YOLO and provides comparable accuracy of object detection. Drone vision is an important part of the internet of drones (IoD) as with increase in demand and feasibility of usage of drones, more, and more services and applications will require data in the visual form to aid or facilitate operations. This chapter puts together some of the state-of-the-art object detection algorithms based on DL which may be useful for drone vision.

The future possibilities of drone vision are limitless. Although this chapter already shows how drones may use advanced DL algorithms for this task, future work can include application of newer object detection algorithms such as Cascade R-CNN [61], detection via class-aware region proposal network (CARPN) [62], Grid-based detection (G-CNN) [63], object detection via Bayesian optimization [64], scale dependent pooling with cascaded rejection classifiers (SDP+CRC) [65], Sub-CNN [66], StuffNet30 [67], networks on convolutional feature maps (NoCs) [68], Hypernet [69], multistage detection with group recursion (MS-GR) [70], inside-outside net [71], etc., with even the possibility of using more optimized base networks for the algorithms discussed in this chapter. Modern-day generative models such as generative adversarial networks (GANs) [72] can also be applied in the workflow of drones with computer vision to generate huge 3D projections over an area for creative visualization and augmented reality (AR). With the successful application of artificial intelligence (AI) through computer vision in drones, we can apply the same to an even bigger industry-aviation. Driven by the success of these algorithms in the IoD, advanced airliner manufacturing companies such as Boeing can integrate vision-based automation in terms of navigation in their future releases.

KEYWORDS

- **computer vision**
- **deep learning**
- **internet of drones**
- **machine learning**
- **neural networks**
- **object detection**

REFERENCES

1. Zhu, P., Wen, L., Bian, X., Ling, H., & Hu, Q., (2018). *Vision Meets Drones: A Challenge.* arXiv preprint arXiv:1804.07437.
2. Campoy, P., Correa, J. F., Mondragón, I., Martínez, C., Olivares, M., Mejías, L., & Artieda, J., (2008). Computer vision onboard UAVs for civilian tasks. *Journal of Intelligent and Robotic Systems, 54*(1–3), 105–135. doi: 10.1007/s10846-008-9256-z.
3. Ren, S., He, K., Girshick, R., & Sun, J., (2015). Faster R-CNN: Towards real-time object detection with region proposal networks. In: *Advances in Neural Information Processing Systems* (pp. 91–99).
4. Liu, W., Anguelov, D., Erhan, D., Szegedy, C., Reed, S., Fu, C. Y., & Berg, A. C., (2016). SSD: Single shot multibox detector. In: *European Conference on Computer Vision* (pp. 21–37).
5. Redmon, J., Divvala, S., Girshick, R., & Farhadi, A., (2016). You only look once: Unified, real-time object detection. In: *Proceedings of the IEEE Conference on Computer Vision and Pattern Recognition* (pp. 779–788).
6. Jiang, H., & Learned-Miller, E., (2017). Face detection with the faster R-CNN. In: *2017 12th IEEE International Conference on Automatic Face & Gesture Recognition (FG 2017)* (pp. 650–657).
7. Womg, A., Shafiee, M. J., Li, F., & Chwyl, B., (2018). Tiny SSD: A tiny single-shot detection deep convolutional neural network for real-time embedded object detection. In: *2018 15th Conference on Computer and Robot Vision (CRV)* (pp. 95–101).
8. Poirson, P., Ammirato, P., Fu, C. Y., Liu, W., Kosecka, J., & Berg, A. C., (2016). Fast single shot detection and pose estimation. In: *2016 Fourth International Conference on 3D Vision (3DV)* (pp. 676–684).

9. Jing, L., Yang, X., & Tian, Y., (2018). Video you only look once: Overall temporal convolutions for action recognition. *Journal of Visual Communication and Image Representation, 52*, 58–65.
10. Raptopoulos, A., (2020). *No Roads? There is a Drone for That.* Retrieved from: https://www.ted.com/talks/andreas_raptopoulos_no_roads_there_s_a_drone_for_that (accessed on 19 November 2021).
11. Amazon.com Inc., (2020). *Amazon Prime Air.* Retrieved from: http://www.amazon.com/b?node=8037720011 (accessed on 19 November 2021).
12. Stewart, J., (2014). *Google Tests Drone Deliveries in Project Wing Trials.* Retrieved from: http://www.bbc.com/news/technology-28964260 (accessed on 19 November 2021).
13. DHL Express, (2014). *DHL Parcelcopter Launches Initial Operations for Research Purposes.* Retrieved from: https://www.dhl.com/discover/content/dam/dhl/downloads/interim/full/dhl-trend-report-uav.pdf (accessed on 30 November 2021).
14. Gharibi, M., Boutaba, R., & Waslander, S. L., (2016). Internet of drones. *IEEE Access, 4*, 1148–1162.
15. He, K., Gkioxari, G., Dollár, P., & Girshick, R., (2017). Mask R-CNN. In: *Proceedings of the IEEE International Conference on Computer Vision* (pp. 2961–2969).
16. Jiang, H., & Learned-Miller, E., (2017). Face detection with the faster R-CNN. In: *2017 12th IEEE International Conference on Automatic Face & Gesture Recognition (FG 2017).* doi:10.1109/fg.2017.82.
17. Yang, S., Luo, P., Loy, C. C., & Tang, X., (2016). Wider face: A face detection benchmark. In: *Proceedings of the IEEE Conference on Computer Vision and Pattern Recognition* (pp. 5525–5533).
18. Simonyan, K., & Zisserman, A., (2014). *Very Deep Convolutional Networks for Large-Scale Image Recognition.* arXiv preprint arXiv:1409.1556.
19. Wang, H., Li, Z., Ji, X., & Wang, Y., (2017). *Face R-CNN.* arXiv preprint arXiv:1706.01061.
20. Wen, Y., Zhang, K., Li, Z., & Qiao, Y., (2016). A discriminative feature learning approach for deep face recognition. In: *European Conference on Computer Vision* (pp. 499–515).
21. Shrivastava, A., Gupta, A., & Girshick, R., (2016). Training region-based object detectors with online hard example mining. In: *Proceedings of the IEEE Conference on Computer Vision and Pattern Recognition* (pp. 761–769).
22. Jain, V., & Learned-Miller, E., (2010). *FDDB: A Benchmark for Face Detection in Unconstrained Settings, 2*(4), 5.
23. Zhang, C., Xu, X., & Tu, D., (2018). *Face Detection Using Improved Faster RCNN.* arXiv preprint arXiv:1802.02142.
24. He, K., Zhang, X., Ren, S., & Sun, J., (2016). Deep residual learning for image recognition. In: *Proceedings of the IEEE Conference on Computer Vision and Pattern Recognition* (pp. 770–778).
25. Yang, W., & Jiachun, Z., (2018). Real-time face detection based on YOLO. In: *2018 1st IEEE International Conference on Knowledge Innovation and Invention (ICKII)* (pp. 221–224).
26. Ghenescu, V., Mihaescu, R. E., Carata, S. V., Ghenescu, M. T., Barnoviciu, E., & Chindea, M., (2018). Face detection and recognition based on general purpose DNN object detector. In: *2018 International Symposium on Electronics and Telecommunications (ISETC)* (pp. 1–4).

27. Jang, Y., Gunes, H., & Patras, I., (2019). Registration-free face-SSD: Single shot analysis of smiles, facial attributes, and affect in the wild. *Computer Vision and Image Understanding, 182,* 17–29.

28. Chen, J. C., Lin, W. A., Zheng, J., & Chellappa, R., (2018). A real-time multi-task single shot face detector. In: *2018 25th IEEE International Conference on Image Processing (ICIP)* (pp. 176–180).

29. Xu, Y., Yu, G., Wang, Y., Wu, X., & Ma, Y., (2017). Car detection from low-altitude UAV imagery with the FASTER R-CNN. *Journal of Advanced Transportation,* 1–10. doi:10.1155/2017/2823617.

30. Barnich, O., & Van, D. M., (2010). ViBe: A universal background subtraction algorithm for video sequences. *IEEE Transactions on Image Processing, 20*(6), 1709–1724.

31. Shastry, A. C., & Schowengerdt, R. A., (2005). Airborne video registration and traffic-flow parameter estimation. *IEEE Transactions on Intelligent Transportation Systems, 6*(4), 391–405.

32. Viola, P., & Jones, M., (2001). Rapid object detection using a boosted cascade of simple features. In: *Proceedings of the 2001 IEEE Computer Society Conference on Computer Vision and Pattern Recognition* (Vol. 1, pp. 1–8).

33. Dalal, N., & Triggs, B., (2005). Histograms of oriented gradients for human detection. In: *2005 IEEE Computer Society Conference on Computer Vision and Pattern Recognition (CVPR'05)* (Vol. 1, pp. 886–893).

34. Müller, J., & Dietmayer, K., (2018). Detecting traffic lights by single shot detection. In: *2018 21st International Conference on Intelligent Transportation Systems (ITSC)* (pp. 266–273).

35. Fregin, A., Müller, J., Kreβel, U., & Dietmayer, K., (2018). The DriveU traffic light dataset: Introduction and comparison with existing datasets. In: *2018 IEEE International Conference on Robotics and Automation (ICRA)* (pp. 3376–3383).

36. Szegedy, C., Vanhoucke, V., Ioffe, S., Shlens, J., & Wojna, Z., (2016). Rethinking the inception architecture for computer vision. In: *Proceedings of the IEEE Conference on Computer Vision and Pattern Recognition* (pp. 2818–2826).

37. Zuo, Z., Yu, K., Zhou, Q., Wang, X., & Li, T., (2017). Traffic signs detection based on faster R-CNN. In: *2017 IEEE 37th International Conference on Distributed Computing Systems Workshops (ICDCSW)* (pp. 286–288).

38. Lin, J. P., & Sun, M. T., (2018). A YOLO-based traffic counting system. In: *2018 Conference on Technologies and Applications of Artificial Intelligence (TAAI)* (pp. 82–85).

39. Huang, R., Pedoeem, J., & Chen, C., (2018). YOLO-LITE: A real-time object detection algorithm optimized for non-GPU computers. In: *2018 IEEE International Conference on Big Data.* doi:10.1109/bigdata.2018.8621865.

40. Ioffe, S., & Szegedy, C., (2015). *Batch Normalization: Accelerating Deep Network Training by Reducing Internal Covariate Shift.* arXiv preprint arXiv:1502.03167.

41. Jaitly, N., Nguyen, P., Senior, A., & Vanhoucke, V., (2012). Application of pretrained deep neural networks to large vocabulary speech recognition. *Proceedings of Interspeech 2012.*

42. Dede, G., & Sazlı, M. H., (2010). Speech recognition with artificial neural networks. *Digital Signal Processing, 20*(3), 763–768.

43. Scanzio, S., Cumani, S., Gemello, R., Mana, F., & Laface, P., (2010). Parallel implementation of artificial neural network training for speech recognition. *Pattern Recognition Letters, 31*(11), 1302–1309.
44. Wijoyo, S., & Wijoyo, S., (2011). Speech recognition using linear predictive coding and artificial neural network for controlling movement of mobile robot. In: *Proceedings of 2011 International Conference on Information and Electronics Engineering (ICIEE 2011)* (pp. 28, 29).
45. Agarwal, M., Jain, N., Kumar, M. M., & Agrawal, H., (2010). Face recognition using eigen faces and artificial neural network. *International Journal of Computer Theory and Engineering, 2*(4), 624.
46. Zhang, M., & Fulcher, J., (1996). Face recognition using artificial neural network group-based adaptive tolerance (GAT) trees. *IEEE Transactions on Neural Networks, 7*(3), 555–567.
47. Réda, A., & Aoued, B., (2004). Artificial neural network-based face recognition. In: *First International Symposium on Control, Communications and Signal Processing* (pp. 439–442).
48. Propp, M., & Samal, A. K., (1992). Artificial neural network architectures for human face detection. In: *Proceedings of the 1992 Artificial Neural Networks in Engineering, ANNIE'92,* (pp. 535–540).
49. Singh, R., Yadav, C. S., Verma, P., & Yadav, V., (2010). Optical character recognition (OCR) for printed Devanagari script using artificial neural network. *International Journal of Computer Science and Communication, 1*(1), 91–95.
50. Amin, A., Al-Sadoun, H., & Fischer, S., (1996). Hand-printed Arabic character recognition system using an artificial network. *Pattern Recognition, 29*(4), 663–675.
51. Mani, N., & Srinivasan, B., (1997). Application of artificial neural network model for optical character recognition. In: *1997 IEEE International Conference on Systems, Man, and Cybernetics. Computational Cybernetics and Simulation* (Vol. 3, pp. 2517–2520).
52. Patil, V., & Shimpi, S., (2011). Handwritten English character recognition using neural network. *Elixir. Comput. Sci. Eng., 41,* 5587–5591.
53. Barve, S., (2012). Optical character recognition using artificial neural network. *International Journal of Advanced Research in Computer Engineering & Technology, 1*(4), 131–133.
54. LeCun, Y., Bottou, L., Bengio, Y., & Haffner, P., (1998). Gradient-based learning applied to document recognition. *Proceedings of the IEEE, 86*(11), 2278–2324.
55. Szegedy, C., Liu, W., Jia, Y., Sermanet, P., Reed, S., Anguelov, D., et al., (2014). *Going Deeper with Convolutions.* arXiv 2014. arXiv preprint arXiv:1409.4842, 1409.
56. Girshick, R., Donahue, J., Darrell, T., & Malik, J., (2014). Rich feature hierarchies for accurate object detection and semantic segmentation. In: *Proceedings of the IEEE Conference on Computer Vision and Pattern Recognition* (pp. 580–587).
57. Girshick, R., (2015). Fast R-CNN. In: *Proceedings of the IEEE International Conference on Computer Vision* (pp. 1440–1448).
58. Uijlings, J. R. R., Van De, S. K. E. A., Gevers, T., & Smeulders, A. W. M., (2013). Selective search for object recognition. *International Journal of Computer Vision, 104*(2), 154–171. doi:10.1007/s11263-013-0620-5.
59. Zeiler, M. D., & Fergus, R., (2014). Visualizing and understanding convolutional networks. In: *European Conference on Computer Vision* (pp. 818–833).

60. Szegedy, C., Reed, S., Erhan, D., Anguelov, D., & Ioffe, S., (2014). *Scalable, High-Quality Object Detection.* arXiv preprint arXiv:1412.1441.
61. Cai, Z., & Vasconcelos, N., (2019). Cascade R-CNN: High quality object detection and instance segmentation. *IEEE Transactions on Pattern Analysis and Machine Intelligence.*
62. Tao, X., Gong, Y., Shi, W., & Cheng, D., (2018). Object detection with class aware region proposal network and focused attention objective. *Pattern Recognition Letters.* doi: 10.1016/j.patrec.2018.09.025.
63. Najibi, M., Rastegari, M., & Davis, L. S., (2016). G-CNN: An iterative grid based object detector. In: *Proceedings of the IEEE Conference on Computer Vision and Pattern Recognition* (pp. 2369–2377).
64. Zhang, Y., Sohn, K., Villegas, R., Pan, G., & Lee, H., (2015). Improving object detection with deep convolutional networks via Bayesian optimization and structured prediction. In: *Proceedings of the IEEE Conference on Computer Vision and Pattern Recognition* (pp. 249–258).
65. Yang, F., Choi, W., & Lin, Y., (2016). Exploit all the layers: Fast and accurate CNN object detector with scale dependent pooling and cascaded rejection classifiers. In: *Proceedings of the IEEE Conference on Computer Vision and Pattern Recognition* (pp. 2129–2137).
66. Xiang, Y., Choi, W., Lin, Y., & Savarese, S., (2017). Subcategory-aware convolutional neural networks for object proposals and detection. In: *2017 IEEE Winter Conference on Applications of Computer Vision (WACV)* (pp. 924–933).
67. Brahmbhatt, S., Christensen, H. I., & Hays, J., (2017). StuffNet: Using 'stuff' to improve object detection. In: *2017 IEEE Winter Conference on Applications of Computer Vision (WACV)* (pp. 934–943).
68. Ren, S., He, K., Girshick, R., Zhang, X., & Sun, J., (2016). Object detection networks on convolutional feature maps. *IEEE Transactions on Pattern Analysis and Machine Intelligence, 39*(7), 1476–1481.
69. Kong, T., Yao, A., Chen, Y., & Sun, F., (2016). Hypernet: Towards accurate region proposal generation and joint object detection. In: *Proceedings of the IEEE Conference on Computer Vision and Pattern Recognition* (pp. 845–853).
70. Li, J., Liang, X., Li, J., Wei, Y., Xu, T., Feng, J., & Yan, S., (2017). Multistage object detection with group recursive learning. *IEEE Transactions on Multimedia, 20*(7), 1645–1655.
71. Bell, S., Lawrence, Z. C., Bala, K., & Girshick, R., (2016). Inside-outside net: Detecting objects in context with skip pooling and recurrent neural networks. In: *Proceedings of the IEEE Conference on Computer Vision and Pattern Recognition* (pp. 2874–2883).
72. GM, H., Gourisaria, M. K., Rautaray, S. S., & Pandey, M., (2020). A comprehensive survey and analysis of generative models in machine learning. *Computer Science Review, 38,* 100285.

CHAPTER 7

Security Analysis of UAV Communication Protocols: Solutions, Prospects, and Encounters

P. PRAVEEN KUMAR,[1] T. ANANTH KUMAR,[1] PAVITHRA MUTHU,[1] RAJMOHAN RAJENDIRANE,[1] and R. DINESH JACKSON SAMUEL[2]

[1]*Department of Computer Science and Engineering, IFET College of Engineering, Tamil Nadu, India, E-mail: ananth.eec@gmail.com (T. A. Kumar)*

[2]*Faculty of Technology, Design, and Environment, Oxford Brookes University, Oxford, United Kingdom*

ABSTRACT

Unmanned aircraft can be piloted independently or remotely by ground crews, and can save lives by putting people out of work in unsafe conditions. Typical UAV applications include monitoring and detection, collaborative search, detection, and tracking, consistent household monitoring, radio tracking, and forest fire surveillance. Despite their enormous utility scope, research, and development of UAV technology have grown exponentially in the last decade. UAVs are typically remote-controlled, which in effect opens paths for cyber-attacks. Popular methods for attacks, minor findings, and related vulnerabilities are discussed in this chapter. Drones are an evolving type of modern IoT devices operating in the sky with complete network communication capabilities. Drones Internet is an infrastructure designed to provide organized access to controlled airspace for unmanned aerial vehicles (UAVs), also called drones. With continuing miniaturization of sensors and

The Internet of Drones: AI Applications for Smart Solutions. Arun Solanki, PhD, Sandhya Tarar, PhD, Simar Preet Singh, PhD & Akash Tayal, PhD (Eds.)

processors and omnipresent wireless networking, drones are seeking several different ways to enrich our lifestyle. Applications of drone technology is discussed here, from on-demand shipping to traffic and wildlife control, facilities monitoring, search, and rescue, agriculture, and cinematography.

7.1 INTRODUCTION

Unmanned aircraft (UAV), commonly referred to as automobiles, are widely used in private, commercial, and military fields. It is estimated that by 2022 there will be 5 million UAVs worldwide, according to qualified journals. Their use includes businesses, such as aerial reconnaissance or photography/ video/scenographies, as well as military uses, for recreational purposes. UAVs are increasingly being used for protection, surveillance, emergency response, and infrastructure inspections [1]. Although this is appropriate, UAVs bear the physical dangers of aircraft just as they bear unmanned frames. Because of the exponentially growing number of UAVs, these risks are bound to occur. Given operational dangers, safety-related dangers are also present. UAVs are typically remotely operated, opening up digital assaults (Figure 7.1) [2].

FIGURE 7.1   Emerging fields in UAV.

Indeed, even very good UAV experts rely on established models of processing that were not built to be exceptionally secure. It has led to a number of sessions in which adversaries say they have learned how antagonistic UAVs are limited by interfering with control signals [3]. These events are not yet accountable for private or corporate UAVs but to rely on known pieces. Strangely, traders rely on the mystery of structure or use as the essential technique to secure the structure. Therefore, assailants can utilize standard "hacking" instruments to "effectively" assume responsibility for a UAV so as to frustrate it finishing its undertakings or, far more detestable, to make harm. Underneath UAV seizing most UAVs gather and store information locally. Frequently put away information is not scrambled and outsiders can effortlessly catch even its transmission (remote telemetry). Strikingly, the impact of advertising new gadgets that are defenseless against "standard" assaults over the Internet is valid for UAVs or automatons too. As an outcome, there is a pressing requirement for improving UAV security [4]. However, more examination is required that efficiently looks at security dangers in current UAV innovation and to characterize for relieving these dangers. This chapter analyzes potential security dangers of UAV correspondence conventions such as AFHDS_2A, DSMX, DSM2 and D16. In detail, sellers depend on recurrence bouncing, range spreading, and key sharing as dynamic safety efforts. Be that as it may, lawfully sellers can just work on ISM groups with an emphasis on the 2.4 GHz band utilizing correspondence moves toward that depend on bundle-based transmission. In this way, they are firmly identified with conventions, for example, IPv6, yet do not utilize safety efforts, for example, cryptography. This makes them helpless against known assaults. We will exhibit this defenselessness by seizing a UAV (drone) utilizing known techniques. We will also give a few thoughts on how the security of UAV conventions can be improved while keeping properties with respect to inertness and throughput.

The abbreviation UAV is utilized to speak to a force-driven, reusable plane or copter that is worked without a human pilot on board [5]. Most UAVs have remote control and correspondence implies. Control by remote correspondence bears the risk of abuse. Be that as it may, research on UAVs is fundamentally centered around self-governing conduct and control. This chapter analyzes and acknowledges the following exploration challenges such as security and monitoring of aerial systems, collision protection and obstacle avoidance, configuration security, high-level control security in hardware and communication [6]. As referenced, remote correspondence with other participating UAVs or potentially the ground is key concerning security angles. The UAV to ground issue regularly addresses, for example, "off the mark of-sight" or "long-run interchanges." Strangely security issues and particularly the security of UAV correspondence conventions are regularly disregarded. Also, general digital assault techniques and on arranged frameworks are then used to recognize potential dangers and vulnerabilities of current UAVs. This chapter is concluded with an investigation report of various security protocols and communication techniques for assuming control over the control of industrially accessible UAVs designs. In detail, we look at and break down a standard UAV correspondence and control convention and innovative methodologies for assaults, minor perceptions, and related security vulnerabilities.

7.2   ARCHITECTURE OF DRONE

In addition, the three main components are either UAV or drone architecture: flight controller, ground control stations (GCS) and communication data-link (CDL). The flight controller was the central processing unit of the drone. The GCS is built using an OLF system, allowing operators to control or track UAVs remotely. GCSs differ in scale, form, and function by drone. Data Links is the wireless communications between the drone and GCS to monitor the flow of information. It depends on the preference of UAVs. Draws may be classified on the basis of the data. Visual sight line (VLOS) [7] Gap enables direct waves to transmit and receive power signals VLOS (Figure 7.2) [8].

FIGURE 7.2   Drone architecture.

The distance across the visual viewing line (BVLOS) enables drones to operate satellite communication [9]. Small UAVs are identified in the communication network survey. The authors have defined different network architectures such as fixed, mobile, and unplanned. There are multiple faults in direct contact and wireless networks. A particular amount of bandwidth is needed to support a high network node density for each UAV. UAV-to-UAV interchanges are not feasible under the immediate connectivity infrastructure as information flow is to be directed to the GCS. An Unmanned aircraft system(UAS)'s transmitter should be had secure system for transmission, UAS installations should provide a fanatical cellular network. Nonetheless, a financial handicap may be the expensive installation of such simple stations.

A UAANET (UAV Ad hoc network) is also used to report the shortcomings of the communication architectures for UAV flights at the time of powerful and fast networking between UAVs and GCS. Comparing with the other networking networks above, UAANETs have many advantages [10]. Due to UAVs' rapid mobility, scalability is ensured to cover a huge area quickly. Furthermore, reliability is increased because a single UAV failure does not impact the entire network. Also, because of multi-hop connectivity, bandwidth is reused more often and thus more efficient. From the security point of view, the lack of a central node in the network decreases the possibility of a security breach. For network security and authentication, UAVS, and GCS will be handling the process.

7.3   ROUTING PROTOCOL CHALLENGES IN UAANET

For UAANETs, specific hierarchical systems are helpful. Separate protocols are topology-based systems including proactive, reactive, and mixed routing. Regional routing can, nevertheless be active in various contexts [11]. Performance evaluation of MANETs for UAANETs based on imitation are discussed below. This critical analysis takes into consideration the Linux component which organizes stack specifications, convention usage, base traffic, features for real-time execution, and a sensible mobility model. The findings of the tests revealed that the UAANET steering convention for AODV matches OLSR and DSR most effectively. As AODV is the most sensitive protocol to change in topology and has low overhead, we have come to the conclusion that AODV is the most appropriate routing protocol for drones. Since many studies have shown that AODV is often overperformed by routing protocols with a more significant number of nodes, its on-demand mechanism allows quicker reaction when UAANET is disconnected [12]. Our usability architecture and copying process certainly influenced the measures; however, we saw similar results. Therefore, the SUAP (Secure UAV unplanned routing protocol) model architecture is implemented in that can be a reliable AODV-supported UAANET routing protocol (Figure 7.3).

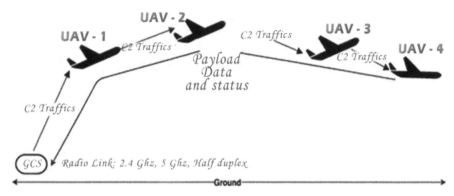

FIGURE 7.3   Architecture of UAANET.

7.3.1 SECURITY CHALLENGES IN UAANET

A routing protocol 's ultimate objective is to find a simple, secure route while maintaining friendly and cooperative network nodes. Nevertheless, this statement is not always true, as UAANET uses insecure wireless media and lacks the fixed infrastructure to recognize nodes (attackers). In fact, the critical amount of command and control (c2) traffic exchanged must ensure its protection at low cost. Be mindful that attackers are also inspired to crack UAANET. For examples, an attacker may aim to traffic control and order the capture of UAVs. It is achieved by eavesdropping and routing of packets. UAANET can typically be challenging to obtain due to its cooperative qualities and uncontrollable atmosphere. In UAANET, a freighter can effectively eavesdrop communications over insecure wireless connections, in comparison to wired networks on which perpetrators use to access to physical infrastructure to carry out any form of attacks. So, few secure UAANET routing protocols from our best understanding have been suggested. Either, a new routing protocol with the protection of its targets or a stable expansion of the existing routing protocol is always created from scratch. The latter solution is used more frequently as the IETF MANET group uses a standardization routing protocol, such as AODV, DSR, and OLSR [13]. Protecting such routing protocols includes assessing and protecting unique to that protocol attacks. The latter method is used to render the SUAP routing protocol.

7.3.2   UAS CERTIFICATION CHALLENGES

UAS is split into several categories in order to qualify for certification, starting with sensors, hardware, and software modules requiring clearance for airworthiness. The benefit of separate research is twofold: first, it allows the developer to design several UAS components (motors, propellers). Furthermore, it only allows the UAS building agent (for Delair Tech), not each component itself, to be included in the device validation process. It eliminates the problem of the UAS user having a license [14]. Remember that only in the country where the designer 's principal place of business is situated will the US be accredited. DGAC will then verify our UAS coordination system at a French national level. While selected requirements for validations and qualification for UAANETs have yet to be proposed by the French professional UAV civil federation 3, frequently the methods and safety standards apply to UAANETs. DO-178 may be a commercial aviation standard to validate aviation applications focusing on model-based implementation and further testing through structured evaluation methods. Please notice that Delair-Tech has already approved that the GCS (the operator) sight DT-18 UAV is taking off. Nevertheless, UAANET architecture shifts the dominant contact system. This calls for a validation and credential replacement process. The robust routing protocol we are designing at present must, therefore, be strictly checked. We noticed out by using professional processes in software creation focused on pattern development [15].

7.4   MODEL DRIVEN DEVELOPMENT OF SECURE UAANET ROUTING PROTOCOL

The most widely-used AODV (AODV-UU) implementation may be implemented with an ASCII text file to check the functional requirements defined by IETF standards. No consistency test between specifications and the ASCII text file is possible. Consequently, UAANET 's direct delivery is not recommended because it can cause serious damage and loss during flight operations. For distributed communication networks, various methods of research have been developed. Nevertheless, some testing approaches are goal-oriented and include choosing a particular reference product property which is possible to be defective. One way to overcome this is by using MDD methods which produce, deploy, and validate code automatically using the high-level model as inputs [10, 16]. As a central component in the development process, MDD appears to use templates. The event cycle is provided with a system layout consisting of block diagrams and state maps, from specifications to simulation testing and integration.

7.4.1 SECURE ROUTING PROTOCOL DESIGN ARCHITECTURE

There are numerous steps to guarantee the AODV 's security. Provide message verification, and use encrypted path discovery and route management packets. Most are vulnerable to wormhole attacks. The wormhole assault involves two conniving criminals. One attacker catches packet at a remote location and replays it through a private high-speed network at some other attacker. This maze between two attackers is the wormhole. The channel's written architecture lets attackers get unaddressed data packets. Because each node encapsulates each packet with cryptographic keys, wormhole attackers breach UAANET security, because there is no way to handle wormhole attacks. This is announced as an easy-to-execute wormhole attack using a high-gain billboard antenna installed at different frequencies, which will test the frequency and collect signals from the GCS and thus the UAVs [18]. However, in our network and intruder model, the following assumption is known: UAVs and GCS come from an equal provider, which lets us exclude security from greedy nodes. A node cannot alter its behavior and still communicate with its neighbors besides being observed or exploited by an enemy; nodes have ample power and network space (i.e., bandwidth) for encryption; all UAVs use all directional antennas.

The limit is r and cannot surpass D_{max} ($r < D_{max}$). D_{max} can be a maximal one-hop range; knots believe an efficient symmetric and asymmetric encryption/decryption, authentication, and hash algorithm. We intend to use RSA and SHA-256 as a hash algorithm for message authentication.

All nodes are synchronized with clocks. It is also feasible to use GPS onboard UAVs and GCS; we presume safe and productive network key management to exchange, maintain, and delete cryptographic keys; confidentiality of the routing packet is not guaranteed. Sure, real-time UAV routing packets. Therefore, although an intruder can eavesdrop the message, their behavior is restricted in passive mode as former information is no longer useful in future. Hash function (H) is only detected by legitimate nodes and is preloaded with keys on a bootstrapper; advanced investigations in the supported Dolev-Yao scenario are included in the attacker capabilities, where an agnostic topology and protocol model is proposed to take a real-world scenario into account [19]. We also noted the actual threats which are data leaks, routing information leakage, deterioration of performance and change in topology (Figure 7.4).

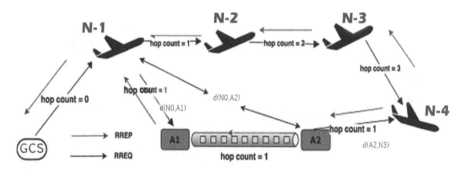

FIGURE 7.4 Illustration of wormhole attack in UAANETs.

The SUAP Routing Protocol is proposed as an AODV Protocol Security Extension, supported public-key authentication, hash chains and regional leashes, to resolve these requirements. For mutable fields and hash chains, this uses digital signatures. A server that produces the message signatures a message using the private key and then verifies the signature of the message using the public key of the sender. Because any hop needs to be increased, the sender cannot sign the hop register. The system with hash chains is used. SUAP is vulnerable to attacks by the wormhole. The relationship between the distance traveled and the hop count frequency is calculated using the variance of spatial leashes. SUAP makes it possible to connect closely every node within the network and to establish communication with its immediate neighbor node. Two different mechanisms are, in reality, protecting themselves from wormhole attacks. Hi, and route error messages are first used. Nodes transmit information to their controllers (one hop) when bootstrapping and senses linking breaks in broadcast mode. To secure these packets from wormhole attacks during transmission, we use a method to examine the association between hop count and packet size. Each node contains its exact location when sending messages. To prevent malicious alteration, message fields (including location) are signed.

7.5   DRONE COMMUNICATION AND SECURITY

Some crucial data such as the position of the UAV, Remaining flight time, Distance, Target location, speed, and several other parameters have to be transferred from drone-to-drone (D2D), drone-to-base station (D2BS) and drone-to-network (D2N) and drone-to-satellite (D2S). The drone uses RF communication to send and receive information from and to the UAV [21].

1. **Drone-To-Drone (D2D):** Unstandardized machine learning (ML) can be applied to the development and automation of an intelligent UAV wireless communications network. D2D networks are now operating as a peer-to-peer networking framework (P2P). This susceptibility to different types of P2P threats, including jamming.
2. **Drone to Base Station:** This networking strategy is built on existing standards focusing on Bluetooth and Wi-Fi 802.11, namely 2.4 GHz and 5 GHz wireless networks. However, many communications between drones are simple and unreliable. They use an authentication scheme that is easily broken, which renders them vulnerable to actively endangering and hostile (man-in-the-middle) attacks.
3. **Drone-to-Network (D2N):** This contact type enables network selection depending on the mandatory security level. It could be included with 3 GHz, 4 GHz, 4G+(LTE) and 5 GHz wireless networking. Having these networks is important.
4. **Drone-to-Satellite:** The form of communication that needs to be provided by GPS to provide real-time coordinates. It allows any drone to return to its original base, within or without control. Satellite contact is deemed secure. Nonetheless, we have high maintenance and expense requirements. Armed forces employ them extensively.

7.5.1   SECURITY IN AERIAL SURVEILLANCE AND TRACKING

Surveillance means close monitoring for the control, power, control, or protection of an individual, a group of people, activities, operations, facilities, buildings, etc. There is a range of different surveillance techniques including GPS tracking, camera detection, outputs, data mining and profiling and biometrical surveillance [22]. The typical stationary design of the camera that is typically manually controlled or attached to a tripod or other structure is generally restricted to conventional tracking. Aerial monitoring can be carried out by helicopter; while it provides the desired result, it also costs a great deal. Unmanaged aircraft systems provide the perfect solution to challenges and restrictions that other monitoring approaches face. Drone tracking offers a simpler, quicker, and cheaper data collection tool and a range of other primary benefits. Drone aircraft can be fitted with night-vision cameras and thermal sensors that have imagers that the human eyes cannot identify. UAVs can quickly cover vast and difficult areas, slash employee numbers and costs and need little room for its operators.

7.5.2   SECURITY IN COLLISION AND OBSTACLE AVOIDANCE

The new guidance legislation requires multiple UAVs to avoid obstacles and provides advice on training reconfiguration. Differential geometry uses sight vector line and relative vector information; therefore, the real-time guidance command is efficient. Recognition mechanism is introduced to decide whether collision avoidance is effective. Lyapunov theorem proves the consistency of the proposed guidelines. Analyzing collision avoidance also uses the full angular rate and recognition capability of the UAV [23]. Unmanned aerial vehicles (UAVs) showed great plant safety potential in agriculture. Within the unstructured agricultural property, unpredictable hazards typically pose a significant flight safety threat. The solutions for the able mentioned problems are given, which consist of deep-learning object recognition, image processing, RGB-D information fusion and task control system (TCS). These techniques improve UAV intelligence and adverse effects of operational protection and performance. Using the fundamental understanding and the distance monitoring, the UAV can be allowed to perceive obstacles and characteristics such as spatial scale, profile, and 3D location. Collision avoidance strategy development process and estimation technique of optimal collision avoidance flight path was detailed based on object detection studies. A series of tests are performed to verify the UAV's environmental awareness capability and the efficacy of autonomous obstacle avoidance. A CNN-based aerial vehicle (UAV) learning program to conquer challenges in new and unstructured environments. To minimize decision latency and increase UAV robustness, an end-to-end two-stage avoidance architecture is planned using a single, forward-facing camera. The prediction approach initially adopts a coevolutionary neural network (CNN)-based model. Three effective processes, depth convolution, group convolution, and channel break, concurrently estimating steering angle and collision likelihood. In the second level, the control mechanism maps the steering angle to an order adjusting UAV's yaw angle. When UAV hits an obstacle, automatic steering avoids collision. Thus, collision risk is calculated as forwarding speed to sustain or escape takeoff. For UAV flight preparation, A multi-strategy obstacle prevention approach based on the rotating vector field is introduced [24]. As inter-UAV control function, distance impact repulsion and directional gravity are used. UAVs are guided outside of the hazard by the Markov random field approach. To block convex polyhedral barriers, a rotating vector field solution is used.

7.6   COMMUNICATION ARCHITECTURES

A connectivity framework describes information flow between GCS and UAV, or between the UAVs. This section includes a few UAV coordination designs.

7.6.1   CENTRALIZED COMMUNICATIONS

A centralized coordinating network is connected to all UAVs with a ground station central node. It is one of the most popular topologies in the network. Every UAV is directly linked to the transmission and reception data of the ground station within this network, and the UAVs are not directly related. The entire network is based on the ground station and two UAVs are routed by a relay from the ground station. Of all UAVs attached directly to the ground station, the command-and-control data transmitted between the ground crew and the UAV is briefly interrupted [25]. The transmitted data between two UAVs are however fairly late, as the data is transmitted through the ground station (Figure 7.5).

Moreover, as long-term contact between the ground station and UAVs is routine, UAVs requires advanced radio transmissions equipment with high transmission capability, which, due to its size and payload limitations, might not be feasible for medium or small UAVs. In reality, the ground station demonstrates UAV network weakness through a single-point failure within the centralized UAV communication system. The whole UAV network is interrupted if the ground station faces problems as it not secure.

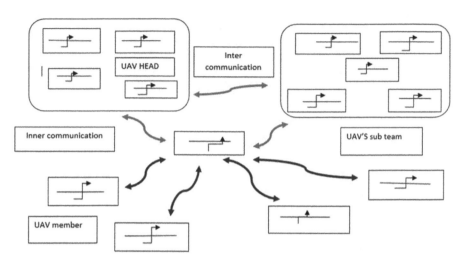

FIGURE 7.5   Centralized/decentralized UAV's-team research allocation scheme.

7.6.2 DECENTRALIZED COMMUNICATIONS

A central node is not necessary for a decentralized communication architecture, and Two UAVs can directly or indirectly communicate. It means that non-ground station details can be redirected via UAV instead of the ground station. Three unified contact systems are discussed below [26]:

1. **UAV Ad Hoc Network (UAANET):** Usually the UAANET has no existing facilities and UAVs are interested in the transmission of data to other UAV networks. Backbone UAV acts as an ad hoc communication network for terrestrial stations and other UAVs. In addition, two antennas are needed for the UAV Gateway, one for communication with other UAVs and the other for communicating with the ground station. The network coverage area of this ad hoc UAV network is substantially reduced as only one UAV is able to link to the ground station. Because many UAVs are thus reasonably close together, the receiver mechanism can be inexpensive and lightweight in a UAV, making it better suited to small to medium-sized UAVs. Nonetheless, mobility habits such as speeds and headings will be identical to ensure network availability. This ad-hoc network design is suitable for networking a cluster of connected UAVs for operations like continuous monitoring. There are also UAVs on

individual planes. Such UAVs also differ between large and small UAVs and payloads. Usually, related UAVs are physically adjacent to multi-type UAV settings and constitute a grouping, while different UAVs are relatively distinct. These UAVs shall form dissimilar classes, depending on their styles. Because of its flow flows and overhead network capacity, it seems to be challenging to create a single UAANET. Multi-type UAV networking, multi-group UAV networking and multi-layer UAV ad hoc networking can be used as communications networks.

2. **UAV Network of Multiple Groups:** UAVs makes up an ad hoc group of UAV networks for backbone UAV links. Intra-group contact (for example, ad hoc UAV network communications are carried out, and inter-group communications (for example, two-group communications) shall be carried out via the respective UAVs and ground stations [27]. A network integrating a centralized UAV network with an ad hoc UAV network may be called a multi-group UAV network architecture. This communication framework is appropriate for the mission of many UAVs of different flight or communication capabilities. Nevertheless, this connectivity framework also droughts robustness owing to its semi-centralized nature.

3. **Ad Hoc Network Multi-Layer UAV:** An ad-hoc multi-layer UAV network is an alternative communications system for networking various heterogeneous UAV communities. Ad hoc multi-layer UAV networks are shown. Inside the structure of the lower layer of the multi-tier UAV ad-hoc network architecture, UAVs create a UAV ad-hoc network. The UAV ad-hoc network's top layer consists of all UAV groups. Just one UAV backbone is directly connected in a multi-layer UAANET to the Earth station. The information exchange between two UAV groups must not be routed via this network's ground station [28]. The ground station handles only information that reduces the measurement and coordination pressure of the ground station substantially. The multi-layer architecture of the ad hoc UAV network is the key to a single UAV network system. A secure ad hoc UAV multilayer network is used because it does not have a single failure point. UAV may be clustered or dispersed. The consolidated UAV network is ground station-based and can be linked to any long-distance UAV. Decentralized communications infrastructure provides expanded coverage through multi-hop communication. For reference, both an ad-hoc UAV and a multi-layer UAV ad-hoc

network are compared using robust network schemes, as all networks have a common fault point and greater access without the need for Internet connections. The easiest way to implement a UAV ad-hoc network is to use a single-party UAV network and one-layer ad-hoc UAV network architecture for networking multiple heterogeneous UAVs. Due to its special UAV operation style, it can be used to favor multi-layer UAV ad-hoc networks with several high-autonomy UAV classes.

7.7   PROMISING COMMUNICATION MODES

The UAV network is a talking drone team. In comparison to one flight device, the whole community of vehicles will fulfill the operating criteria of different missions. That network drone can perform various tasks and implement different network topologies to fulfill specific task requirements. The Drone team must broaden the reach of the project. The drones also carry sensitive national security details or activities [29]. During the execution cycle, Drones talked. If malicious or adversaries hack in the UAV squad, they may affect mission execution or delete critical data via contact that may pose a significant risk to national security. Communications safety is central to rowing strategy. Iraqi protesters secretly captured the U.S. using $26 COTS. Real-time video signals that give Predator drones information to avoid or track U.S. military operations. A year later, militant groups used Sky Grabber to catch a drone video from U.S. helicopters via U.S. military satellites via unencoded aircraft. As the UAV network is used for military and civilian purposes, it also includes sensitive information which an attacker can access and send for specific security attacks, including GPS Spoofing or Wi-Fi, to catch drones and obtain the necessary intelligence.

Furthermore, since the UAV network has a complex topology and versatile connectivity, it is particularly vulnerable to external attacks, such as adverse weather and sparse communication networks, resulting in protective vulnerabilities that cannot usually be resolved due to loss of packet information. Two popular algorithms used here are AES and SM4. As we all know, AES is commonly used and has four distinct cryptographic operating modes in various cases. AES well-being late struggled to meet UAV network health requirements for interaction with device invention. The SM4 algorithm can be a national secret classification algorithm and a low-power chip encryption algorithm. The algorithm has unique packet length specifications, thus the

128-bit key. SM4's encryption and decryption algorithms have an equivalent structure, and implementation requires only the encryption algorithm. The SM4 algorithm has a design-friendly, structurally defined, a stable and efficient algorithm that meets an easy-to-use communication mechanism criterion. The latest stable data encryption and decoding measurement cannot meet UAV 's lightweight performance and protection requirements [30].

On the one hand, because encryption and decryption are too long, the real-time performance of the system or function is limited, and there is little room for drone training and significant repercussions. However, unless the network interruption of the intended communication system is unacceptable, the ciphertext cannot usually be read, and the correct plaintext information is not obtained. Research on compact communication systems that endorse transmitting power is, therefore of great importance for the safety of communication in the UAV network. An attacker can use network authentication methods or techniques to evaluate mr message for legitimate information like drone detection—attack email interception. Ip interception attack captures confidential network information without interference.

1. **Fake ID Attack:** The attacker gets the legitimate drone ID in the UAV network by intercepting an authentication message mr between the UAV nodes. Thus, attacker F will use this network link ID. Thus, the hacker appears to be a legitimate drone_A for linking to the UAV network and communicating with UAV nodes and intercepting messages across the network or sending false information messages.

2. **Replay Attack:** The attacker will return the message received from the host to accomplish the device spoofing purpose and to receive the sender's authentication message that mostly is used to authenticate the identity and to break the accuracy of the authentication. The assault could be a C-described drone or ground facility. C can intercept the A-B message via a secure line. C masks A to B according to the path of line and B misunderstood C to be and sends the answer message.

7.7.1   SECURITY ATTACKS IN UAV

Today, recording all our canals and plans is increasingly essential, which are catch stunning with unexpected scenes [31]. Due to drones' rapid mobility, this is always possible, an action camera needs only to be seen and these moments saved. This approach poses a challenge for many airspace customers. In the boom of autonomous vehicles and therefore growing technological advances, one or more of those drives flying over a neighborhood are frequently expected, causing irritation to residents and even property damage. As a trendy item, manufacturers put cheaper models increasingly back, without listening to information such as product health. The safety aspect of a number of different models is examined to understand how well protection is handled by signals which are sent through WLAN and how the multi-layer safety system is used, which is one of the most relevant functional safety frameworks so far—multi-layer protection network solution. The same strategy for shielding commercial drones should be used. We focused on wireless internet-based drones. The protection issue with these unmanned aircraft is that they are operating wireless devices. For computers, navigation devices and the wireless network, safety should be considered.

7.7.2   MULTI-LAYER SECURITY FRAMEWORK

Some of the most practical methods of defense to date are in-depth defense. This is a multi-layer solution to network security. It is assumed that when securing commercial drones, the same approach will be taken. Wi-Fi-based drones are introduced. The health challenge of these unmanned aerial systems (UAS) is that they are flying autonomous robots. Device, navigation system and wireless network security should also be considered [32].

1. **Security Attack Model:** This threat model is based on three simple attacks such as Denial-of-Service (DoS), Buffer Overflow, and ARP Cache Poisoning. This model was developed by drone penetration testing and three vulnerabilities. These three security attacks will interrupt drone's in-flight actions. The drone used and related drones are possibly vulnerable to other security attacks; however, this was not apparent during this study. In this chapter, protection multi-layer security architecture against such intrusion attacks will be restricted. Store and use inputs to prototype the drone's IP address and MAC username. Controller's IP and MAC address are checked. The access point will now delete any ARP response packets containing controller/drone IP or MAC address. This will ensure drone and controller honesty interconnect. In other words, OSI layers 2 and 3 notify the access point to record the drone's IP address and MAC and hence the controller and ensure that no computer inside the network can falsify them.
2. **Issues in Controlling Drone Communication:** Multi-layer security framework may result directly from penetration testing. The

issue, backed by the drone's safety evaluation, is regulating drone interconnection. In other terms, processes, and protocols were seen to interconnect power, so the drone is dangerous. This issue is also alleviated by adding a watchdog timer limit to control the CPU 's time while the browser is not being used; filtering all data sent to the drone system, and drone access point anti-spoofing systems [17]. The suggested monitor time is intended to defend against DoS attacks and will incorporate the OS domain. All π and μ will be focused on computational tests to guarantee that non-navigational procedures cannot access the CPU and prevent DoS attacks. The suggested filtering protection file works inside the network environment but is able to disrupt and terminate non-conforming operations. This ensures that no mechanism on the embedded drone network that accepts input data directly can allow load data over αA. The suggested filtering feature is illustrated. The character length specified in αA needs additional drone testing. Additionally, the access point's OSI network and link layers enforce the anti-spoofing security scheme and delete ARP responses with fake IP or MAC addresses. This would ensure that any ARP Cache Poison Attack controllers fail to damage the network. The Drone command helps the controller enter its control ports, copying the drone's ARP when the access point detects the data.

A system should be established with the suggested techniques, including protection on active and open drone services. The authentication system could fail and, thus, the services will be exposed. Therefore, besides authentication algorithms, Buffer Overload, DoS, and ARP Cache Poison attacks must be secured. This will mitigate possible damage allowing specific individuals to take advantage of such weaknesses. Because of the work carried out and therefore the tests carried out, it was possible to verify that these errors are still successful and obviously not the developers' concern, being a commercial type drone is not meant to have a particular application as is presumably granted to a military-style drone. In any case, the frame event should shield drones and successively users from external attacks, making use of these tools more widespread. In future work, it will be important to require details on applying the program suggested in this chapter and to test that such attacks can be mitigated by limiting access to power. As there is currently no technology involving securing these drones and related data is not required. It is hoped that producers of different business drones will take these vulnerabilities into account and develop appropriate applications to escape them.

7.8 FUTURISTIC RESEARCH AREAS

Unmanaged aerial vehicles may be small, remote-controlled aircraft, commonly known as drones. We predict battlefield, destruction, and death when we speak of drones. It may be because soldiers often use drones to drop bombs and shoot down enemy targets. They are known as ARMY drones. However, for this destructive intent, many possible drones exist. They are primarily used for civilian and industrial purposes to carry smaller goods to hard-to-reach places. They include fast pictures and real-time videos.

7.8.1 MEDICINAL DRONES

The roads in developing nations and mountain, desert or forest areas that need long-distance travel are impassable. For medical supplies, including vaccines and medications, the lack of road access is critical. So far, air travel like a helicopter is the only option, but it is costly and not available to patients or the healthcare system. Drones' performance in the fields of ecology and environment makes us believe they can also be used as medical couriers in the public health sector (Figure 7.6).

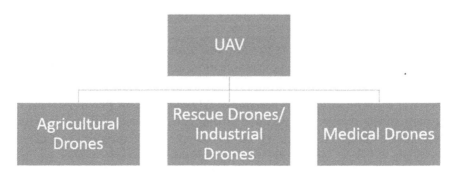

FIGURE 7.6   Future trends-UAV's.

Drones have tremendous public health benefits. They are used for transporting blood, specimens, and biological agents such as vaccines and reducing travel times to remote locations. They can be used to relieve disasters and save lives. Although there are limitations and risks in Drones' operation, development, and science advancements will solve them. Drone delivery and drone delivery ports can be constructed in the vicinity of health systems. Public security and privacy evaluations must be carried out before public health scaling on drones. More drone safety studies, including drone crashes and the reason for crashes, are required. Health awareness is also essential for people to alleviate drone apprehension.

7.8.2   AGRICULTURAL DRONES

By offering farmers three comprehensive views, the drone's interest can be understood. First, air monitoring crops may lead to the disclosure of trends showing irrigation, soil variability and fungal infestations. Second, the drone uses the remote satellite sensing system to detect crop growth by comparing multiple satellite images. Second, aeronautical cameras can capture several spectral images with visual and infrared information, revealing the difference between depressed and healthy plants, which cannot be seen with naked eyes. Agricultural drones may boost the crops and contribute to an insight into the technique of disease management via imaging and sensors. These will also provide assistance in tracking irrigation and water sources by forecasting water quality by glaciers. Agricultural drones can assist farmers in transforming farming to a higher level.

7.8.3   RESCUE SCENARIOS

UAVs have substantial capacity to view vast and remote locations, can relay image and sensor data from remote places more quickly than traditional ways, and UAVs can help emergency troops understand things and locate injured persons or persons in need of assistance without the risk of injury. This is also particularly relevant when the environment becomes difficult to navigate or may endanger rescuers. Those involve the evacuation of avalanches, explosions, flooding, and pollution (e.g., atomic disasters). UAVs are not only used for transmitting visual information in these situations but can also transmit other sensor data including temperature, air quality or radioactivity. In the case of natural disasters, UAVs may provide communication links to inaccessible people or even provide urgently-needed resources. When people are participating in physical activities, such as running or walking, health conditions can change rapidly, and the injured person is often able to provide remote assistance. In addition, people can get lost in outdoor sports and have to be taken home. UAVs may support this. It is also the case when children go out, or pets elope while playing. People lose their friends or children in amusement parks or at significant events. UAVs, like computer vision, can allow people to locate in crowds. Not only for outdoor situations but even for fairs, shopping centers and also in smaller buildings, UAVs may be used in the event of a fireplace, is this challenge to move. UAVs may be used by firefighters to examine items in a home without entering the house. Before sending firefighters, missing persons may be found here. In addition, information obtained by thermal imaging and air quality sensors can help firefighters operate quicker and safer.

7.9   CONCLUSION AND FUTURE SCOPE

This chapter provides a summary and a detailed discussion on UAV architecture and communication protocols security issues. The flaws in UAV design in the drone protection field are completely exploited. The UAANET model can overcome these limitations, and promising advances have been demonstrated in the securing of high-density node contact in UAV components. In UAANET architecture, first protocol routing problems and security issues are exposed. A UAANET Routing Protocol powered by the model is then addressed to avoid active and passive attacks on UAV communication. Extending security ideology, different systems for drone communication are used to classify safety conflicts in monitoring and tracking applications. Later, various UAV systems communication models are subjugated to successful modes of communication. Eventually, several potential paths and applications for study are envisaged.

KEYWORDS

- **drone**
- **protocols**
- **task control system**
- **unmanned aerial vehicle**
- **visual sight line**
- **wireless**

REFERENCES

1. Luppicini, R., & Arthur, S., (2016). A techno ethical review of commercial drone use in the context of governance, ethics, and privacy. *Technology in Society, 46*, 109–119.
2. Rao, B., Ashwin, G. G., & Romana, M., (2016). The societal impact of commercial drones. *Technology in Society,45*, 83–90.
3. Zeng, Y., Rui, Z., & Teng, J. L., (2016). *Wireless Communications with Unmanned Aerial Vehicles: Opportunities and Challenges, 54*(5), 36–42. IEEE Communications Magazine.
4. Yoon, K., Daejun, P., Yujin, Y., Kyounghee, K., Szu, K. Y., & Myles, R., (2017). Security authentication system using encrypted channel on UAV network. In: *2017 First IEEE International Conference on Robotic Computing (IRC)* (pp. 393–398). IEEE.
5. Kini, A., J., (2018). *Implementation of a Trusted I/O Processor on a Nascent SoC-FPGA Based Flight Controller for Unmanned Aerial Systems.* PhD diss., Virginia Tech.
6. Campbell, S., Wasif, N., & George, W. I., (2012). A review on improving the autonomy of unmanned surface vehicles through intelligent collision avoidance manoeuvres. *Annual Reviews in Control, 36*(2), 267–283.
7. Karpenko, S., Ivan, K., Alexander, M., Boris, M., & Dmitry, N., (2015). Visual navigation of the UAVs on the basis of 3D natural landmarks. In: *Eighth International Conference on Machine Vision (ICMV 2015)* (Vol. 9875, p. 98751I). International society for optics and photonics.
8. Ozger, M., Michal, V., & Cicek, C., (2018). Towards beyond visual line of sight piloting of UAVs with ultrareliable low latency communication. In: *2018 IEEE Global Communications Conference (GLOBECOM)* (pp. 1–6). IEEE.
9. Zmarz, A., Mirosław, R., Maciej, D., Izabela, K., Korczak-Abshire, M., & Katarzyna, J. C., (2018). Application of UAV BVLOS remote sensing data for multi-faceted analysis of Antarctic ecosystem. *Remote Sensing of Environment, 217*, 375–388.
10. Jean-Aimé, M., Mohamed, S. B. M., & Nicolas, L., (2016). Extended verification of secure UAANET routing protocol. In: *2016 IEEE/AIAA 35th Digital Avionics Systems Conference (DASC)* (pp. 1–16). IEEE.
11. Oubbati, O. S., Abderrahmane, L., Fen, Z., Mesut, G., Nasreddine, L., & Mohamed, B. Y., (2017). Intelligent UAV-assisted routing protocol for urban VANETs. *Computer Communications, 107*, 93–111.
12. Jean-Aimé, M., Gilles, R., & Nicolas, L., (2015). Emulation-based performance evaluation of routing protocols for UAANETS. In: *International Workshop on Communication Technologies for Vehicles* (pp. 227–240). Springer, Cham.
13. Macker, J. P., & Scott, C. M., (1999). Mobile ad hoc networking and the IETF. *ACM SIGMOBILE Mobile Computing and Communications Review, 3*(1), 11–13.
14. Ramalingam, K., Roy, K., & Chris, N., (2011). Integration of unmanned aircraft system (UAS) in non-segregated airspace: A complex system of systems problem. In: *2011 IEEE International Systems Conference* (pp. 448–455). IEEE.
15. Jean-Aimé, M., Mohamed, S. B. M., & Nicolas, L., (2016). Joint model-driven design and real experiment-based validation for a secure UAV ad hoc network routing protocol. In: *2016 Integrated Communications Navigation and Surveillance (ICNS)* (pp. 1E2-1-1E2-16, doi: 10.1109/ICNSURV.2016.7486324). IEEE.

16. Jean-Aimé, M., Mohamed, S. B. M., & Nicolas, L., (2015). Secure routing protocol design for UAV ad hoc networks. In: *2015 IEEE/AIAA 34th Digital Avionics Systems Conference (DASC)* (pp. 4A5–1). IEEE.
17. Mozaffari, M., Walid, S., Mehdi, B., Young-Han, N., & Mérouane, D., (2019). A tutorial on UAVs for wireless networks: Applications, challenges, and open problems. *IEEE Communications Surveys & Tutorials, 21*(3), 2334–2360.
18. Fletcher, S. D. A., Norman, P. J., Galloway, S. J., & Burt, G. M., (2008). *Evaluation of Overvoltage Protection Requirements for a DC UAV Electrical Network.* No. 2008-01-2900. SAE Technical Paper.
19. Qiu, Y., & Maode, M., (2016). A mutual authentication and key establishment scheme for m2m communication in 6lowpan networks. *IEEE transactions on industrial informatics, 12*(6), 2074–2085.
20. Jean-Aimé, M., Mohamed, S. B. M., & Nicolas, L., (2019). Performance evaluation of a new secure routing protocol for UAV Ad hoc Network. In: *2019 IEEE/AIAA 38th Digital Avionics Systems Conference (DASC)* (pp. 1–10). IEEE.
21. He, D., Sammy, C., & Mohsen, G., (2017). *Drone-Assisted Public Safety Networks: The Security Aspect, 55*(8), 218–223. IEEE Communications Magazine.
22. Kamate, S., & Nuri, Y., (2015). Application of object detection and tracking techniques for unmanned aerial vehicles. *Procedia Computer Science, 61,* 436–441.
23. Wang, X., Vivek, Y., & Balakrishnan, S. N., (2007). Cooperative UAV formation flying with obstacle/collision avoidance. *IEEE Transactions on Control Systems Technology, 15*(4), 672–679.
24. Nassar, A., Karim, A., Reda E., & Mohamed, E., (2018). A deep CNN-based framework for enhanced aerial imagery registration with applications to UAV geolocalization. In: *Proceedings of the IEEE Conference on Computer Vision and Pattern Recognition Workshops* (pp. 1513–1523).
25. Scherer, J., Saeed, Y., Samira, H., Evsen, Y., Torsten, A., Asif, K., Vladimir, V., et al., (2015). An autonomous multi-UAV system for search and rescue. In: *Proceedings of the First Workshop on Micro Aerial Vehicle Networks, Systems, and Applications for Civilian Use* (pp. 33–38).
26. Dionne, D., & Camille, A. R., (2007). Multi-UAV decentralized task allocation with intermittent communications: The DTC algorithm. In: *2007 American Control Conference* (pp. 5406–5411). IEEE.
27. Wu, J., Liangkai, Z., Liang, Z., Al-Dubai, A., Lewis, M., & Geyong, M., (2019). A multi-UAV clustering strategy for reducing insecure communication range. *Computer Networks, 158,* 132–142.
28. Li, J., Yifeng, Z., & Louise, L., (2013). Communication architectures and protocols for networking unmanned aerial vehicles. In: *2013 IEEE Globecom Workshops (GC Wkshps)* (pp. 1415–1420). IEEE.
29. Liu, D., Yuhua, X., Jinlong, W., Jin, C., Kailing, Y., Qihui, W., & Alagan, A., (2020). *Opportunistic UAV Utilization in Wireless Networks: Motivations, Applications, and Challenges, 58*(5), 62–68. IEEE Communications Magazine.
30. Li, T., Jianfeng, M., Xindi, M., Chenyang, G., He, W., Chengyan, M., Jing, Y., Di, L., & Jiawei, Z., (2019). Lightweight secure communication mechanism towards UAV networks. In: *2019 IEEE Globecom Workshops (GC Wkshps)* (pp. 1–6). IEEE.

31. Sedjelmaci, H., Sidi, M. S., & Nirwan, A., (2017). A hierarchical detection and response system to enhance security against lethal cyber-attacks in UAV networks. *IEEE Transactions on Systems, Man, and Cybernetics: Systems, 48*(9), 1594–1606.

32. Hooper, M., Yifan, T., Runxuan, Z., Bin, C., Adrian, P. L., Lanier, W., William, H. R., & Wlajimir, A., (2016). Securing commercial WIFI-based UAVs from common security attacks. In: *MILCOM 2016–2016 IEEE Military Communications Conference* (pp. 1213–1218). IEEE.

PART III

DRONES IN THE MACHINE LEARNING ENVIRONMENT

CHAPTER 8

Challenges and Opportunities of Machine Learning and Deep Learning Techniques for the Internet of Drones

ROSHAN LAL,[1] SANDHYA TARAR,[1] and
NAVEEN CHILAMKURTI SMIEEE[2]

[1]Department of Information and Communication Technology, Gautam Buddha University, Uttar Pradesh, India, E-mails: rchhokar@amity.edu (R. Lal), tarar.sandhya@gmail.com (S. Tarar)

[2]Department of Computer Science and IT, La Trobe University, Melbourne, Australia, E-mail: n.chilamkurti@latrobe.edu.au

ABSTRACT

The term IoD originates from internet of things (IoT), drones replace things. The term IoD is introduced in the autonomous flying robot or machine to make drones smarter. The internet of drones (IoD) is a layered framework control building organized generally for arranging the passage of unmanned aeronautical vehicles to controlled airspace and giving course organization between zones suggested as center points. Artificial intelligence (AI) is a term used straightforwardly in the drone business: anyway, Now human-made think AI is used in drones to make this flying machine increasingly splendid.

In collaboration of drones to IoD, it will make, the existing sectors intelligent and smarter to work. Many studies have been utilized the power of supervised, semi-supervised, unsupervised machine learning (ML) and deep learning (DL) algorithms and the IoD to extract useful information from

The Internet of Drones: AI Applications for Smart Solutions. Arun Solanki, PhD, Sandhya Tarar, PhD, Simar Preet Singh, PhD & Akash Tayal, PhD (Eds.)

various sector where we can use application of IoD. ML, DL and IoD are the only two fields which when both applied together are able to give solutions to solve the real-world problems. This study focuses on the significant research in the area of smart drones. We will find the various algorithms of ML, deepfake, GAN, and DL. We also compare the performance of various algorithms in terms of computational performance which will make more smart drones. So, findings of this study are helpful for various application field of drones.

8.1 INTRODUCTION: INTERNET OF DRONES (IoD)

Internet of things (IoT) had taken over many aspects of our day-to-day life and its applications are implemented in many fields such as healthcare, agriculture, transportation, healthcare. However, a major drawback when it comes to such intelligent systems is that the rate of energy being drained is very high. Same goes for drones, which are an emerging new form of IoT. In the coming future, scientist predict that drones will be used in many spheres of life including the day-to-day package deliveries to tracking down criminals. Platform of cognitive drones, also known as "Flying IoT" focuses on the machine learning (ML) application called object detection. This is a significant step towards automatically recognizing and tracking objects (Figure 8.1).

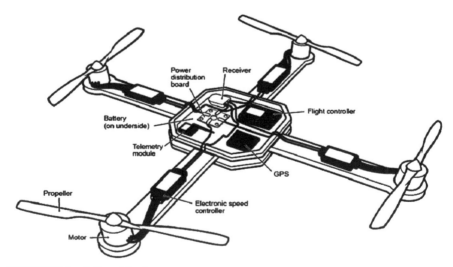

FIGURE 8.1   IoD architecture.

Source: Reprinted from Ref. [63]. https://creativecommons.org/licenses/by/4.0/

Object detection is a tedious and computationally intensive task. A large amount of energy is taken up by the drone while tracking, tracing, and detecting a person. For short distances, this is not a problem, but for long distances it becomes impossible due to the increasing workload which is significantly affecting the flight. Computing drones in real time essentially needs a number of ALUs per processor and a bigger size of chip. This results in an increase in the heatsink and makes the device bulky. All this affects the take-off weight of the drone, subsequently decreasing the flight time [1–4].

An alternative architecture that would ease off the workload of the drone would be to introduce a sensor cloud system to the existing architecture. If the computationally heavy tasks such as data collection and analysis is taken care by the edge devices in the cloud architecture, it would reduce the workload. It will provide greater efficiency and speed to the drone. This is where internet of drones (IoD) comes into play [5, 6].

8.1.1 IoD NETWORKS

There are three layers in the IoD network. Each layer has its own functionality which helps in achieving the required structure (Figure 8.2).

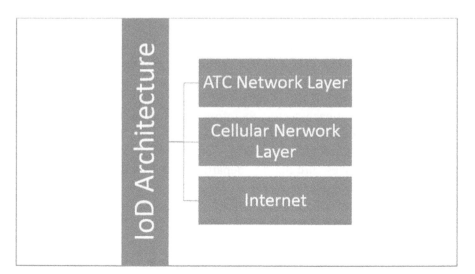

FIGURE 8.2 IoD network.

1. Air Traffic Control Network Layer: This layer uses the airspace and efficiently maintains a collision free course for the drone. It is an integral part of the architecture. The components of A.T.C. are discussed while keeping the A.T.C. structure in United States as the focus, however similar structure is followed all over the world. The airspace in United States is divided into 21 zones, which are also called centers. Each center is further divided into sectors. Every zone is also divided in 50 miles (in diameter) portions known as the terminal radar approach control (TRACON) where consists a number of airports (Figure 8.3).

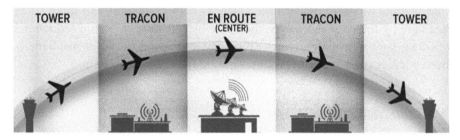

FIGURE 8.3   A.T.C components.

The traffic control divisions are divided into the ATCSCC, ARTCC, TRACON, ATCT, and F.S.S.

i.   Air Traffic Control System Command Center (ATCSCC): It looks after the air traffic. It controls and manages the traffic within the centers. Some problems that might take place are bad weather, inoperative runways and traffic overload.

ii.   Air Route Traffic Control Center (ARTCC): It manages the traffic in each sector; therefore, each center has one ARTCC. However, it does not manage the traffic for TRACON airspace.

iii.   Terminal Radar Approach Control (TRACON): It manages and makes sure that the departure and arrival of the aircraft is smooth within its space.

iv.   Air Traffic Control Tower (ATCT): These towers are located at each airport and handle the ground traffic. They also manage take-off and landing.

v.   Flight Service Station (F.S.S.): F.S.S. is responsible for assisting during emergencies. It coordinates operations where search and rescue are required (Figure 8.4).

FIGURE 8.4   Traffic control division.

There are differences between IoD and A.T.C. ATCSCC cannot be used in case of drones. This is because drones share a certain amount of airspace at a time. Therefore, tasks that use centralized entities like load prediction and load assignment which are taken care by ATCSS are not possible. Separation of volume of flights must be done in an autonomous manner with the help of drones. It would give better and efficient results than human intervention [7, 8].

2.   Cellular Network Layer: The cellular network consists of honeycomb structured cells. The cells are hexagonal in nature so as to cover most of the area of signals. These signals are directed to a mobile with the help of a base station (BS) which are further connected to a mobile telecommunications switching office (MTSO). To avoid the interference of signals of different cells, each station uses a different frequency. MTSO is a central entity which is responsible for the purpose of assigning B.S. to each mobile unit in a particular cell. It also performs handoff. However, handoff is challenging as it is not known if the call will end once the device changes cells. There may be an occurrence where this happens due to the unavailability of channels in the new cell. A similar situation may exist in the case of drones. But in IoD, the source and the destination of the drone is known. This helps in better utilization of resources [9–12].

3.   Internet: The internet architecture has five layers (Figure 8.5): application layer; transport layer; internet layer; link layer; and physical layer.

IoD has similarities with the internet as both networks perform the task of routing. However, as the internet's time scale is smaller. As IoD is computationally more able, it can calculate optimal routes.

A difference between the two is that in case of the internet, the overloading packets in the system can be dropped as resending them is not expensive. However, in IoD this is not economically feasible.

Application
Transport
Internet
Link
Physical

FIGURE 8.5   Internet architecture.

8.2   ADVANTAGES OF DRONES USING MOBILE TECHNOLOGIES

Drones are capable of reaching places that are inaccessible to humans with minimum causalities. As discussed above, wireless connectivity is required by the drone to perform operations and tasks.

Using mobile networks can be used to take low altitude drone tasks into global airspace. Mobile networks are already running successfully and therefore they provide economically feasible solutions on a nationwide scale.

The drone utility in today's time is expanding and growing at a rapid scale, thereby being extensively used in industries and research. The quality connectivity provided by the mobile networks provide a base for the applications of drones.

Drones which are a part of IoD are not only used for tracking and recognizing but also for geo-fencing. Moreover, the mobile network technologies are constantly evolving and therefore offer a greater scope for IoD (Figure 8.6) [13–16].

8.3   ADVANTAGES OF INTERNET OF DRONES (IoD)

Due to the technological advancement in the society unmanned aerial vehicles (UAV) or drones are being used excessively-from small-scale remote-control toys for kids to powerful military systems. Research committee has jotted down a number of characteristics that have been appealing.

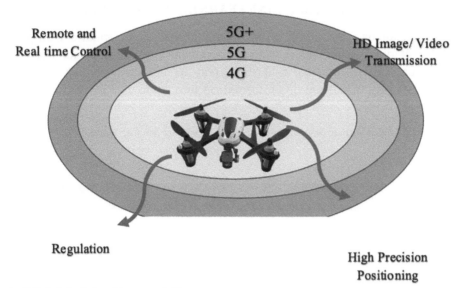

Remote and
Real time Control

5G+
5G
4G

HD Image/ Video
Transmission

Regulation

High Precision
Positioning

FIGURE 8.6   IoD with types of G's.

Drones are the best suited for navigation purposes over uneven terrain. Adding to this, the aerial view offered by UAV gives a new perspective while discovering geographical points which are not easy to reach by humans. Drones are smaller, lighter, economically feasible and do not require a pilot onboard unlike aircrafts. This allows us to carry out experiments and military operations without putting human life in risk. The smaller size makes the drones fuel efficient and helps us dodging casualties to a minimal [16].

8.4   DISADVANTAGES OF INTERNET OF DRONES (IoD)

Keeping aside the advantages of the drones, there are shortcomings of drones as well. They do not work efficiently in adverse weather conditions. Due to the multiple propellers in the drones architecture, there is a significant amount of noise that is produced. There may also be times when there is a crash of the drone due a hardware malfunctioning or a software glitch. Moreover, the battery power of the drone is limited. Most drones have features of visual and audio aid such as cameras and voice recorders, privacy becomes a very important issue [18, 19].

8.5　INTRODUCTION: MACHINE LEARNING (ML) ALGORITHMS

In the earlier half century there was an outbreak of information, ample amount of data which was expanding exponentially and the data was of no use unless and until we find some way to analyze that data or find the correct way of putting this information into some meaningful insights (Figure 8.7).

"We are drowning in information and starving for knowledge"

—John Naisbitt

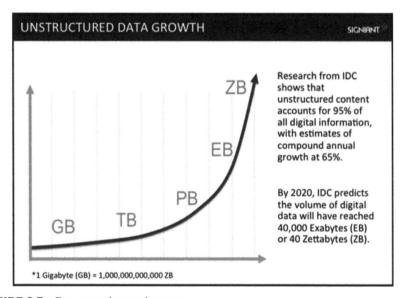

FIGURE 8.7　Data growth over the years.

Source: Reprinted from https://www.signiant.com/resources/tech-articles/the-historical-growth-of-data-why-we-need-a-faster-transfer-solution-for-large-data-sets/.

　　Now in order bring outcomes from that information unknowingly we interacted with one of the biggest types of learning known as ML. Now machine can also learn [21].

8.5.1　HISTORY OF MACHINE LEARNING (ML)

The history of Machine leaning evolves from 1950 when Alan Turing creates "Turing test" to 2015 "when amazon launches its first own ML platform" and till date it is evolving with its great impacts.

"The Analytical Engine weaves algebraic patterns just as the Jacquard weaves flowers and leaves"

—Ada Lovelace

The foremost computer programmer and the founder of computing quoted that anything the world could think of can be well described and proved by MATH. Quite essentially, we can say that a mathematical representation can be used to create the relation between the representation of phenomenon. Our foremost engineer Ada Lovelace states that the machines are enough capable to understand the world without the assistance of human.

It took two centuries, there are some core objectives which are quite critical in the concepts of ML. Irrespective of the clause, the obtained knowledge can be graph plotted as data points. Then comes the concept and algorithms of "machine learning" which helps to find the mathematical expressions and relations which are hidden with the authentic piece of knowledge.

8.5.2   PROBABILITY THEORY

"Probability is orderly opinion… inference from data is nothing other than the revision of such opinion in the light of relevant new information"

—Thomas Bayes.

The mathematician he originates the plans of propagating probability theory into ML algorithms [22, 23]. The world is filled with probability. All our actions are uncertain they come up with different possibilities. Bayesian probability refers to computing the uncertainty of an event on which ML is based. As a result of which, we consider probabilities based on the information available, rather than trials [24].

Let us consider an example of a Cricket Match, instead of taking the Number of matches won by India against Australia. The Bayesian considers to-the-point information of them like their current forms, starting teams [24]. The core advantages of taking Bayesian approach are that the rare possibilities can be in-sight while making the decision based on their relevant Features and reasoning [25].

8.5.3 STATISTICAL THEORY

Statistical theory is a system for ML drawing from the fields of insights and practical analysis. Statistical theory manages the issue of finding a prescient capacity dependent on information. Statistical theory has prompted effective applications in fields such as, P.C. vision, discourse acknowledgment, and bioinformatics [25].

The objectives of learning are understanding and expectation. Learning falls into numerous classifications, including regulated learning, solo learning, web-based learning, and support learning. From the viewpoint of statistical learning hypothesis, administered learning is best understood. Supervised learning (SL) includes learning from a preparation set of information. Each point in the preparation is an info yield pair, where the information maps to a yield. The learning issue comprises of construing the capacity that maps between the information and the yield, with the end goal that the educated capacity can be utilized to foresee the yield from future information [25].

8.5.4 MACHINE LEARNING (ML)

"Computers are able to see, hear, and learn. Welcome to the future."

—*Dave Waters [26]*

A computer is able to learn from experience without being specifically programed. ML (a subset of AI) is a branch of computer science which has evolved from the study of data sciences and data analytics. ML is basically the invention of algorithm from the type of data being analyzed (Figure 8.8) [27].

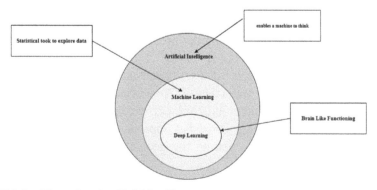

FIGURE 8.8 Hierarchy of artificial intelligence.

ML is the process of learning from its past experiences. ML is to produce better outcomes and precise predictions. It is the popular technique to predict the future, classify it, and make the suitable necessary decisions (Figure 8.9) [28].

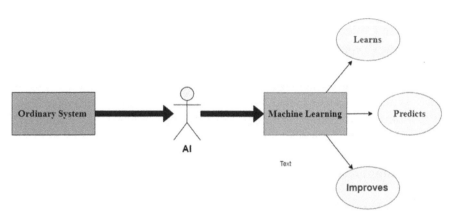

FIGURE 8.9   Uses of machine learning.

As there is ample amount of information which is expanding exponentially and in order to get meaningful insights from the data we are in need of some learnings or set of rules or algorithms that can provide us beneficial results from this data. ML has created great impact on several industries like healthcare, banking, management sectors, medicinal, and much more. ML is being used all over from the electronic gadgets you use or the applications that are a piece of your regular daily existence are controlled by amazing algorithms from ML.

8.5.5   ESSENTIAL KEYWORDS

1.   **Dataset:** A collection of data which contains the essential features for solving a problem.
2.   **Features:** Essential chunks of data which makes the problem understandable for us. They are stored into the algorithms of ML.
3.   **Model:** It is the mathematical representation of model parameters for each and every input parameter for prediction or the class and action for the regression, classification, and reinforcement techniques. For

example, "a decision tree algorithm would be trained and produce a decision tree model."

8.5.6 MECHANISM OF MACHINE LEARNING (ML)

1. **Data Collection:** Collection of data for the algorithm.
2. **Data Preparation:** Editing and preparing the data in the favorable pattern with the essential characteristics and the reduction in the dimensionality.
3. **Training:** Can be called "fitting stage," it is the stage where the algorithm of Machine Learning gains knowledge from the broadcasted data prepared after data processing.
4. **Evaluation:** Testing is performed to check the performance.
5. **Tuning:** Here the original performance is maximized.

8.5.7 TRADITIONAL PROGRAMMING VS. MACHINE LEARNING (ML)

Basically, traditional programming is defined as process where user inputs the data and programming rules which results in the output (Figure 8.10).

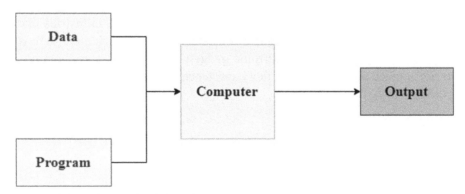

FIGURE 8.10   Conventional algorithm.

ML is a technique is quite different from traditional programming where it uses the user input along with the result of the program and then predicts the rules of program associated with it (Figure 8.11) [1].

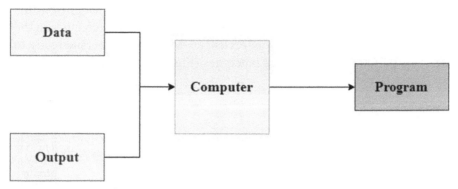

FIGURE 8.11 Machine learning algorithm.

Machines go through a process of learning, testing on different algorithms, rules, and based on their performance they conclude to the ML [1].

8.5.8 HOW MACHINE LEARNING (ML) WORKS?

With the increment in the data there is a requirement for having a framework that can deal with this huge heap of information. ML models like deep learning (DL) allows the vast majority of data to be handled with accurate predictions. ML has reformed the manner in which we see the data or information being provided and the results can increase out of it (Figure 8.12) [29].

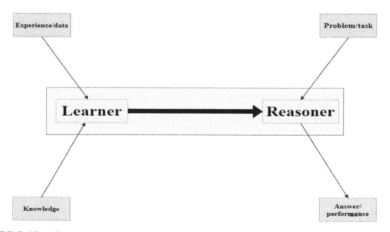

FIGURE 8.12 Components of machine learning algorithm.

ML models learns from the useful patterns that are extracted from the data to perform suture predictions. Whenever there are some changes in the data, machine learn from its past experiences. ML makes machine ready to adapt new patterns to show better outcomes (Figure 8.13) [7].

FIGURE 8.13   Steps for implementing machine learning algorithm.

Some of the related disciplines to ML are shown in Figure 8.14.

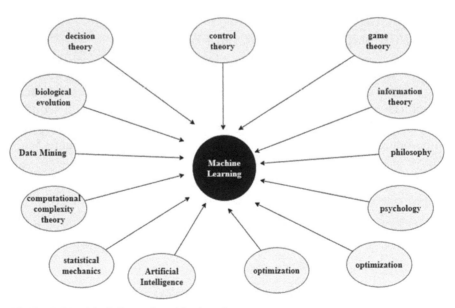

FIGURE 8.14   Disciplines of machine learning.

8.5.9   KEY ELEMENTS OF MACHINE LEARNING (ML) (FIGURE 8.15)

1.   **Representation:** How to represent knowledge?
Examples:

     i.   Decision trees;
    ii.   Model ensembles;
   iii.   Sets of rules;
   iv.   Neural networks;
    v.   Instances;
   vi.   Graphical models;
  vii.   Support vector machines.

2. Evaluation: The way to evaluate candidate programs (hypotheses).

Examples:
     i.   Accuracy;
    ii.   Prediction and recall;
   iii.   Squared error;
   iv.   Likelihood;
    v.   Posterior probability;
   vi.   Cost, margin;
  vii.   Entropy k-l divergence.

3. Optimization: The way candidate programs are generated.

Examples:
     i.   Combinatorial optimization;
    ii.   Convex optimization;
   iii.   Constraint optimization.

FIGURE 8.15   Elements of machine learning.

8.5.10 MACHINE LEARNING (ML) APPROACHES

The focal point of the field is realizing, that is, obtaining abilities or information as a matter of fact. Most generally, this implies combining helpful ideas from authentic information.

All things considered, there is a wide range of kinds of discovering that you may experience as a professional in the field of ML: from entire fields of study to explicit methods.

The focal point of ML is "learning" and there are numerous types of learning you may experience as an expert. These learning techniques gives a way to analyze the information provided and then process it through the respective algorithms.

8.5.11   SIMPLE LINEAR REGRESSION ALGORITHM

Simple linear regression is great learning for data analytical process. It works with two variables out of which one is the dependent variable and the other one is an independent variable, and finds the relationship between the two as a linear function. The following model function is considered and worked upon-

$$y = b_0 + b_1x \tag{1}$$

where; b_0 is the slope; and b_1 is the y-intercept.

> **Advantages of Simple Linear Regression:**

- Speed of the machine is fast;
- Machine does not require complex calculations;
- Machine runs very fast when the amount of data is huge;
- Data is labeled in this type;
- It is simple and easy to apply and understand;
- The computational cost is low;
- Space complexity is very low;
- It is a high latency algorithm;
- It is good interpreter;
- It provides ground for very complex learning algorithms.

> **Disadvantages of Simple Linear Regression:**

- Machine does not work on non-linear data;
- First it has to be checked whether the data is linear or not linear. If it is simple linear, then it may be implemented;
- Machine require training data set and test data set;
- Machine assumes data is normally distributed in real;
- Machine over simplifies many real-world problems.

8.5.12   RANDOM FOREST ALGORITHM

Random forest algorithm is a SL process which can be used for both regression as well as classification [30]. It is basically used for classification tasks.

As we know that a forest is made up of trees and very robust forest. Random forest algorithm creates decision tree on sample data and then gets the prediction from each tree and lastly selects the best solution. It is ensemble method which is better than a single decision tree. Random forest algorithm reduces the over-fitting as compared to single decision tree [30].

➢ **Advantages of Random Forest Algorithm:**

- Random forest algorithm retains good accuracy;
- Random forest algorithm does not scaling of data;
- Random forest algorithm has very less variance;
- Random forest algorithm reduces the overfitting;
- Random forest algorithm retains good accuracy;
- Random forest algorithm works very well on large scale data;
- Random forest algorithm possesses high accuracy;
- Random forest algorithm is more flexible rather than decision tree.

➢ **Disadvantages of Random Forest Algorithm:**

- Random forest algorithm has very complex complexity;
- Random forest algorithm has time consuming process than decision tree;
- Random forest algorithm has very hard construction;
- Random forest algorithm requires more computational resources;
- Random forest algorithm takes more time in prediction with other algorithm.

8.5.13   LOGISTIC REGRESSION ALGORITHM

In logistic regression, the concept of the probability of odds is used, which lies between 1 and 0. Some real-life examples exist in situations where we have a student who can either pass or fail, a lottery that can either we won or lost, a person who is potentially sick or healthy. The cost function in the linear equation is taken as the mean squared error (M.S.E.) [31]. If the same is used in logistic regression algorithm, with the help of gradient descent (GD) one can find the global minima of the function will converge if the function is convex in nature.

➢ **Advantages of Logistic Regression:**

- Implementation is very easy;

- Efficiency increases once the data is trained;
- It can easily overfit in high dimensional datasets;
- It performs well when the datasets that are linear in nature;
- It is highly interpretable.

➢ **Disadvantages of Logistic Regression:**

- Prediction of discrete functions can be done using this type of regression;
- The dependent variable is bounded and restricted to the discrete set of numbers;
- In Logistic regression if the number of features is more than the number of observations, it is not used because it may be lead to induce overfitting in the model;
- It is the linear assumption between the dependent variables and the independent variables. In the real world, the data is rarely linearly differentiable.

8.5.14   KNN (K-NEAREST NEIGHBORS)

K-nearest neighbors (KNN) is a simple algorithm of ML used to solve the classification problems. KNN means K-nearest neighbors, where K is the number of neighbors in KNN algorithm does not require learning during training period [32]. KNN takes data sets during real time predictions. KNN is a very fast algorithm rather than others like SVM, linear, and Logistic regression, etc. KNN algorithm require no training before predictions. So, adding of new data cannot be impact on the accuracy of KNN. KNN algorithm is very easy to implement require only two parameters the value of K and the distance function.

➢ **Advantages of KNN Algorithm:**

- KNN algorithm has no training period of data sets so it is called lazy learner or instance-based learning;
- KNN is very faster algorithm rather than others;
- Adding of new data may not impact on accuracy;
- KNN is an easy to implement due to only two parameters;
- KNN is very effective algorithm on small data set rather than others.

➢ **Disadvantages of KNN Algorithm:**

- KNN algorithm does not work very well on large data set;

- KNN algorithm has low performance due to the cost of distance between the new point;
- KNN does give optimum results with high dimensional datasets;
- KNN requires feature scaling before being applied to any data;
- KNN is sensitive to noisy data, missing values, and outliers;
- KNN need to manually to remove outliers.

8.5.15 *Decision Tree Algorithm*

As the name suggests, this algorithm works with the help of trees which are basically flowchart like structure with leaf nodes. In the tree, every leaf node depicts a label class whereas the nodes inside (internal nodes) are test on the attributes or features. Decision tree is an algorithm of ML which is used to solve both types of problems like regression and classification [33]. Decision tree leads to overfitting the data which is overcome by random forest algorithm. Decision tree algorithm's output can be interpreted by humans in a very simple way.

> **Advantages of Decision Tree Algorithm:**

- It can be used both regression and classification;
- It is very easy to implement and simple to understand;
- It can handle variables of both continuous nature and categorical nature;
- It can handle missing values automatically;
- It is robust to outliers;
- It takes less training period as compared to random forest;
- It uses if-else statements which are very easy to understand;
- It can handle non-linear parameters efficiently.

> **Disadvantages of Decision Tree Algorithm:**

- It leads overfitting data;
- It leads wrong predictions;
- It has very high variance due to overfitting of the data set;
- It is very Unstable due to adding new point;
- It has a little bit of data noise that can make it unstable which leads to wrong predictions;
- It is not used with large data sets, if data size is large then one single tree can grow complex and leads to overfitting.

8.5.16   SUPPORT VECTOR MACHINE (SVM)

This ML algorithm can be used for both classification and regression, however is used mostly to classify data. SVM was introduced in the 1960s and was further refined in the 1990s [34]. This algorithm aims at plotting data as a refined point in an n-dimensional plane such that the coordinate of the point is equal to the value of the features. SVM transforms the non-linear data into linear data and used to find the best and optimal results of the problems [35].

> **Advantages of Support Vector Machine (SVM):**

  - It has a good generalization with the help of which we can prevent over-fitting to take place;
  - SVM handles the non-linear data in an efficient manner;
  - It follows and uses the Kernel method;
  - Classification and regression problems can both be solved using SVM;
  - SVM has more model stability.

> **Disadvantages of Support Vector Machine (SVM):**

  - SVM has a difficult to choose an appropriate kernel function;
  - SVM require more extensive memory;
  - SVM has an algorithmic complexity;
  - SVM require feature scaling;
  - SVM takes a long time for training data;
  - SVM is more difficult to understand;
  - SVM is interpret by human.

8.5.17   NAÏVE BAYES

A Naive Bayes classifier is based on the probabilistic ordering of ML models. The heart of the classifier is the theory of Bayes.

$$P(A/B) = P(B/A)\ P(A)/P(B) \qquad (2)$$

Using the hypothesis from Bayes, we can find the probability of an occurrence of A after B occurred. The proof here is B and the hypothesis is A [36]. It is meant to be free of metrics / highlights. It is the proximity between one specific variable and the other. It is either referred to as being faithful or naïve.

➢ **Advantages of Naïve Bayes:**

- Naïve Bayes gives better performance than other models if the predictors are independent of each other;
- As compared to other models, Naïve Bayes theorem requires a smaller amount of training data therefore training period is less;
- Naïve Bayes can handle multicategory tasks;
- It is suitable for incremental training of data;
- It is easy to implement;
- It is relatively easy to understand, is relatively simple and used for classification of text;
- It explains the results easily.

➢ **Disadvantages of Naïve Bayes:**

- In real life problems, it is very rare to have predictors that are independent of each other;
- In this algorithm, there is a need to calculate the prior probability;
- It is sensitive to the form of input data.

8.5.18   UNSUPERVISED LEARNING ALGORITHM

Unsupervised learning algorithm is a major category of ML algorithm that uses previously undetected patterns in a dataset that has no pre-existing labels, taking in account a minimum of human supervision [37]. In contrast to SL algorithm, it uses of human-labeled data. It is a self-organization that allows modeling of probability. Unsupervised learning algorithm gains through the observation, also finding structures in the data set simultaneously. Once the machine has been given a dataset then it automatically starts finding the relations and in the data set with the help of clustering analysis.

Some of the following examples of unsupervised algorithm are:

- K-Means algorithm;
- Apriori algorithm;
- Clustering algorithm.

8.5.19   K-MEANS ALGORITHM

K-means algorithm is an iterative method that creates subgroups by predefined K distinct clusters. Each point belongs to one group without any overlapping. With the help of centroids, the data points in the clusters are clubbed which are similar and closer to the other intra-cluster points simultaneously keeping the different clusters as far away from other clusters as possible [38]. The algorithm works as follows:

- Number of clusters (K) that will be used are specified;
- Initial centroids are selected by randomly picking K data points from the dataset;
- Sum of mean distance is computed between the centroids and all data points;
- Each point is put under the closest calculated centroid and hence the closest cluster;
- Centroid is updated by assigning the new value as the average of the data points present in the cluster;
- Steps (c) to (e) are repeated until there is no change in the assignment of the clusters of the data points. Hence it is an iterative algorithm.

➢ **Advantages of K-Means:**

- It is easy to implement;
- It scales to larger data sets;
- There will be a definite convergence;
- It easily adapts to new examples.

➢ **Disadvantages of K-Means:**

- The value of K is difficult to predict and if often chosen manually;
- The results are obtained for different parameters;
- It does not work effectively with clusters of different density and size;
- Centroids are effected by outliers and therefore should be considered to be removed or clipped before making sub-groups.

8.5.20 APRIORI ALGORITHM

This unsupervised learning algorithm is often used for data mining and applies the downward closure property. According to this property, in a subset of data items, the frequent item sets occurring in the data set are frequent, however the converse may or may not hold true [39]. It identifies individual items that occur frequently in the transactional databases. There are two parameters that are used in this algorithm, namely being "support" and "confidence." The prior is associated with the frequency of the item's occurrence whereas confidence refers to the conditional probability [40]. The main steps of the algorithm are:

- Candidate set is generated, i.e., support of each item in the database is calculated;
- Candidate set is updated by eliminating or removing the items that have a lower support than the given threshold;
- Frequent item sets are joined and the steps 1 and 2 are repeated until no new itemsets can be joined in the final set.

➢ **Advantages of Apriori Algorithm:**

- It is easy to understand;
- It is easy to implement;
- The results are easy to communicate to the end user as they are intuitive in nature;
- Various scans are created for candidate sets;
- It can easily be parallelized.

➢ **Disadvantages of Apriori Algorithm:**

- It is slow in comparison and the bottleneck is candidate generation in this algorithm;
- It requires a large number of database scans;
- It assumes transaction database is memory resident;
- Performance time is more.

8.5.21   CLUSTERING ALGORITHM

Clustering is a data analysis technique. It aims at identifying and creating subgroups of the given data and on the basis of the given attributes of the sample space [37–39]. The differentiation and division of the datapoints is on the basis of similarity between the points.

➢ **Advantages of Clustering:**

- It is relatively easy to implement;
- It is easy to understand.

➢ **Disadvantages of Clustering:**

- In this algorithm, it is difficult to detect if the dataset is homogeneous or contains clusters;

- Variables cannot be ranked by their contribution in heterogeneity of dataset;
- Special patterns (zone of mixing, noise, outliers, etc.) cannot be detected;
- Inability to estimate correct number of clusters.

8.5.22   *SEMI-SUPERVISED LEARNING ALGORITHM*

Semi-supervised learning algorithm is a process of combining supervised approach and Unsupervised approach [40]. Semi-supervised learning algorithm is used when less amount of labeled data is used with a greater amount of unlabeled data while training data set. It is a simple understand and easy to implement on the given data set.

> **Advantages of Semi-Supervised Learning:**

- It is helpful when labeled data is hard to get or expensive;
- It can be used in N.L.P;
- It has a lot of medical applications;
- Unlabeled data is cheaper than labeled data.

> **Disadvantages of Semi-Supervised Learning:**

- It makes a lot of assumptions like the smoothness, clusters, manifolds, etc.;
- Some techniques require very specific setup.

8.5.23   *REINFORCEMENT LEARNING ALGORITHM*

Reinforcement learning algorithm is an algorithm in which we use the ability of an agent to interact with the environment, which leads us to find the best outcome [39, 40]. It is based on the concept that is followed is the hit and trial approach. The agent is rewarded with a point for a correct answer, i.e., plus point or penalized with a point for a wrong answer, i.e., minus point. On the grounds of the reward points gained by the machine, it retrains itself. Once it has been trained, it becomes ready to predict new data that is given to it. Some examples of reinforcement learning algorithm are:

- Markov decision process;

- Q-learning;
- Chess game.

8.5.24   CHARACTERISTICS OF REINFORCEMENT LEARNING ALGORITHM

- Determinization of the subsequent data that will be received will due to the agent's actions;
- There is no concept of a supervisor, only a real number or reward signal exist;
- Feedback is always delayed;
- Time plays an important role;
- Sequential decision making in terms of reward points.

8.5.25   REINFORCEMENT LEARNING ALGORITHM APPLICATIONS

There are the various applications of reinforcement learning algorithm:
- Airspace control system;
- Robot motion control system;
- Data processing;
- Planning for business strategy;
- Robotics for industrial automation;
- Reward points for gaming.

➢ **Advantages of Reinforcement Learning Algorithm:**

- It can be used to solve complex problems;
- It gives long term results;
- Errors can be corrected during the training process;
- If there is absence of dataset, it can learn from experience.

➢ **Disadvantages of Reinforcement Learning Algorithm:**

- This type of learning can be non-stationary;
- It may lead to an overload of states;
- Realistic environments may have partial observability;
- Parameters can be effect on the speed of learning;
- Reward design which should be involved.

8.6   INTRODUCTION TO NEURAL NETWORKS

The neural nets provide a way to implement ML, where a computer learns about a specific instruction to be performed be it classification, prediction, generation, etc., by analyzing past data. The data fed to the ML model is labeled beforehand, and cleaned and processed for any redundancy and missing values [41, 42].

The neural net architecture is inspired from the human brain, where there are hundreds or in some situations, tens of thousands of processing nodes arranged within spider web configuration where each node is connected to every other node in the next layer. The trend for today's neural networks is to be put in layers of nodes and forward-feed in nature, what that means is that the data moves in just forward direction, providing an output ate the last layer.

Each incoming connection to the node is assigned a number identified as weight associated to that node, when an input is passed to the node it gets multiplied by this weight before getting added to the resultant product for the layer. Now if the resultant is smaller than the set threshold it is not passed to the next layer, otherwise it passed to every node of next layer. This passing of resultant is known as firing of neuron.

In the initial phase of training neural nets are set with random values of weights and thresholds. The data for training is fed to the bottommost layer, known as the input layer, where each node corresponds to a unique parameter/feature of the dataset. These are then multiplied and summed up in a complex way, depending on their activation function and end up at the output layer radically transformed. The initial values od weights and threshold are adjusted throughout the training process based on the error in yield.

Perceptron, the first trainable neural network, was presented by the Cornell University professor Frank Rosenblatt in 1957. The basic structure for the perceptron was very similar to modern neural net, besides the fact that it just had a single layer of activation function-based nodes wedged between an input and an output layer [42].

Functioning of a perceptron involves taking in several feature selections as inputs from dataset and producing a class/ prediction as an output (Figure 8.16).

Consider X1, X2, X3 as some features, so for different data these can range from just a few to hundreds of features. Meaning an input layer can have neurons/nodes ranging from just one or two to hundreds. Rosenblatt associated weights such as w1, w2, w3…., having real values ranging from

(–0.5, 0.5) generally indicating the significance of respective inputs. Output can range from [0,1], determined by the weighted sum of the associated weights along with their inputs ($\sum_j w_j x_j$) with respect to the threshold values. Similar to weights threshold values are real number with initial range (–0.5, 0.5) [42, 43].

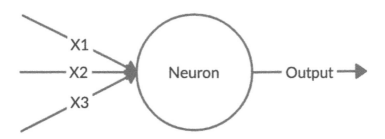

FIGURE 8.16   A single perceptron similar to neuron in human neural network.

Algebraic expression for the above is as follows:

$$Output = \begin{cases} 0 \; if \; \sum_j w_j x_j \leq threshold \\ 1 \; if \; \sum_j w_j x_j \geq threshold \end{cases}$$

And that is it for the functioning of a perceptron!

To make things simple for understanding a perceptron. The condition $\sum j w_j x_j >$ threshold being clunky, can be made elegant by making two notational changes. First, changing $\sum j w_j x_j$ to be represented as a dot product, i.e., $w \cdot x \equiv \sum j w_j x_j$, where w and x are representative vectors symbolizing the weight and inputs respectively. Second, moving the threshold to the other side of the inequality, in order to be replaced but a new term that changed the way we perceived threshold known as bias, $b \equiv$ –threshold. Bias can be understood as how much the machine is reluctant to being changed or simply set in its ways. With bias the above expressions can be rewritten as follows:

$$Output = \begin{cases} 0 \; if \; w.x + b \leq 0 \\ 1 \; if \; w.x + b > 0 \end{cases}$$

The bias is a measure symbolizing the ease by the which a neuron can be fired. A perceptron with a significantly large bias would be easily fired and the machine would be quick to adapt to change. While a perceptron with a relatively small bias would be much more harder to trigger and machine being reluctant to change, thus we have to set the bias in somewhere between the two as one can lead to overfitting and the other to underfitting, either of which is not something we require. Introduction of bias made a small change in how we describe a perceptron, but down the line it leads to further notational simplifications [44].

8.6.1 DEEP NEURAL NETWORKS

A deep neural network can be distinguished from an everyday single-layer perceptron based neural network by its depth. What we mean by depth is the number of node layers that must be used to process the data before it can be recovered from the other side as a useful value as a class/prediction.

Early model of neural networks takes for example the first perceptron's were shallow, having one input and one output layer, or in some cases had a single hidden layer in between. To be classified as DL model a minimum of three layers along with an input and an output layer was considered mandatory.

Throughout the deep-learning process, as we move deep into the neural nets more and more complex features can be recognized and each layer is responsible for a certain set of feature selection. This is because if the fact that the feature from the previous layers is carried to the future layers, resulting in complex combinations which are difficult to perceive from a human eye.

Deep neural networks were capable of feature extractions without any human help, contrary to the earlier ML models. Feature extraction is a function which can take hundreds and thousands of data scientists' years to complete and they need years of experience. DL developed as a way to cheat the threshold of limited expertise and time.

During the process of training on unlabeled data, the layers try to select features that can be used to reconstruct the input, try to minimize distinction between the output and the input data. Take for example Restricted Boltzmann machines (RBMs), whose purpose is reconstructions in this manner [43–45].

During this process, the neural nets start identifying patterns and utilize these patterns for getting optimal results. Further they learn to understand the significance of these features and start drawing conclusions from what these patterns represent.

In case of a neural network trained on a labeled data can work on unstructured data providing those models with more inputs than conventional models. This methodology leads to higher performance, more inputs to train on, more data to learn from and are more likely to be accurate. This gives DL the ability to process and learn from huge quantities of unlabeled data giving it a distinct advantage over previous algorithms.

8.6.2 CONVOLUTIONAL NEURAL NETWORK (CNN)

Primary use of convolutional networks is to classify, cluster images by similarity, or perform object recognition within scenes. The ability of convolutional networks in image recognition explains why the world looks up to the capabilities of DL. The field of computer vision developed after the development of CNN has an obvious application for robotics, security, self-driving cars, drones, medical diagnoses, and treatments for the visually impaired.

Convolutional systems can similarly perform business-situated undertaking, Convolution originates from a Latin word *convolvere,* "to convolve" signifying to roll together. For numerical purposes, a convolution is the essential estimates of how much two capacities cover as one disregards the other. Think about a convolution as a method of blending two capacities by increasing them [44, 45].

In image analysis, the most basic and foremost function is to analyze and break down the input image, the second being the application of a filter. The purpose of filter is to pick up signal and features that are required from the image. Then the result is the multiplication of the two features, which in case of an image is the overlap of the two feature images to form the final picture. In case of convolutional network, we are taking large numbers of filters that are passed over one another with different strengths, each picking a unique signal. In the initial phase of processing, One can imagine them as passing a horizontal, vertical, and diagonal line filter to create a map of the edges in the image. By learning different portions of a feature space, convolutional net allows for easily scalable and robust feature engineering [45–47].

> **Application of CNN:**
> - Computer vision;
> - Recognition of faces;
> - Label generation for images;
> - Classification of images;
> - Recognition of action;
> - Pose estimation;

- Analysis of documents;
- Natural language processing;
- Speech recognition;
- Text classification;
- Radiology: Classification, segmentation, detection.

➢   **Advantages and Disadvantages of CNN:**

SL. No.	Advantages	Disadvantages
1.	Good accuracy in image recognition	Overfitting and Underfitting
2.	Feature learning	Parameter to memory requirement
3.	Weight sharing	Non-expressive learning
4.	Performance better than conventional neural networks	Non-expressive logics
5.	Feature extractors	Hyper tuning required
6.	Pre training can be done, thus saving time when required	Computationally expensive

8.6.3   RECURRENT NEURAL NETWORK

Recurrent neural networks (RNN), a type of artificial neural network (ANNs) developed to analyze patterns in chain of data. This is applicable in numerical times series data emanating from sensors, stock markets and government agencies. An example of RNN is LSTMs, the differentiating factor between the two, RNNs and LSTMs from other neural networks is that they consider time and the order of occurrence into account, and the results of the past experiences (RNN) and past error factors(LSTM) from the last neuron are passed onto the next neuron along with the output.

In order to understand recurrent neural nets, we have to have a basic understanding of feedforward nets. The basis for their naming is by the way they channel information through the nodes of the network. This involves a series of mathematical operations performed at each and every node of the network. In the traditional feedforward networks the data is passed straight through from one layer to another to the last, while a RNN cycles the information through a loop (Figure 8.17) [46, 47].

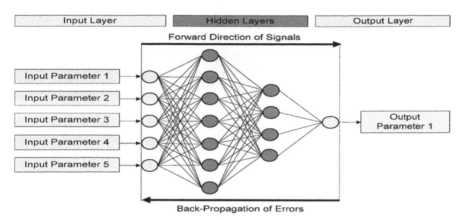

FIGURE 8.17   Forward feed neural network.

Recurrent networks not only take just the input as they appear but also what they perceived it as in the previous neuron. The main distinction between a feedforward and a recurrent network is a feedback loop connected to the previous nodes, consuming their own outputs as inputs in the future nodes. Recurrent networks are considered to have memory, thus becoming a huge memory dump during training.

Adding memory is for a reason: The information present is used to perform tasks that are not possible through feedforward networks. This sequential information is retained through the recurrent network's concealed state, which spans over a prolong period of time as it cascades forward to affect the processing of each new example. Correlations between events separated over many steps are called "long-term dependencies," as the event down the line in time depends on/ may be function of events that occurred before (Figure 8.18). The mathematical way to implement this weight sharing methodology is:

$$h_t = \varnothing\left(Wx_t + Uh_{t-1}\right)$$

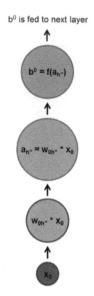

b^0 is fed to next layer

$b^0 = f(a_{h''})$

$a_{h''} = w_{0h''} * x_0$

$w_{0h''} * x_0$

x_0

FIGURE 8.18   Working of a recurrent neural network.

➢ **Application of RNN:**
  • Text summarization;
  • Document generation;
  • Report generation;
  • Conversational interfaces and chatbots;
  • Machine translation;
  • Image recognition and description generation;
  • Speech recognition;
  • Text classification;
  • Sentiment analysis;
  • Fraud and spam detection;
  • Stock price prediction.

➢   **Advantages and Disadvantages of RNN:**

SL. No.	Advantages	Disadvantages
1.	Captures sequential information	Vanishing and exploding gradient problem
2.	Parameter sharing	Training an RNN is a very difficult task
3.	RNN used with convolutional layers to extend the powerful pixel neighborhood for computing	Cannot process very long sequence for tanh and rely functions

8.6.4 GENERATIVE ADVERSARIAL NETWORK (GAN)

Generative adversarial networks (GANs) are algorithms designed for utilization of two neural systems, setting one in opposition to the next, for manufacturing new examples of information that can be misinterpreted for genuine information. They can be applied in the field of image, video, and voice generation. GANs' potential for both great and evil is colossal, on the grounds that they can figure out how to imitate any appropriation of information. That is, GANs can be educated to make universes frightfully like our own in any area: pictures, music, discourse, exposition. They are robot craftsmen one might say, and their yield is amazing-strong even. In any case, they can likewise be utilized to create counterfeit media content, and are the innovation supporting Deep fakes [48, 49].

To understand GANs, we need to understand how a generative algorithm works in order to so, contrasting them with discriminative algorithms is necessary. Discriminative algorithms have the purpose of classifying input data.

The designed neural network comprises of a generator algorithm with the purpose of generating new data instances, the other a discriminator that verifies the output instance for originality/authenticity, so it can be placed in the original dataset. As the generator is creating new, synthetic images that it passes to the discriminator. The hope is that the generated data be authentic, even though they are fake (Figure 8.19) [50, 51].

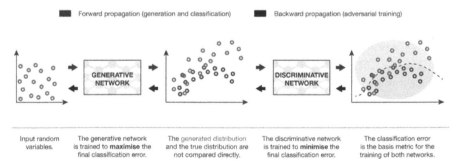

FIGURE 8.19 Working of a generative adversarial network.

Source: Adapted from Ref. [65].

> **Applications of GAN:**
> - Generate image datasets;
> - Generate human photographs;

- Generate lifelike photographs;
- Generate anime characters;
- Caption generation;
- Description to picture generation;
- Face frontal view generation;
- Photos to emojis;
- Face aging;
- Photo blending;
- Video prediction;
- 3D object generation.

8.7 INTRODUCTION TO DEEP FAKE

Deep fake is a kind of man-made brainpower used to make persuading pictures, sound, and video tricks. The term, which portrays both the innovation and the subsequent fake substance, is a portmanteau of DL and fake.

Deepfake substance is made by utilizing two contending AI calculations-one is known as the generator and the other is known as the discriminator. The generator, which makes the fake mixed media content, requests that the discriminator decide if the substance is genuine or fake [52, 53].

Together, the generator and discriminator structure something many refer to as a generative adversarial system (GAN). Each time the discriminator precisely recognizes a substance as being manufactured, it furnishes the generator with significant data about how to improve the following deep fake. The initial phase in building up a GAN is to recognize the ideal yield and make a preparation dataset for the generator. When the generator starts making a satisfactory degree of yield, video clasps can be taken care of to the discriminator.

As the generator shows signs of improvement at making fake video cuts, the discriminator improves at spotting them. On the other hand, as the discriminator shows signs of improvement at spotting fake video, the generator shows signs of improvement at making them. Up to this point, video content has been progressively hard to adjust in any considerable manner. Since deepfake AI are made through AI nonetheless, they do not require the significant aptitude that it would take to make a practical video in any case. Lamentably, this implies pretty much anybody can make a deep fake to advance their picked plan. One risk is that individuals will fully trust such recordings; another is that individuals will quit confiding in the legitimacy of any video content whatsoever [54, 55].

8.7.1 HOW DOES DEEPFAKE WORK?

A Deepfake video appears as though a unique substance having the individual doing an activity or talking on a point. And keeping in mind that making such fake recordings, numerous pictures of the focused-on individual from various edges is utilized to superimposed on unique face.

Pictures are undermined with faces and other body parts, to make it look unique, while making recordings, the voice is likewise cloned with focused individual utilizing the AI-empowered instruments to disentangle the procedure and match the lips moving as indicated by words verbally expressed. The Deepfake recordings influences life of famous characters in our general public. Government officials, entertainers or on-screen character, different big names, and remarkable characters from corporate world. Computer based intelligence and AI based apparatuses are utilized to produce such substance. In any case, with the assistance of Deepfake location administrations assists with recognizing such fake recordings accurately.

What is more, Deepfake works like making an exciting report about well-known characters that individuals love to watch and furthermore share with others or incorporate into their tattles. Also, deepfake pornography recordings get more consideration and welcomes all the more hitting on grown-up destinations expanding on the web their guests rely on such entryways [56–58].

8.7.1.1 DEEPFAKES DISADVANTAGES

Or maybe profiting anybody, this AI-based innovation has detriments influencing various gatherings of our general public. Aside from making fake news and purposeful publicity deepfake is significantly utilized for retribution pornography to slander the striking famous people.

When fake recordings turn into a web sensation individual accept at first, and continue imparting to others causes the focused-on individual to become humiliate observing such unordinary acts. Until and except if official explanation of the focused on big name not comes, numerous individuals begin thinking making, that makes their life troublesome, particularly when they are scrutinize by their fans on different stages like online networking and so forth [59, 60].

8.7.1.2   DEEPFAKES ADVANTAGES

However, it is unsafe for the general public yet it additionally has barely any focal points like making a remarkable consideration among the online crowd making the website page mainstream on the web crawler, as increasingly number of individuals begin looking on such suggestive subjects.

What is more, not many V.I.P.s who are not known to everybody likewise get well known for the time being, as individuals begin looking through finding out about them, when they do, what is there calling and increasingly about their own experience and current notoriety in the market [60–62].

Another genuine favorable position of deepfake is, it causes us to get mindful about such fake things and we ought not have confidence in all that we see around us. When we find that it is fake, we learn and next time when such substance comes through comparative sources, we set aside some effort to accept or do some examination to validate the news.

8.7.2   DEEPFAKE FUTURE PROSPECTIVE

Deepfakes have started to dissolve trust of individuals in media substance as observing them is not, at this point similar with having confidence in them. They could make pain and negative impacts those focused on, uplift disinformation and detest discourse, and even could invigorate political strain, excite general society, savagery, or war. This is particularly basic these days as the innovations for making deepfakes are progressively receptive what is more, internet-based life stages can spread those fake substance rapidly. Here and there deepfakes do not have to be spread to enormous crowd to cause adverse impacts. Individuals who make deepfakes with pernicious reason just need to convey them to target crowds as a component of their harm procedure without utilizing web-based social networking. For instance, this methodology can be used by insight administrations attempting to impact choices made by notable individuals, for example, lawmakers, prompting national and universal security dangers. Getting the deepfake disturbing issue, research network has concentrated on creating deepfake recognition calculations and various outcomes have been accounted for. This chapter has surveyed the best-in-class techniques what is more, an outline of normal methodologies. It is perceptible that a fight between those AI who utilize propelled AI. To make deepfakes with the individuals who put forth attempt to distinguish deepfakes is developing.

Deepfakes' quality has been expanding and the exhibition of identification techniques should be improved as needs be. The motivation is that what is broken by AI can be fixed by AI too. Another exploration bearing is to coordinate location techniques into dissemination stages, for example, social media to expand its viability in managing the across-the-board effect of deepfakes. The screening or separating component utilizing powerful recognition techniques can be executed on these stages to facilitate the deepfakes recognition. Lawful prerequisites can be made for tech organizations who own these stages to evacuate deepfakes rapidly to decrease its effects. What is more, watermarking apparatuses can likewise be coordinated into gadgets that individuals use to make advanced substance to make permanent metadata for putting away creativity subtleties, for example, time, and area of mixed media substance just as their untampered attachment. This incorporation is hard to execute however an answer for this could be the utilization of the troublesome blockchain innovation. The blockchain has been utilized successfully in numerous zones and there are not many examinations up until now tending to the deepfake discovery issues dependent on this innovation. As it can make a chain of one of a kind unchangeable square of metadata, it is an extraordinary instrument for advanced provenance arrangement. The joining of blockchain innovations to this issue has exhibited certain outcomes yet this examination bearing is a long way from develop. Then again, utilizing discovery strategies to spot deepfakes is vital, yet understanding the genuine aim of individuals distributing deepfakes is considerably progressively significant. This requires the judgment of clients dependent on social setting in which deepfake is found, for example who dispersed it and the thing they said about it. This is basic as deepfakes are getting increasingly more photorealistic and it is exceptionally foreseen that discovery programming will fall behind deepfake creation innovation. An investigation on social setting of deepfakes to help clients in such judgment is in this way worth performing. Recordings and photographic have been broadly utilized as confirmations in police examination and equity cases. They might be presented as confirmations in a courtroom by advanced media, crime scene investigation specialists who have foundation in proper law requirement and involvement with gathering, looking at and dissecting advanced AI. The advancement of AI and AI innovations may have been utilized to change these computerized substances and consequently the specialists' sentiments may not be sufficient to verify these confirmations on the grounds that even specialists cannot perceive controlled substance. This angle needs to consider in courts these days when pictures and recordings are utilized as confirmations to convict culprits in light of the presence of a wide scope of computerized control techniques. The computerized media crime scene investigation results along these lines must be end up being substantial and solid before they can be utilized in courts. This requires cautious documentation for each progression of the crime scene investigation procedure and until the outcomes are reached. AI and AI calculations can be utilized to help the assurance of the genuineness of computerized media and have gotten exact and dependable outcomes, for example yet the vast majority of these calculations are unexplainable. This makes an enormous obstacle for the utilizations of AI in crime scene investigation issues in light of the fact that not just the legal sciences specialists in many cases do not have aptitude in calculations, yet the experts likewise cannot clarify the outcomes appropriately as a large portion of these calculations are discovery models. This is progressively basic as the latest models with the most precise outcomes depend on DL strategies comprising of AI and neural system boundaries. Reasonable AI in vision along these lines is an exploration bearing that is expected to advance and favorable circumstances of AI in computerized media crime scene investigation.

8.7.3 DISCUSSIONS AND CONCLUSION

Nowadays, IoD is used for autonomous flying robot and machine to make more effective devices. The IoD offers nonexclusive kinds of help for various machine applications. Drones used to discover the data using the camera and sensors, which is later inspected to isolate vital information to apply for a specific explanation. Drones have two areas: the plane and a control structure. Multiple drones are furnished with top tier advancement that uses human-made cognizance, a piece of programming building that has some aptitude in making sharp machines that think and act as individuals. Man-caused awareness can be used to make machines that perform tasks more viably than individuals, allowing them to work in undesirable conditions 24 hours of the day all through the whole year with no late morning breaks, days off, or paid get-away. This has quite recently incited regardless of unfathomable achievements in domains like space travel and medicine and could be the reaction to overall issues like world longing and nuclear holocausts.

As machines develop the likelihood to improve the world, specialists represent a considerable request: Nowadays, there are many applications of IOD in various sectors in typical situation like pandemic or COVID-19. By using supervised, semi-supervised and unsupervised machine learning and deep learning algorithms we can make more intelligent and smarter system like drone to solve the real world problems in such pandemic situation. Sometimes ML and DL models, most of the algorithms either do not provide high accuracy or sometimes provide misleading results which can be a future threat in various fields. This study focuses on the significant research in the area of smart drones. Drones may play very important role during the COVID-19 pandemic and in the future also.

Drones can improve the life of the people during any pandemic situation in the world. The aim of the smart drones to reduces the shortcomings which are existing in the current scenario of the COVID-19 pandemic situation in the whole world. So, ML, DL, GAN, and Deepfake techniques are used to make the smart drone which can work more efficiently and effectively in the real-world situation. In this book chapter we will find the role of smart drones and its applications using various techniques of ML, DL, GAN, and deepfake. We will find the various algorithms of ML, deepfake, GAN, and DL. By using these techniques, we can improve the advanced feature od IoD so that we can design smart drone and they can work in various fields in most efficiently and effectively. There are various challenges and opportunities of ML, DL, GAN, and Deepfake to deploy the functioning of smart drone. This study is helpful for various application field of drones which may improve the life style of people during such pandemic situation in the world like COVID-19.

KEYWORDS

- **artificial intelligence**
- **deep learning**
- **deepfake**
- **generative adversarial network**
- **internet of drones**
- **machine learning**
- **terminal radar approach control**

REFERENCES

1. Goldman, S., (2017). *Drones: Reporting for Work*. Online: http://www.goldmansachs. com/our-thinking/technology-driving-innovation/drones/ (accessed on 19 November 2021).
2. Gharibi, M., Boutaba, R., & Waslander, S. L., (2016). Internet of drones. *IEEE Access, 4*, 1148–1162.
3. Zeng, Y., Zhang, R., & Lim, T. J., (2016). *Wireless Communications with Unmanned Aerial Vehicles: Opportunities and Challenges, 54*(5), 36–42. IEEE Communications Magazine.
4. He, D., Chan, S., & Guizani, M., (2017). *Drone-Assisted Public Safety Networks: The Security Aspect, 55*(8), 218–223. IEEE Communications Magazine.
5. Lin, X., Wiren, R., Euler, S., Sadam, A., Maattanen, H. L., Muruganathan, S. D., Gao, S., et al., (2018). *Mobile Networks Connected Drones: Field Trials, Simulations, and Design Insights*. Submitted to IEEE Communications Magazine, Available at http:// arxiv.org/abs/1801.10508 (accessed on 19 November 2021).
6. Chandrasekharan, S., Gomez, K., Al-Hourani, A., et al., (2016). *Designing and Implementing Future Aerial Communication Networks, 54*(5), 26–34. IEEE Communications Magazine.
7. Lin, X., Yajnanarayana, V., Muruganathan, S. D., Gao, S., Asplund, H., Maattanen, H. L., Bergström, M., et al., (2017*). The Sky is Not the Limit: LTE for Unmanned Aerial Vehicles*. To appear in IEEE Communications Magazine. Available at: https://arxiv.org/ abs/1707.07534 (accessed on 19 November 2021).
8. RP-170779, (2017). *Study on Enhanced LTE Support for Aerial Vehicles*. NTT DOCOMO, Ericsson. Online: http://investors.fedex.com/files/doc_downloads/ statistical/FedEx-Q1-FY15-Stat-Book_v001_t195uu.pdf (accessed on 19 November 2021).
9. 3GPP TR 36.777, *Enhanced LTE Support for Aerial Vehicles*. Online: http://investors. fedex.com/files/doc_downloads/statistical/FedEx-Q1-FY15-Stat-Book_v001_t195uu. pdf (accessed on 19 November 2021).
10. GSMA, (2017). *Mobile Spectrum for Unmanned Aerial Vehicles*. GSMA public policy position. White paper. Online: https://www.gsma.com/spectrum/wp-content/ uploads/2017/10/Mobile-spectrum-for-Unmanned-Aerial-Vehicles.pdf (accessed on 19 November 2021).
11. Lin, X., Bergman, J., Gunnarsson, F., Liberg, O., Razavi, S. M., Razaghi, H. S., Ryden, H., & Sui, Y., (2017). *Positioning for the Internet of Things: A 3GPP Perspective, 55*(12), 179–185. IEEE Communications Magazine.
12. GSMA, (2018). *Mobile-Enabled Unmanned Aircraft*. White paper. Online: https://www. gsma.com/iot/wp-content/uploads/2018/02/Mobile-Enabled-Unmanned-Aircraft-web. pdf (accessed on 19 November 2021).
13. Yajnanarayana, V., Wang, Y. P. E., Gao, S., Muruganathan, S., & Lin, X., (2018). *Interference Mitigation Methods for Unmanned Aerial Vehicles Served by Cellular Networks*. Submitted to IEEE 5G Word Forum. Available at: https://arxiv.org/ abs/1802.00223 (accessed on 19 November 2021).
14. Lin, S., Kong, L., Gao, Q., Khan, M. K., Zhong, Z., Jin, X., & Zeng, P., (2017). Advanced dynamic channel access strategy in spectrum sharing 5G system. *IEEE Wireless Communications, 24*(5), 74–80.

15. Li, S., Zhang, N., Lin, S., Kong, L., Katangur, A., Khan, M. K., Ni, M., & Zhu, G., (2018). Joint admission control and resource allocation in edge computing for internet of things. *IEEE Network, 32*(1), 72–79.

16. D'Andrea, R., (2014). 'Guest editorial can drones deliver?' *IEEE Trans. Autom. Sci. Eng., 11*(3), 647–648.

17. Gross, D., (2013). *Amazon's Drone Delivery: How Would it Work?* [Online]. Available: http://www.cnn.com/2013/12/02/tech/innovation/amazon-drones-questions/ (accessed on 19 November 2021).

18. FedEx Corporation, (2015). *Q1 Fiscal 2015 Statistics.* [Online]. Available: http://investors.fedex.com/files/doc_downloads/statistical/FedEx-Q1-FY15-Stat-Book_v001_t195uu.pdf (accessed on 19 November 2021).

19. Raptopoulos, (2013). *No Roads? There is a Drone for That.* [Online]. Available: https://www.ted.com/talks/andreas_raptopoulos_no_roads_there_s_a_drone_for_that (accessed on 19 November 2021).

20. Fortino, G., et al., (2012). Enabling effective programming and flexible management of efficient body sensor network applications. *IEEE Transactions on Human-Machine Systems, 43*(1), 115–133.

21. Hinselmann, G., et al., (2011). Large-scale learning of structure– activity relationships using a linear support vector machine and problem-specific metrics. *Journal of chemical information and Modeling, 51*(2), 203–213.

22. Visa, S., et al., (2011). *Confusion Matrix-based Feature Selection* (pp. 120–127). MAICS 710.

23. Szepesvári, C., (2010). Algorithms for reinforcement learning. *Synthesis Lectures on Artificial Intelligence and Machine Learning, 4*(1), 1–103.

24. Kaufman, L., & Peter, J. R., (2009). *Finding Groups in Data: An Introduction to Cluster Analysis, 344.* John Wiley & Sons.

25. Zhu, X., & Andrew, B. G., (2009). Introduction to semi-supervised learning. *Synthesis Lectures on Artificial Intelligence and Machine Learning, 3*(1), 1–130.

26. Watson, P., (2008). Naive Bayes classification using 2D pharmacophore feature triplet vectors. *Journal of Chemical Information and Modeling, 48*(1), 166–178.

27. Iyengar, S., et al., (2008). A framework for creating healthcare monitoring applications using wireless body sensor networks. *Proceedings of the ICST 3rd International Conference on Body Area Networks.*

28. Leung, K. M., (2007). *Naive Bayesian Classifier* (pp. 123–156). Polytechnic University Department of Computer Science/Finance and Risk Engineering.

29. Konovalov, D. A., et al., (2007). Benchmarking of QSAR models for blood-brain barrier permeation. *Journal of Chemical Information and Modeling, 47*(4), 1648–1656.

30. Igel, C., & Marc, T., (2005). A no-free-lunch theorem for non-uniform distributions of target functions. *Journal of Mathematical Modelling and Algorithms, 3*(4), 313–322.

31. Raileanu, L. E., & Kilian, S., (2004). Theoretical comparison between the Gini index and information gain criteria. *Annals of Mathematics and Artificial Intelligence, 41*(1), 77–93.

32. Votano, J. R., et al., (2004). Three new consensus QSAR models for the prediction of Ames genotoxicity. *Mutagenesis, 19*(5), 365–377.

33. Lemon, S. C., et al., (2003). Classification and regression tree analysis in public health: Methodological review and comparison with logistic regression. *Annals of behavioral Medicine, 26*(3), 172–181.

34. Rizvi, S. J., & Jayant, R. H., (2002). Maintaining data privacy in association rule mining. *VLDB'02: Proceedings of the 28ᵗʰ International Conference on Very Large Databases.* Morgan Kaufmann.
35. Kauffman, G. W., & Peter, C. J., (2001). QSAR and k-nearest neighbor classification analysis of selective cyclooxygenase-2 inhibitors using topologically-based numerical descriptors. *Journal of Chemical Information and Computer Sciences, 41*(6), 1553–1560.
36. Eysenbach, G., (2001). What is e-health? *Journal of Medical Internet Research, 3*(2), e20, Apr-Jun.
37. Lewis, R. J., (2000). An introduction to classification and regression tree (CART) analysis. *Annual Meeting of the Society for Academic Emergency Medicine in San Francisco* (Vol. 14). California.
38. Chadha, A., Iyer, B. R., Messatfa, H., & Yi, J. (2000). U.S. Patent No. 6,032,146. Washington, DC: U.S. Patent and Trademark Office.
39. Guha, S., Rastogi, R., & Shim, K. (1998). CURE: An efficient clustering algorithm for large databases. ACM Sigmod Record, 27(2), 73–84.
40. Kaelbling, L. P., Littman, M. L., & Moore, A. W. (1996). Reinforcement learning: A survey. Journal of Artificial Intelligence Research, 4, 237–285.
41. Bhandare, A., Bhide, M., Gokhale, P., & Chandavarkar, R. (2016). Applications of convolutional neural networks. International Journal of Computer Science and Information Technologies, 7(5), 2206–2215.
42. Tabian, I., Fu, H., & Sharif Khodaei, Z. (2019). A convolutional neural network for impact detection and characterization of complex composite structures. Sensors, 19(22), 4933.
43. Camuñas-Mesa, L., & Linares-Barranco, B., & Serrano-Gotarredona, T., (2019). Neuromorphic spiking neural networks and their memristor-CMOS hardware implementations. *Materials, 12*, 2745. 10.3390/ma12172745.
44. Asteris, P., Roussis, P., & Douvika, M., (2017). Feed-forward neural network prediction of the mechanical properties of sand Crete materials. *Sensors, 17*(6), 1344. doi: 10.3390/s17061344.
45. https://imgur.com/kpZBDfV (accessed on 19 November 2021).
46. http://news.mit.edu/2017/explained-neural-networks-deep-learning-0414 (accessed on 19 November 2021).
47. http://neuralnetworksanddeeplearning.com/chap1.html (accessed on 19 November 2021).
48. Kaelbling, L. P., Littman, M. L., & Moore, A. W. (1996). Reinforcement learning: A survey. Journal of Artificial Intelligence Research, 4, 237–285.
49. Avgeraki, K. (2021). Skin lesion analysis towards melanoma detection from dermoscopic images using Convolutional Neural Networks (Master's thesis, Πανεπιστήμιο Πειραιώς).
50. Adams, J., Murphy, E., Sutor, J., & Dodd, A. (2021). Assessing the Qualities of Synthetic Visual Data Production. In 2021 9th International Conference on Information and Education Technology (ICIET) (pp. 452–455). IEEE.
51. https://machinelearningmastery.com/impressive-applications-of-generative-adversarial-networks/ (accessed on 19 November 2021).
52. Badrinarayanan, V., Kendall, A., & Cipolla, R., (2017). Segnet: A deep convolutional encoder-decoder architecture for image segmentation. *IEEE Transactions on Pattern Analysis and Machine Intelligence, 39*(12), 2481–2495.

53. Guo, Y., Jiao, L., Wang, S., Wang, S., & Liu, F., (2017). Fuzzy sparse autoencoder framework for single image per person face recognition. *IEEE Transactions on Cybernetics, 48*(8), 2402–2415.
54. Tewari, A., Zollhoefer, M., Bernard, F., Garrido, P., Kim, H., Perez, P., & Theobalt, C., (2018). High-fidelity monocular face reconstruction based on an unsupervised model-based face autoencoder. *IEEE Transactions on Pattern Analysis and Machine Intelligence.* doi: 10.1109/TPAMI.2018.2876842.
55. Yang, W., Hui, C., Chen, Z., Xue, J. H., & Liao, Q., (2019). FV-GAN: Finger vein representation using generative adversarial networks. *IEEE Transactions on Information Forensics and Security, 14*(9), 2512–2524.
56. Liu, F., Jiao, L., & Tang, X., (2019). Task-oriented GAN for PolSAR image classification and clustering. *IEEE Transactions on Neural Networks and Learning Systems, 30*(9), 2707–2719.
57. Cao, J., Hu, Y., Yu, B., He, R., & Sun, Z., (2019). 3D aided duet GANs for multi-view face image synthesis. *IEEE Transactions on Information Forensics and Security, 14*(8), 2028–2042.
58. Zhang, H., Xu, T., Li, H., Zhang, S., Wang, X., Huang, X., & Metaxas, D. N., (2019). StackGAN++: Realistic image synthesis with stacked generative adversarial networks. *IEEE Transactions on Pattern Analysis and Machine Intelligence, 41*(8), 1947–1962.
59. Lyu, S., (2018). *Detecting Deepfake Videos in the Blink of an Eye.* Retrieved from: http://theconversation.com/detecting-deepfake-videos-in-the-blink-of-an-eye-101072 (accessed on 19 November 2021).
60. Bloomberg, (2018). *How Faking Videos Became Easy and Why That's so Scary.* Retrieved from: https://fortune.com/2018/09/11/deep-fakes-obama-video/ (accessed on 19 November 2021).
61. Chesney, R., & Citron, D., (2019). Deepfakes and the new disinformation war: The coming age of post-truth geopolitics. *Foreign Affairs, 98*, 147.
62. Tucker, P., (2019). *The Newest AI-Enabled Weapon: Deep-Faking Photos of the Earth.* Retrieved from: https://www.defenseone.com/technology/2019/03/next-phase-ai-deep-faking-whole-worldand-china-ahead/155944/ (accessed on 19 November 2021).
63. Pothuganti, Karunakar & Jariso, Mesfin & Kale, Pradeep. (2017). A Review on Geo Mapping with Unmanned Aerial Vehicles. International Journal of Innovative Research in Computer and Communication Engineering. 3297.
64. Tabian, Iuliana, Hailing Fu, and Zahra Sharif Khodaei. 2019. "A Convolutional Neural Network for Impact Detection and Characterization of Complex Composite Structures" Sensors 19, no. 22: 4933. https://doi.org/10.3390/s19224933
65. Rocca, J. Understanding Generative Adversarial Networks (GANs) Building, step by step, the reasoning that leads to GANs. Toward Data Science, Jan 7, 2019. https://towardsdatascience.com/understanding-generative-adversarial-networks-gans-cd6e4651a29

CHAPTER 9

Machine Learning and Deep Learning Algorithms for IoD

JAGJIT SINGH DHATTERWAL,[1] KULDEEP SINGH KASWAN,[2]
VIVEK JAGLAN,[3] and AANCHAL VIJ[2]

[1]Department of Artificial Intelligence & Data Science,
Koneru Lakshmaiah Education Foundation, Vaddeswaram, AP, India,
E-mail: jagjits247@gmail.com

[2]School of Computing Science and Engineering, Galgotias University,
Greater Noida, Uttar Pradesh, India, E-mails: kaswankuldeep@gmail.com
(K. S. Kaswan), aanchal.vij@galgotiasuniversity.edu.in (A. Vij)

[3]Department of CSE, Graphic Era Hill University, Dehradun, Uttarakhand,
India, E-mail: jaglanvivek@gmail.com

ABSTRACT

In the world's fast emerging web, everybody, and all is connected to the internet and it is part of the life of the people. The new development for secure operations in commercial and public use behind the Internet of Drone (IoD) poses connectivity and technological problems in real-life areas. The use of multi-drone in which the tasks for each drone are delegated and the values from the data center transmitted to the cloud provide an unreachable area of data management techniques. In today's drone, the cameras can quickly process images in each frame. The effective drone data analysis processes for the tracking, evaluation, and assessment of results are defined to evaluate the data for key decisions. The main resources include: aerial grade image, cloud analysis 3D points, better, and effective connectivity and teamwork, quicker resolution of problems for the pre-visualization of several concepts. The data

The Internet of Drones: AI Applications for Smart Solutions. Arun Solanki, PhD, Sandhya Tarar, PhD,
Simar Preet Singh, PhD & Akash Tayal, PhD (Eds.)

viewed and viewed by drones must still be collected and transferred to the Cloud data center. The visibility graph provided by the 3D Path Planning algorithm for the shorter route and the sum of data obtained is determined by means of the visibility graph technique. SPF learning algorithm that lets every scenario progress rapidly. Based on image recognition successful decision to preserve the existence and properties of human beings should be made to secure the cloud with bandwidth control for data processing. This focus is on endlessly scaled computational power with an algorithm for web analysis which ends the automation of processes. The model's efficiency is calculated using the specified resource types and the number of nodes used. Recently, deep learning (DL) has produced outstanding results to address a Various UAV function in the awareness, preparation, position, and control fields. Its excellent learning capabilities; it is quite useful for a variety of individual UAV implementations from the diverse data collected in actual environments. In addition, unmanned air vehicles (UAV), for a range of civilian activities, are currently commonly used in health, surveillance, and disaster relief applications, including postal distribution or warehouse management. In this aspect, the cost-effective and scalable approach is to attach unmanned aerial (UAV) terrestrial RAN (Radio Access Network) to it. In order to successfully operate these UAV-based RANs (U-RANs), some problems need therefore to be solved. As such networks are extremely dynamic and heterogeneous, model-based modeling methods, mostly depending on constraints and expectations, are heavily limited in real-world scenarios. In fact, it is a highly specialized job to build a series of acceptable protocols for these U-RANs. This chapter explores extensively the latest recorded uses and implementations of profound learning and machine learning (ML) for UAVs, and their efficiency and limitations as well. A thorough description of the core methods of DL and ML will also be used for realistic and efficient approaches. By concentrating on supervised and reinforcing learning strategies and, lastly, by describe the main challenges for DL in UAV based applications, we address why, where, and which ML approaches are helpful in constructing U-RANs.

9.1   INTRODUCTION

The usage of Unmanned Air Vehicles (UAVs) has recently increased [15–20]. These aircraft are able to build their own course to move from preferred places, and also to avoid fixed obstacles in their way, while working entirely autonomously. A variety of scholars have suggested separate route design algorithms. Strategies such as alternative fields [1] and interactive analysis methods such as A* and D* [2, 3], as well as evolutionary technologies in algorithm (EA) [4, 5]. Nevertheless, UAVs have historically been limited to certain regions where there are no other aircraft outside the jurisdiction of the UAV authority. If we want to exclude UAVs from such constraints and push them into general areas, the UAV requires to be able to build a route that not only eliminates static barriers, but can also tackle shifting obstructions (other vehicles) effectively. If the design issue is modified from set to changing barriers, it would shift the issue to fluid ('Don't reach this area') ('Don't reach this field at this moment, given the capacity to alter the position at that point'). It also changes from a deterministic issue to a stochastic one, which is equally important. For preparing a route to remove a moveable barrier, we will at any future period avoid the direction of the obstruction (other vehicle). But the potential position of the barrier (other vehicle) can be only inferred in most non-trivial situations. This motion will be associated with some uncertainty. A highly successful pilot manager should explicitly tackle the mixture of the expected hazard movement, local instability and how this complexity evolves with time. The degree of ambiguity about another vehicle's potential position can differ due to the complexity of the individual circumstance. A commercial aircraft that meets or does not comply with a flight schedule has a significant degree of vulnerability relative to an unregulated personal aircraft. A clever, adversarial aircraft will be much more unpredictable for a combat situation. All these circumstances may be called the same general question, but with different degrees of insecurity development.

9.2 RELATED WORK

Sequential approach in single-layer solutions is one of the types primarily used for the identification of intervention. Sequential action detection is carried out by examining the sequential features that apply to the function in each frame of a picture. Therefore, before action detection, a function vector is derived from an image sequence. The function matrix is then evaluated to figure out how the characteristics interact with the association for action recognition. There are usually two forms of systematic strategies in consideration of human behavior, i.e., exemplary, and state strategies. Exemplary methods operate by matching a human action series prototype with the variance picture inputs. Subsequently, the input structure should be understood while the sequence prototype varies least. This approach will also allow identical activity to be recognized with changes in poses and rates. Darrell et al. [8] established a dynamic time warping (DTW) algorithm, together with the exemplar technique, which can optimum synchronization of two nonlinear sequences with polynomial numerical variations. A visual-base behavior identification is proposed in the works of Fogel et al. [4]. The size of the reference picture was not sufficient at the time of the application, because the average height was 30 pixels wide. To contend with this, the human two-dimensional optical flows are measured using a spatial discrepancy in the picture frames to assess the space time amount of people. The two-dimensional optical flows are then transformed to a spatial / temporal motion description of distorted acceleration channels. Finally, the k-nearest procedure is used for the space time motion classification to categorize series. State model methods interpret human behavior in a statistical fashion as a control machine. The state framework is structured to take advantage of space-times functionality in a video series between the space applications. Therefore, based on a spatial input function, Statistical equipment to produce spatial and time structure is given for the state program. Specifically, a qualified mathematical A spatial-temporary sequence is required for the model.

The probability series is determined when the spatial variation between the image sequences is measured. For human activity recognition, approaches such as the Hidden Markov (HMM) model and a dynamic Bayesian network (DBN) are commonly used. Likewise, human behavior is viewed by HMM and DBN as a secret condition. In every picture clip, it is suggested human action reflects a society, which results in the likelihood of change between countries. These possibilities lead to human behavior and therefore human actions may be identified as functioning through function of Yamato et al. [9] across the level of the state model's production likelihood.

For recent studies, the System paradigm is used in a more effective profound thinking process of concurrent approaches. Research reported by Montes et al. [10] demonstrates the usage of spatial functions CNN. Similar to DMM, RNN produces the likelihood of DMM production, which is also known as a spatial-time function. The Methods The seen can be seen in today's high-resolution feedback images. Li et al. [11] have showed that new optical flows have been incorporated in works in Montes et al. to increase the detection efficiency further.

9.3   MACHINE LEARNING (ML) ALGORITHM (FIGURE 9.1)

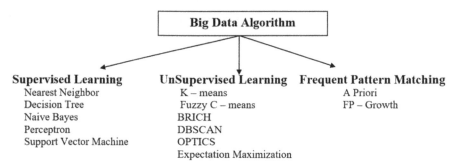

FIGURE 9.1   Machine learning framework.

9.3.1   SUPERVISED LEARNING (SL)

Overall, supervised learning (SL) [21–25] takes place when the input and output variables of a system are presented with the intention to learn how they are arranged together. The goal is to produce enough map to forecast the output when new input is given. This is an iterative process and the algorithm is updated or requested every time the prediction is made, until an appropriate level of output is achieved.

9.3.1.1   NEAREST NEIGHBOR

The KNN is a form of machine learning (ML) algorithm [26–30], which can be used both for multiclass classification and for probabilistic dimensionality reduction. Nonetheless, it is used primarily for statistical analysis problems in the industry. The following two characteristics will characterize KNN:

1. **Lazy Algorithm for Learning:** KNN has little expert experience and using all the information during identification as an algorithm of lazy reporting.
2. **Algorithm of Nonparametric Analysis:** KNN is also an algorithm of non-parametric learning because it assumes nothing about the basic data.

➢ **Algorithm KNN:** In order to forecast new data points values, K-nearest neighbors (KNN) algorithm is based on 'feature similarity' that means the new data point will be given a value based on how closely it coincides with training points. With the aid of the next steps, we can understand his work:

- **Step 1:** We need a dataset to implement an algorithm. Therefore, we have to load the training and test data during the first step of KNN.
- **Step 2:** Next, the K value must be chosen, i.e., the closest data points. K may be a whole. K may be any whole.
- **Step 3:** The following is performed for every point in the test results:
  ○ Measure the distance between test data and each trainings row using the method: Euclidean, Manhattan or Hamming. Euclidean is the most common method for distance calculation;
  ○ Sort them in ascending order now, based on the distance interest;
  ○ Next, from the sorted array, it will select the top K rows;
  ○ Now, a class based on the most common type of rows will be assigned to the test point.
- **Step 4:** End.

➢ **KNN in Drone:** Using a remoteness method, the range of the drone from the secure zone is determined. The Ultra-sound wave takes time throughout the first calculation and time the Ultra-sound wave takes during the second test, using specifications in this procedure. The values on the basis of which the algorithm calculates the drone wavelength and also estimates drone path.

9.3.1.2   DECISION TRESS

The algorithm decision tree relates to the controlled deep neural network. Algorithms may be used to overcome correlation and rating issues, as opposed to several other algorithms. The main idea for the usage of the decision tree is to construct a learning approach that can be utilized for trying to forecast the class or properties of resulted in different decision rules derived from previous data (training data). With tree representation, the decision tree algorithm attempts to solve the problem Every tree internal node is an attribute and each leaf node is the class label.

The popular attribute selection measures:

1. **Information Gain:** We attempt to approximate the value details by using the information gain as a criterion. To calculate, we will use certain knowledge theories concepts to allow sense of the unpredictability or uncertainty of a randomized input variables.
2. **Gini Index:** A Gini index is a calculation of how frequently an object that is randomly picked is wrongly identified. It means that you would choose an attribute with lower Gini index.

➢ **Assumptions of Decision Tree:**
  - The full range of routines is at the start called the kernel;
  - Tend to be categorical for practical standards. When the concepts are constant, they must be simulated before building the model;
  - Attribute-based reports are shared sequentially;
  - By using a mathematical approach, attributes such as tree root or internal node are ordered.

➢ **Decision Tree Algorithm:**
  - **Step 1:** Place the appropriate value for the schema on the root tree.
  - **Step 2:** Sub-sets split the activity. To include details of the similar meaning for an element inside every subset, subsets must be created.
  - **Step 3:** Upon each sub-set, repeat step 1 and step 2 until all branches of the tree have leaved nodes.

➢ **DTA in Drone:** During the operating region, a small configured DT drone is beneficial to best decide on BDI.

9.3.1.3 NAÏVE BAYES

Naive Bayes is a simple method to construct classifiers: model class labels, represented as vectors of functional values, that assign class labels to problem instances, where certain finite systems are derived from class labels. There is only one family of algorithms based on a popular principle: Both naive classificatory of the Bayes conclude that the importance of one function in the light of the facts class variable, is irrelevant to Some other characteristic value. For example, a fruit can also be called an apple if it is small, red, and around 10 cm in diameter. A cynical Bayes Classifier treats these features as separate and contributes to the assumption that this fruit is an apple irrespective of any possible color, roundness, and diameter associations.

➢ **Algorithm:**
- **Step 1:** Read the T data set for training.
- **Step 2:** This measures the average and normal deviation of the regression model in every ranking.
- **Step 3:** Repeat:
  - o The probability of fi can be determined using the gauss density equal for every class;
  - o Before all predictor variables are calculated $(f_1, f_2, f_3, \ldots, f_4)$.
- **Step 4:** Calculate the likelihood for each class.
- **Step 5:** Get the greatest likelihood.

➢ **Usage of Naïve Bayes in Drone:** A parameter estimation algorithm for adaptation back-to-home sensing algorithm that integrates naive classification of Bayes with binary search to adjust take off and land sensors for adaption back-to-home.

9.3.1.4 PERCEPTRON

The goal of this learning problem is to make predictions of future data with correct labels for model training. Classification to predict class labeling is a common problem for SL. A linear classifier which categorizes the perceptron as an algorithm of classification, which relies on a linear forecast function. Its forecasts are based on a combination of weights and vectors. Two groups for the classification of training data are proposed in the linear classification.

➢ **Components of a Perceptron:**
- Input perceptron;
- Weights perceptron;
- Bias perceptron;
- Activation/step function perceptron;
- Weighted summation perceptron.

➢ **Steps of Perceptron:**
- Feed the features of the model to be trained in the first layer as an input;
- Both weights and inputs are increased-increasing weight and input results are multiplied;
- To change the output function, the Bias value will be added;
- The activation function is shown with this value (the form of activation function depends on the necessity);
- After the last stage, the value obtained is the output.

➢ **Algorithm of Perceptron:** Our objectives are to evaluate the W vector that can identify our data in a perfect manner. I 'm trying to get to the algorithm directly. This is the following:
- **Step 1:** Initialize w randomly.
- **Step 2:** While! convergence do

Pick random x Ɛ P U N;

If x Ɛ P and w.x<0 then

w = w+x;

end

If x Ɛ N and w.x>=0 then

w = w-x;
- **Step 3:** End.

➢ **Perceptron in Drone:** Template is being used to prepare an artificial neural network (ANN) multi-layer perceptron (MLP) to correctly identify a drone that MLP can also be used for returning to the impeller measurements of the drone, such as the duration of the wings and the velocity of spinning.

9.3.1.5 SUPPORT VECTOR MACHINE (SVM)

Learning a hyperplane in linear SVM is achieved by a transformation of a linear algebra that is not covered by this introduction to SVM, but through the use of the internal product of two or more of those comments rather than by observations itself. An important insight is that the linear SVM can be rephrased. The internal product between the two vectors is the sum of the multiplication of each couple of input values [2, 3, 5, 6] is $2 \times 5 + 3 \times 6$ or 28.

➢ **Algorithm:**
- **Step 1:** Start;

- **Step 2:** Input the dataset;
- **Step 3:** Classify the dataset;
- **Step 4:** Add four kernel functions to the SVM Machine Learning (linear, binomial, fractal dimension and radial) (RFB);
- **Step 5:** Join the hyperplane;
- **Step 6:** When consistency and integrity cannot be obtained so continue to Step 4;
- **Step 7:** End.

9.3.2 UNSUPERVISED LEARNING

9.3.2.1 CLUSTERING

Clusters are one of the most common methods used to analyze exploratory data in order to obtain an intuition about Data layout. Code form. It is necessary to create the function of identifying data subgroups. Such that in the same subgroup (cluster) graphs are quite similar where the data points are quite different in the different clusters.

9.3.2.2 K-MEANS

K-means an iterative algorithm is intended to segment the datasets into a fixed, separate, and overlapping subgroup (clusters). This is meant to include details in the intrinsic cluster as close as possible while still holding the clusters as distinct (far). This sets the data points for a cluster in order to minimize the Measured width from the middle of the group data sets (arithmetic sum of all pieces of data contributing to the group). The larger the variation of the samples is, the greater the uniformity (similarity) of the data points.

➢ **Algorithm:**
  - **Step 1:** Choose 'c' cluster centers randomly.
  - **Step 2:** Distance between each dataset and cluster core can be computed.
  - **Step 3:** Assign the cluster center data point with Total Cluster Center spread to all Cluster Centers.
  - **Step 4:** Using the current cluster core to recalculate:

$$Vi = (1 / Ci \sum_{j=1}^{Ci} \binom{n}{k} Xi$$

Where the number of observations in the group is 'ci.'
- **Stage 5:** Calculate the gap per data point from current collected cluster centers.
- **Stage 6:** Start, repeated from step 3, when no data sets are reassigned.

➢ **K-Means in Drone:** The suggested algorithm calculates the minimum shipping times, which uses K mean clusters to identify release positions, and a genetic algorithm to resolve the lorry path as an issue for journeying salespersons in drones.

➢ **Fuzzy C-Means:** This algorithm is based on the distance between the cluster center and the database, specifying membership for every cluster core-related data frame. The closest the knowledge to the cluster core is to the cluster center. Every data point will of course be the same as one membership summation. Membership and cluster centers will be modified after each iteration.

➢ **Algorithm:**
- **Step 1:** Randomly select 'c' cluster centers.
- **Step 2:** Calculate the fuzzy membership 'μ_{ij}' using:

$$\mu_{ij} = 1 / \sum_{k=1}^{c} (d_{ij} / d_{ik})^{(2/m-1)}$$

- **Step 3:** Compute the fuzzy centers 'v_j' using:

$$V_j = (\sum_{i=1}^{n} (\mu_{ij})^m x_i) / (\sum_{i=1}^{n} (\mu_{ij})^m), \forall_j = 1, 2, ---, c$$

- **Step 4:** Repeat step (2) and (3) until the minimum 'J' value is achieved or $\|U^{(k+1)} - U^{(k)}\| < \beta$.

where; 'k' is the iteration step; 'β' is the termination criterion between [0, 1]; 'U' = $(\mu_h)_{n \times c}$ is the fuzzy membership matrix; 'J' is the objective function.

➢ **Fuzzy C Mean in Drone:** A two collection of fuzzy laws are used to handle the changeover of the headings (plane perpendicular control system) and the second to govern the height turn (vertical avion controller) of the airliner for the application of the fuzzy conceptual management framework for autonomous UAV autopilots.

9.3.2.3   BIRCH (HIERARCHICAL MODELS)

- The clustering process is scalable.
- For large sets of data planned.
- Only a data scan is needed.
- Based on the CF (Clustering Feature) notation of a CF chain.
- CF tree is a height-balancing tree that stores hierarchical clustering characteristics.
- The data point cluster is identified by three (N, LS, SS) numbers, where; N is the number of items in the sub cluster; LS is the linear sum of the points; SS is the sum of squared of the points.

➤ **Algorithm:**
  - **Step 1:** Memory load data. Create a CF tree, search Database and load memory data. Reconstruction of the tree from the tree structure memory is depleted.
  - **Step 2:** Data from condensation. Resize the data collection to construct a smaller CF tree. Delete any outlier's possible condensation.
  - **Step 3:** Clustering regional. Using current cluster algorithms on CF entries (e.g., K-means, HC).
  - **Step 4:** Optimization of the cluster enhancement. Fixes CF trees problem, where different leaf entries can be allocated equal valued data points.

➤ **BRICH in Drone:** An overall selection is generated and provided by the hierarchical tree topology-based wireless drone network.

9.3.2.4   DBSCAN

Clustering is an unsupervised method of learning, splitting the data points into many separate bunches or classes so that the data points in the same classes have identical characteristics and in different groups different characteristics.

➢ **DBSCAN Algorithm Parameters:**
- **Steps:** Define the data point neighborhood, i.e., whether it is less than or equal to 'eps' the distance between two points is seen as neighbors, as friends. If the significance of eps is too small, significant portions of the data are referred to as outliers.
- **MinPts:** Represents the collection of neighborhoods in the episode. Lengthening the dataset, the greater MinPts value should be picked.

➢ **DBSCAN Algorithm:**
- **Step 1:** Choose a point P arbitrary.
- **Step 2:** Retrieve from w.r.t. Eps and MinPts all points reachable by density.
- **Step 3:** A cluster is created when P is a core point.
- **Step 4:** If P is a border point, P, and DBSCAN are visiting the next point of the database without density.
- **Step 5:** Continue the process until all of the points have been processed.

➢ **DBSCAN Algorithm in Drone:** DBSCAN was introduced to determine the placement of the drones into the centers of species clustered by utilizing a Density-based clustering algorithm.

9.3.2.5 OPTICS (DENSITY-BASED MODELS)

Clustering structure (OPTICS) recognition orders are an algorithm for density-based determination. The database points are (linearly) arranged to be neighbors in the ordering of space closest points. Therefore, for each point, a special distance representing the density that is required for the cluster is stored in order to have the two points on the same cluster.

➢ **Algorithm:**
- **Step 1:** Begin with an arbitrary object as the current object from the input database p.
- **Step 2:** It recalls the E-neighborhood of p, defines the core distance, and sets the distance to reach without specification.
- **Step 3:** The actual object, p, is then entered in the output.
- **Step 4:** If p is not an item of interest.

- **Step 5:** In OrderSeeds list (or input database if OrderSeeds is zero, OPTICS simply shifts into the next component.

➤ **OPTICS (Density Based Models) in Drone:** DBSCAN was introduced to determine the placement of the drones into the centers of species clustered by utilizing a Density-based clustering algorithm.

9.3.2.6 *EXPECTATION-MAXIMIZATION (DISTRIBUTED MODELS)*

While the MLE and EM can both find the "best fit" parameters, the model is very different in their way of finding. MLE first collects all the data and then uses them to build the most probable model. EM first estimates the parameters for the missing data, then changes the model to match the assumptions and the data observed.

The basic steps for the algorithm are:

- An initial guess is made for the parameters of the model and a distribution of probability is established. The "E-Step" for the "Planned" distribution is sometimes referred to;
- Recent data are entered into the pattern;
- In order to include new data, the distribution of likelihood from the E-step is modified. Often the "M-step" is named;
- Steps 2 to 4 are repeated before equilibrium is reached (i.e., a distribution that does not move from E to M-step).

➤ **Algorithm:**
- Consider a set of starting parameters given the incomplete data;
- Expectation step (E-step): Using the data from the dataset observed, estimate (assume) the missing data values;
- Maximizing step (M-step): To update these parameters, complete data generated following the expectation (E) step is used;
- Repeat step 2 and step 3 until convergence.

➤ **Frequent Pattern Matching:** Frequent patterns are item sets, subsections or sub structuring that appear not less than a user-specified threshold in a frequency range. For instance, a number of items like milk and bread that often appear together in a dataset of transactions are a common set of items.

> **A Priori Algorithm:** For frequent itemset mining, Apriori algorithm was the first algorithm. The R Agarwal and R Srikant were subsequently improved and it was called Apriori. The two measures "join" and "prune" are used to reduce the search space. This algorithm the identification of the most common products is an iterative approach.

> **Steps Apriori Algorithm:**

- **Join Step:** By linking each element to it itself, you create (K+1) element from K-itemsets.
- **Prune Step:** The count is being checked for each object in the database. When the applicant does not provide enough funding, the submission is considered irregular and therefore withdrawn. It is required to raising the scale of the claimant posts.

> **Algorithm:**

- **Step 1:** Scan and compare the transaction data base for items.
- **Step 2:** A set of candidates can be generated. Or using Apriori to prune rare papers from the package.
- **Step 3:** Check the transaction database to compare and obtain the help of candidate element set in the found collection.
- **Step 4:** The candidate set = NULL.
- **Step 5:** All nonempty subsets are created for each frequent itemset.
- **Step 6:** Output and help are specified for every nonempty subset.

> **Expectation-Maximization (EM) in Drone:** Expectative-maximization (EM) algorithm classified into four groups: lane, lawn, structures, and tree branches 3D aerial sensor dispersed height results.

9.4 DEEP LEARNING (DL) ALGORITHM

9.4.1 CLASSIFICATION OF NEURAL NETWORK

The following forms of neural networks may be categorized.

- Neural network feedforward;
- Neural network recurrent (RNN);
- Neural network radial base function;
- Kohonen individual organization neural network awareness;
- Neural network modular.

The knowledge flows from input to output layer in a single direction with a neural network feedforward [31–36] (if any). There are no loopbacks or loops. Figure 9.2(a) illustrates a different form of Integration of an input network with neural multi-layer properties and functions along the potential direction with transition calculated. Z is the convolutional layer number, and y is highly nonlinear for X. activation on increasing strata. W reflects the weight of the two units in the next rows, and b reflects the unit's bias interest. The RNN processing units shape a loop, like neural feedforward networks. The layer output becomes the entrance towards the next node, normally the only network component and node performance, is the data that generates a feedback loop on itself. This enables the network to note and use it to affect the existing performance.

> **Neural Network Feedforward in Drone:** enhanced deep feed forward neural network (EDFFNN) in order to monitor and analyze the traffic of the UAV network to detect the intrusions with maximum detection rate of 94.4%.
>
> One important effect of this distinction is that RNN will use an input sequence and produce a series of output values without the feedforward neural network, making it particularly useful for applications involving time-phased processing of imputation [37–40] such as speech recognition, video detection frame by context, etc. The unrolling of an RNN in time as seen in Figure 9.2(b). For example, if a three-word sentence series is an entry, then every word correlate to a layer and hence the network is unfolded or exposed three times into a three-layer RNN. Here is a graphical model explanation: x is the time t entry. U, W are the studied parameters which all measures share. Otis time t performance. St is at t time and can be determined accordingly, where f (e.g., ReLU) is the activation function.

$$S_t = (Ux_t + Ws_{t-1})$$

> **Neural Network Recurrent (RNN) In drone:** More complex tasks, we propose the use of recurrent neural networks (RNN) instead and successfully train an LSTM (Long-Short Term Memory) network for controlling UAVs.
>
> Radial origin neural network method is used for grouping, approximation method, problem of estimation of time series, etc. The input, secret, and output layers are used. The secret layer has a radial base

function (through Gaussian) and A server center for rising node. The Connection knows how to assign the middle entry and integrates radial base function output with weight parameters to label or decrease the performance level [12].

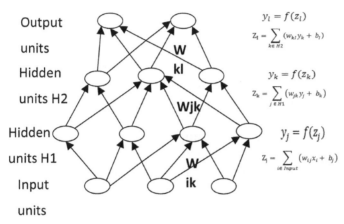

FIGURE 9.2(A)   Feedforward NN [6].

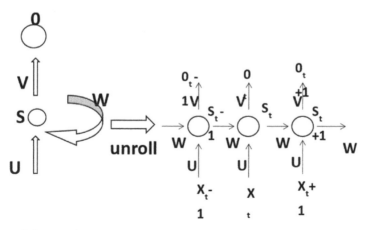

FIGURE 9.2(B)   Unrolling RNN [6].

➢ **Neural Network Radial Base Function in Drone:** The satisfactory performance of the proposed intelligent controller is validated based on the computer-based simulation results of a benchmark

UAS with system uncertainties and disturbances, such as wind gusts disturbance.

The network model is self-organized by Kohonen neural networks into input data with unregulated learning. There are two completely connected layers, i.e., input, and output layer. A two-dimensional grid is the display sheet. The activation feature is not usable and Weighing is the output layer node attributes (position). The distance Euclidian from the input data is calculated in relation to the weights in any output layer node. In order to get them closer to the input info, the weight of the nearest node and its neighbors are actualized with the following formula [13, 41, 42, 44, 45, 48].

➢ **Kohonen Individual Organization Neural Network in Drone:** The processing of the images an ANN technique called Kohonen SOM (self-organized map) will be used in drone.

$$(t + 1) = w_i(t) + \alpha(t)n_{j*_i}(x(t) - w_i(t))$$

where; x(t) is a t-time input value; wi(t) is an ith at t-time and α-channel is an ith-jth neighborhood value.

Modular neural network breaks broad networks into individual modules for neural networks. Then, as part of a common performance from the whole network [14, 43, 46, 47, 49, 50], the smaller networks execute similar functions. The DNNs are usually implemented as follows:

- Difficult autoencoders;
- CNNs or ConvNets convolution neural networks;
- Boltzmann remote machines (RBMs);
- Short-term (LSTM) deep memory.

Autoencoders are neural networks that acquire characteristics or encoding from a single set of data to minimize dimensionality. Sparse Autoencoder is an automobile encoder variant in which some devices generate a value below Nero and do not fire, or are damaged. High CNN is multi-layered (pixel values in the image) of group combinations that communicate with the feedback and create the desired outcome extraction function. This is used in visual recognition programs by CNN, devices recommenders and NLP. RBM is used to understand the distribution of possibilities in the data collection. The backpropagation is used for processing in both such networks.

Backpropagation usages a gradient decrease in order to reduce the mistake, adjusting the weights to increasing weight according to the partial derivative of the mistake. Models for neural networks may be separated into two different groups too:

- Neural networks of discriminative; and
- Neural networks of generative.

Discriminatory model is a to the context in which data travels from the input layer to the output layer via the secret layers. They are used for grading and retrenchment problems in controlled preparation. In the other side, generative structures Down the top and in reverse directions the collected data migrate. We find ourselves in unrestricted pre- and delivery problems. Since x and the corresponding y tags are viewed in the context of a discriminatory model, the probability density p(y) is explanatory; that is, the likelihood y viewed x is assigned, while a generative model knows the joint probability p(x) from which P(y) is estimated [15].

In general, discriminative strategies are followed if the labeled data is accessible to ensure effective testing, even if generative strategies are not applicable to the labeled data [16].

Three forms of preparation may be usually categorized:

- Supervised;
- Unsupervised;
- Semi-supervised.

The controlled learning involves named network data creation while the non-controlled learning comprises of no labeled data collection, and therefore no feedback-based the analysis. Neural networks are preprocessed by designs like RBMs in unattended learning and could be further tailored employing standard monitoring learning algorithms afterwards. It is then used for the identification of trends or classifications in test data collection. With its overwhelming quantity and range of data, broad data moved the envelope even farther for deep research. There is no clear agreement as to whether controlled learning is better than no controlled education, contrary to our intuitive inclination. We all are well-founded and usage examples. All frameworks show in Table 9.1 basically use in UAV.

TABLE 9.1   Deep Learning Libraries' Frameworks

Framework	Institution	License	1st Release
caffe	Berkeley research	AI BSD/free	2015
Microsoft cognitive toolkit	Microsoft	MIT license free	2016
Gluon	Aws microsoft	And open source	2017
Keras	Individual author	MIT license free	2015
MXNet	Apache software foundation	Apache 2.0 free	2015
Tensor flow	Google brain	Apache 2.0 free	2015
Theano	University of Montral	BSD/free	2008
Torch	Rohan Collobert *et al.*	BSD/free	2002
Py troch	Facebook	BSD/free	2016
Chainer	Preferred networks	BSD/free	2015
Deep learning 4j	Adam Gibson *et al.*	Apache 2.0 free	2014

9.4.2   DNN ARCHITECTURES

There are many levels of nodes in the deep neural network. Growing systems were built to fix issues in various fields or applications. For example, CNN is usually used for machine viewing And Graphic and RNN identification for statistical analysis issues is typically used below are three of deep neural networks' most popular architectures.

1. **Neural Network of Convolution:** CNN is a neural network of choice for machine vision (image recognition) and video processing centered on a human visual cortex. CNN includes a set of concentration layers and sub-samples accompanied A totally connected layer (e.g., SoftMax) and normalizer layer. The layer may also be used in many fields, such in NLP and drug discovery, etc. Figure 9.3 reveals a common 7-layered architecture for visual identification, the LeNet-5 CNN developed by LeCun et al. [28].

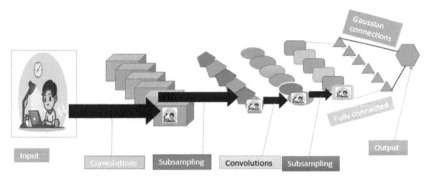

FIGURE 9.3   CNN layer architecture [28].

➢ **Algorithm of CNN Back Propagation:**
Initialization weights to randomly generated value.
Set learning rate to a small value (positive).
Iteration n = 1; **Begin**
for n< max iteration OR Cost function criteria met, **do**
for image x_1 to x_i, **do**
- Forward propagate through convolution, pooling;
- Derive Cost Function value for the image;
- Calculate error term $\delta^{(l)}$ with respect to weights for each.
Note that the error gets propagated from layer to layer:
  ○ Fully connected layer;
  ○ Pooling layer;
  ○ Convolution layer.
- Calculate gradient $\nabla w_k^{(l)}$ and $\nabla_{bk}^{(l)}$ for weights $\nabla w_k^{(l)}$ and $\nabla_{bk}^{(l)}$

Gradient calculated in the following sequence:

  ○ Convolution layer;
  ○ Pooling layer;
  ○ Fully connected layer.
- Update weights:

$$w_{ji}^{(l)}w_{ji}^{(l)} + \Delta w_{ji}^{(l)}$$

- Update bias:

$$b_j^{(l)}b_j^{(l)} + \Delta b_j^{(l)}$$

Here is the popular CNN architectural variety and implementation.

1. **AlexNet:** CNN has been configured to operate parallel GPUs in Nvidia.
2. **Start:** Google-built powerful CNN.
3. **ResNet:** Microsoft's exceptionally broad residual network. It was given first spot in the ILSVRC 2015 ImageNet dataset competition.
4. **VGG:** Really broad CNN for large-scale identification of pictures.
5. **DCGAN:** Proposed by Radford, Metz, and Chintala [33] large coevolutionary networks with generative adversarial. This is used to know the hierarchy of function representations in input artifacts without surveillance.

➤ **Neural Network of Convolution in Drone:** Explores the trade-offs involved in the development of a single-shot object detector based on deep convolutional neural networks (CNNs) that can enable UAVs to perform vehicle detection under a resource constrained environment such as in a UAV.

9.4.3 NEURAL NETWORK OF AUTOENCODER

A neural network that uses Autoencoder [51, 60, 61, 65, 67, 68] unattended algorithms are described in an input dataset for reduction of dimensionality and the regeneration of the original database. The algorithm for learning is based on background propagation.

The principle of principal component analysis (PCA) is generalized by autoencoders. As shown in Figure 9.4, multidimensional data is transformed by a PCA into a linear image. This figure illustrates how a linear vector with PCA is able to reduce the input data in 2D. On the other hand, autoencoders can go even further and start producing vibrational images.

Figure 9.5 reveals RBM task sensor frames of single layer used for pre-training, accompanied by unrolling [36, 40, 42, 43, 46]. The unrolls the RBM bundles in the codec row. The encrypted block is then unrolled and the network is done with back package [36, 44, 45, 47, 49].

Figure 9.6 demonstrates how auto-encoders can which the input data element and know how to replicate it in the output layer. This figure reveals to gain reliability and performance modeling than current approach for orthogonal decomposition (POD) of distributed parameter reduction (DPSs), Wang et al. [37] have implemented a deep self-coder successfully comprising stacks.

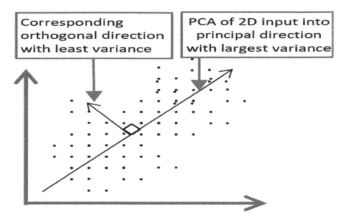

FIGURE 9.4 PCA linear representation.

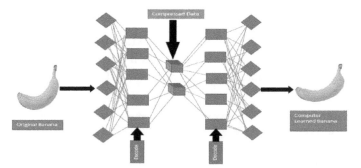

FIGURE 9.5 Autoencoder training stage [36].

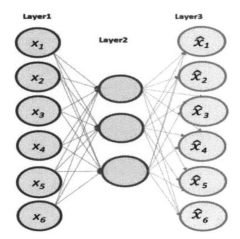

FIGURE 9.6 Nodes of autoencoder.

> ➤ **Neural Network of Autoencoder in Drone:** The deep convolutional denoising autoencoder is widely utilized to extract the target sound source in monaural audio source separation.

9.4.4 RESTRICTED BOLTZMANN MACHINE (RBM)

Boltzmann Limited Machine is an ANN that can be used create variational generative models using unlabeled data using unattended learning algorithm [39]. The aim is to train the network to maximize the likelihood of a variable (for example product or log) in the measurable units such that the data may be reconstructed probabilistically.

RBM's exposed layer and secret layer, as seen in Figure 9.7, are two-layer networking. The unit in the apparent stratum is linked to all units in the secret stratum without the same stratum connections.

> ➤ **Usage of Restricted Boltzmann Machine (RBM) in Drone:** Deep Boltzmann machine (DBM) as a very efficient approach for extracting the structure of explored data in UAV.

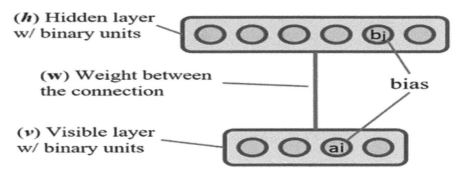

FIGURE 9.7   Machine of restricted Boltzmann machine.

9.4.5 LONG SHORT-TERM MEMORY (LSTM)

LSTM was first introduced by Hochreiter et al. in 1997 [41] and is a Recurrent Neural Network extension. In comparison with the feed forward architectures mentioned earlier, LSTM that Retention and preparation for study with a memory [53–58, 64] or growing knowledge may be preserved on existing policies.

The LSTM is made up of memory cell states that flow signal when controlled with input, forget, and exit windows, as seen in the example in Figure 9.8. Such doors monitor all that is processed, read, and written on the container. Through its speech processing services [42] LSTM is used by Google, Apple, and Amazon.

➢ **Long Short-Term Memory (LSTM) in Drone:** the use of RNN instead and successfully trains an LSTM (Long-Short Term Memory) network for controlling UAVs.

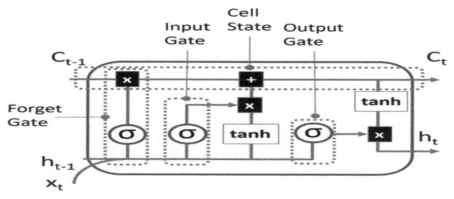

FIGURE 9.8   Block of LSTM.

9.4.6   COMPARISON OF DNN NETWORKS

Table 9.2 offers concise description of the different DNN systems and contrasts. The explanations given in Table 9.2 are not meant to be comprehensive of architectures, programs databases and DL control frames.

9.4.7   TRAINING ALGORITHMS

The important part of the DL is a learning algorithm. A shallow network is distinguished from the deep neural network. The more layers it gets, the deeper it gets. A unique dimension or attribute may be sensed for any base.
    Some of the most famous algorithms for formation are:

1. **Gradient Descent (GD):** For most ML and in-depth learning, GD is the fundamental concept. Centered on the Newton Algorithm

TABLE 9.2 Comparison of DNN Network

Network Type	Network Architecture Model	Training Type	Training Algorithm	Implementation Sample	Common Application	Famous Dataset Sample	Structure of Deep Learning
CNN	Discriminative	Supervised	Gradient Descent based Backpropagation	Deep CNN	Image Classification	MNIST	TensorFlow, Caffe, Torch
Residual Network	Discriminative	Supervised	Gradient Descent based Backpropagation	Deep DenseNet	Dimensionality Reduction; Encoding	Image net	TensorFlow, Keras
Autoencoder	Generative	Unsupervised	Backpropagation	Sparse Autoencoders, Variational Autoencoders		MNIST	TensorFlow, Keras
Adversarial Networks	Generative & Discriminative	Unsupervised	Backpropagation	Generative Adversarial Network	Image Improvement,	CIFAR10	TensorFlow, Keras
RBM	Generative & Discriminative	Unsupervised	Backpropagation	Deep RNN, Neural Machine Traslation	NLP	MNIST	Tensorflow, Microsoft Cognitive tool
Recurrent Neural Network	LSTM, Discriminative	Supervised	Backpropagation	Deep RNN, Recurrent Unit	NLP	MNIST	Tensorflow, PyTorch
Radial Basis Function NN	RBF Network	Discriminative	Supervised k-means and clustering	Radial Basis Function NN	Iris	Time series data set predicition	Tensor Flow
NN	Generative	Unsupervised	Competitive Learning	Kohonen Self Organizing NN	Dimensionality Reduction	SPAMbase	TensorFlow

principle, it determines a 2D function 's roots (or null value). We arbitrarily pick a point in the turn and move the x-axis is zero at the specified point previous to the y-axis measure, i.e., function or x, depending on negative or positive attribute derivatives or slope magnitude.

The error derivatives shown in Figure 9.9 are the weighted overall summation of the fault derivative products with respect to the unit outputs in the layer above.

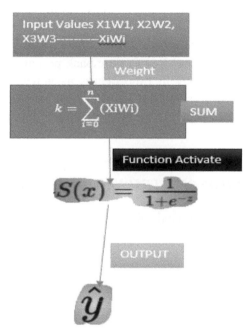

FIGURE 9.9   NN error calculation [6].

> ➢ **Gradient Descent in Drone:** A GD algorithm with adjustable parameter for attitude estimation is developed, aiming at the attitude measurement for small unmanned aerial vehicle (UAV) in real-time flight conditions.

2. **Stochastic Gradient Descent (SGD):** The most famous variant and application of the GD is the stochastic gradient descent (SGD). During the GD, all measurements in the exercise dataset are analyzed before updating weight details. Updates are introduced in SGD after a small ton of n tests have been executed. Because we change the

weights in SGD rather than in GD, we will converge even more easily to a global minimum.

3. **Momentum:** The velocity v vector described in physics is an exponentially decreasing average of the continuum utilizing the principle of momentum [48]. This avoids costly fall in the wrong way.

4. **Levenberg-Marquardt Algorithm:** The algorithm for Levenberg-Marquardt (LMA) is used mainly to address non-linear small square problems, such as curve adaptation. We attempt to match in a minimal square question a certain data point in the system with the least amount of the error squares between the real data points of the system and points.

5. **Backpropagation Through Time:** The typical approach to build the recurrent neural network is time back propagation (BPTT). The unrolling of the RNN in time occurs as a feedforward network. However, in contrast to the feed-forward network, the unrolling RNN has A set of accurate weighting factor is the time field learning for every level.

9.4.8 COMPARISON OF DEEP LEARNING (DL) ALGORITHMS

Table 9.3 summarizes the typical DL algorithms and compares them. The benefits and drawbacks and methods for overcoming the weaknesses are discussed. The most popular kind of training is GD training. Time back-packing is a backpacking suited to the recurrent neural network.

TABLE 9.3   Distinction between Deep Learning Algorithms

Algorithm	Advantages	Disadvantages	Techniques to Address Disadvantages
Deep learning in batch gradient descent	Ranges well following enhancements	It requires a while for masses to accumulate after the whole datasets transfer	Descent of the micro batch gradient
Deep learning in stochastic gradient descent	Ranges well following enhancements	Noise level of inaccuracy because it has been computed in each data set;	Descent of the Mini-Batch Gradient; Shuffle details single time
Deep learning in back propagation through time	Works more efficiently than metaheuristics (e.g., genetic algorithm)	Difficult to use when integrating digitally, as the whole-time sequence must be included	Take some time rather than whole time

TABLE 9.3   *(Continued)*

Algorithm	Advantages	Disadvantages	Techniques to Address Disadvantages
Deep learning in contrastive divergence	You may construct simulations from the collection of data input; better technological;	Hard to prepare	Get the Monte Carlo Markov Chain review
Deep learning in evolutionary algorithms	You may construct simulations from the collection of data input; better technological;	This requires energy to move as different variations are checked	Use GPUs and cloud
Deep learning in reinforcement learning (Q-learning)	Could align research and creation	Rewards are highly uncommon in some situations	Job back from the state without compensation

9.5   DRONE AS THE NEW "FLYING IoT"

Drones like men can automatedly identify and video performers, monitor offenders and send products straight to your doorstep. Like for a smart computer, such ML will therefore consume resources, so a recent study reveals that a drone processing charge is moved even further to the sensor cloud infrastructure to sustain power consumption and drones.

"Drones are a modern generation of IoT systems soaring into the sky and providing complete network communication. Intelligent drones with cognitive computing know-how require the capabilities of identifying and monitoring artifacts automatically to relieve the consumer of the boring role of managing them (Figure 9.10).

The challenge is to build a drone that is clever enough to know its target in a multitude and strong enough to monitor its target over long distances.

"To combine output and utilization of resources is the task of object sensing on drones. In order to act effectively, the leader must identify an entity many times a second, otherwise he can lose himself. Even for simple low-power CPUs, or even GPUs, this real time efficiency criterion is difficult to fulfill Deep desktop or server-level processors are needed for complex multi-class modeling including co evolutionary neural networks (CNNs). Therefore, incredibly low-power processors without expertise in hardware are not enough, however the advanced equipment adds complexities of the architecture and unrepeatable development costs.

FIGURE 9.10   Drone (flying IoT).

9.5.1   ML AND DL BASED APPLICATION OF IoD

1. **Drones for Operations of Disaster Relief:** However, the searching drones and emergency drones will provide foods, water, cables, lifestyle, safety equipment, surfboards or life preservations as shown in the image. For instance, metropolitan Drones  begins to develop drones, which instantaneously inflate life jackets.

2. **Drones First Aid:** For example, if 2 men walk on a road but one falls, a hypothetical will play a role. In situations like this, sudden cardiac arrest (SCA) is a potential life-threatening source. This relies on the use of an AED or Automated External heart

monitor for restarting the pulse. While this apparatus is sometimes located in office buildings' first aid situations, it is not placed in a corner of the lane.

3. **Drones Use in Agricultural Applications:** Drones in smart agriculture: Cropped spraying is an apparent IoT drone framework for agriculture for fertilization and pest management.

   i. **Sowing Seeds in Field:** Seeding seeds, and including nutrients for their maintenance, may become a standard activity over time, as drones are fast and precise in carrying out this function and can reduce seeding prices by 85%.

   ii. **Field Soil Analysis:** Drones could provide additional insight quickly in the cropping season, according to an article in the Technology Review, by generating exact 3-D graphs that allow farmers to schedule their crop root trends as well as fertilizer application and water management.

4. **Drones for Parcels, Materials, or Freight Delivery:** A broad variety of cases of distribution are being investigated and subjected to regulatory requirements on admission to the industry with such applications. Possibly, the distribution of pizzas and other fast meals is not far away until the drone is distributed for orders.

5. **Drones in Military Operations:** For combat activities, helicopters are prominent, so that may be the most popular example of unmanned aerial vehicles. Monitoring is clearly a significant topic for combat operations. A lightweight drone that can spy on other sky powers.

6. **Drones for Sporting and Entertainment:** The usage of aerial imagery and video is now becoming one of the most popular for commercial drones, for sports broadcasting. The usage of aerial photography and video is now  one of the most common for sports and entertainment, commercial drones.

7. **Drone Racing:** Drone racing is becoming a common activity, and hobby and championship drone operators are now open. FPV drone racing is a sport category in which competitors fly drones (usually smaller radio-controlled aircraft or  quad copters) fitted with cameras, with head-mounted displays that screen the view of the drones from live stream cameras.

9.6 MACHINE LEARNING (ML) IN DRONE

URAN design techniques with example data or past expertise may be a strong tool for ML [15, 17, 18, 37]. Algorithms of ML may categorize as a tracked, unattended, and reinforced. In SL, a projection from input and the output can be automatically generated, the right values (labels) being given by a teacher during the training process. There are several criteria to be viewed as input data testing for SL in very specific situations with multiple UAVs.

In the case of SL, the features and variables corresponding to the specified network interference can be used in an ML box and the Based on the given situation, the loss function may be defined. No instructor in unattended learning and the algorithm requires, without previous instruction, to autonomously discover correlations, trends, and variations in the accessible data. For certain instances, though, there may not be a sufficient number of training results.

A reward-driven feature may usually be configured depending on the situation, for example, that visibility or the several clients as seen in Table 9.4.

TABLE 9.4   Elements of Reinforcement Learning in U-RANs

Elements of RL	Speed of Drone	Drone Location
Action in RL	Alteration speed of drone	Alteration in location
State in RL	Recent speed of drone	Recent location of drone
Reward in RL	Raised total throughput	Raised coverage

9.6.1   RADIO RESOURCE ALLOCATION

The distribution of radio resources (RRA) for U-RANs involves power control, dynamic spectrum, beamforming design, back shaft management, cache management and system management of resources. Such tasks can be collectively handled by the ML approaches in Figure 9.11. The key user trends which can be inserted in the ML box that take care of as different variables of user behavior (e.g., network traffic requires, latency tolerances and reliability), such as space locations and coordination specifications.

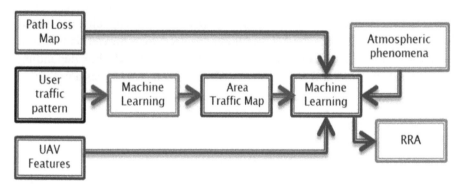

FIGURE 9.11   Block diagram of RRA based ML.

9.6.2   RELAYS DESIGN OF COLLECTORS

For IoT implementations, UAVs can be used as distributors and transmissions [66, 67]. With respect to Figure 9.12, at period t 0 and velocity v the UAV begins to fly to the source (S) to Select the details in t = d/2 (first move or hop). The second one, from t = d, flies to the target (D) and passes data to T = 3d/2 in the buffer. Such a handheld relay is clear technique will reduce the connection gap for either of the two bends as you compare to static relays with fixed relays in a given location.

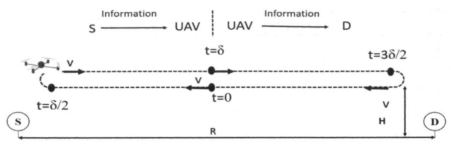

FIGURE 9.12   Design of UAV collector relay.

9.6.3   TYPE OF UAV CHOOSING

The key power costs for UAVs with contact capability maintenance of air vehicle and wireless transmitters and receivers (i.e., floating, and flying). The radio interface energy may be influenced by different UAVs such as distance, wing form, motion characteristics and scale [53, 63, 66, 67, 69]. Therefore, it is critical for users with specific communication criteria to choose the form of UAV's (i.e., time float, speed spread, capacity, download / load speed maximum). The ML methods are used to achieve this mission. The above listed functionality can then be inserted into an ML box depending on the data collection and can be done offline. Figure 9.13 gives a graphical image of this operation.

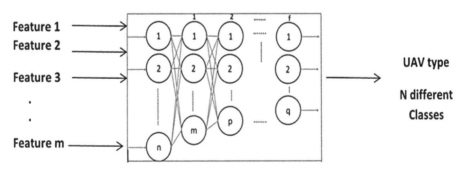

FIGURE 9.13   UAV choose best path for monitoring.

9.6.4   SELECTED SEVERAL UAVS ACTING AS BSS

Although UAVs are being used as BSs for the field operator, selecting the amount and location of UAVs substantially affects the transmitting capacity of the specific UAVs, the coverage region and Receiver SINR [1, 3, 7, 20, 63]. UAVs can also be jointly configured for number and locations difficult to be carried out in several interesting situations, especially if UAVs have to be combined with ground BSs, because of URAN's dynamic network requirements. Of that reason, they are often separately optimized [4, 10, 19, 20, 34, 35, 39, 49]: the first move is to find the best number of UAVs and, for that amount, to get optimum UAV locations.

Consider an example in which many UAVs use the same user-filed data and users often provide device-to-device (D2D) connectivity connections to make it simpler to access a file (see Figure 9.14). The optimal number of UAVs will in this situation be achieved with RL and SL processes.

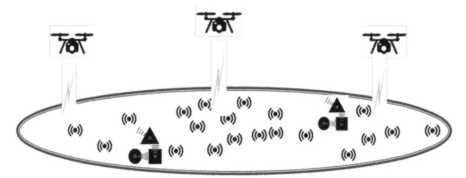

FIGURE 9.14   Multiple UAV share same file.

A more cost, compensation, and penalty is the amount of consumer inability to access the file in which the introduction or decrease of a UAV will then be called an operation and where the So the condition is the several of UAVs. Figure 9.15 shows This operation's flowchart. A variety of situations as a first step are specified with RL with a given loss function as the number of UAVs obtained.

This indicates a situation where several UAVs are given to various users. In Figure 9.14, the upper side picture indicates a situation where 7 UAV (UAVs 1, 2, 3, 3, A, B, C, etc.), are operated by various users. Every UAV is attached to a curve-bound field. The boundaries of these regions are changed when consumer activity changes and the position of the UAV as in the bottom plot of Figure 9.16 are subsequently revised. The partition is designed to optimize the coverage and reduce the UAV's flying time and contact power costs.

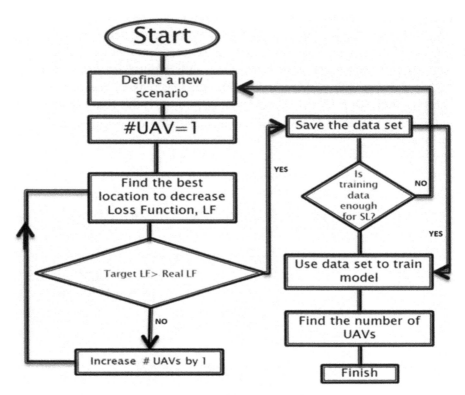

FIGURE 9.15   Combined RL and SL searching number of UAVs.

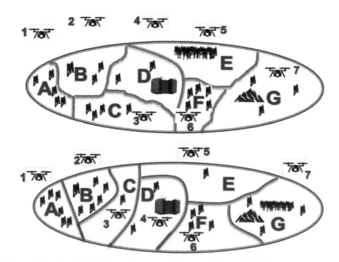

FIGURE 9.16   Changing UAV location according to demanded image.

9.6.5 STRUCTURE OF A MOBILE CLOUD

The data which can be collected and processed in the cloud can be analyzed using ml (see Figure 9.17). A variety of clouds, called mobile clouds or clouds, may be put into a range of UAVs, clouds, and BSs. The main problems are the gathering, storage, and analysis of massive data, the unified balancing (e.g., the mobile cloud), and the deployment of paradigms in the U-RAN (e.g., the mobile edge computer). The cloud can have various cloud computing features. For example, the cloudlet will have an optimal pace or position to meet the computing needs of different users in the region.

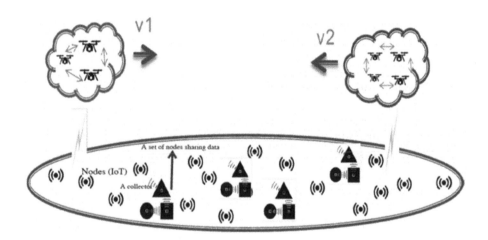

FIGURE 9.17 UAV to UAV collecting data through cloud.

9.7 PRINCIPLE WORKING OF MACHINE LEARNING (ML) IN DRONE

Using the ML concepts the challenges of the current genetic algorithm can be made more precise. Recognition of a visual face is natural and the consequence is apparent. This is different with artificial intelligence (AI). The process flows are shown in Figures 9.18 and 9.19. It is demonstrated in the convolution steps:

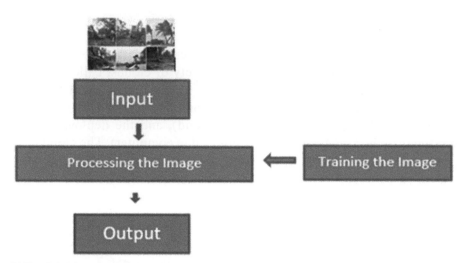

FIGURE 9.18   Machine learning process.

The telemetry data analysis functions like this:

- **Phase 1:** Break the photos into tiles.
- **Phase 2:** Provide the process to each image tile.
- **Phase 3:** Store the result in a new array for any picture tile.
- **Phase 4:** Sampling down.
- **Phase 5:** Last decision making.

Table 9.5. 15,000 Machine Learning images have been evaluated (6,000 are human images, while 9,000 images are not human).

TABLE 9.5   Evaluation of Human Image [53]

	Predicted "Human"	Predicted "Not a Human"
Drone click image of human	5,560	440
Drone no click image of human	162	8,838

In order to access images, the network is completely linked, transformed, and maxed. These steps can be repeated with more convolution layers many times during the taking of samples in real time. Such moves aim to produce a common end goal. Through the method, 97% precision can be achieved (Table 9.6).

TABLE 9.6 Data Collected by Drone

No. of Layers	Neurons per Layer	Dense Layer Activation	Network Optimizer	Accuracy Data Captured
2	768	Elu by drone	Adamax network optimizer	56.03
2	512	Elu by drone	Adamax network optimizer	55.65
3	314	Elu by drone	Adamax network optimizer	52.12

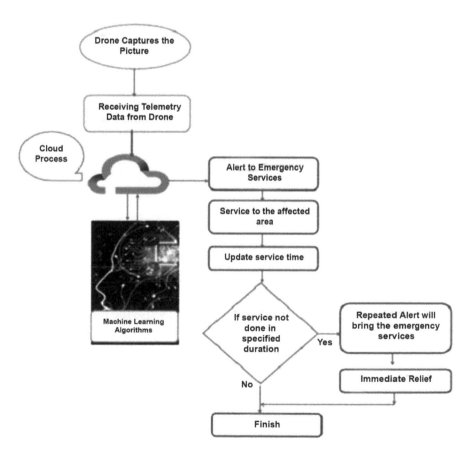

FIGURE 9.19 Flowchart of image received.

Elu is the linear device used for the triggering of the system's dense sheet. The exactness of the data collected is roughly 56% while the number of layers is in the range of 2.

9.7.1 DRONE DATA AND CLOUD COMPUTING

The drone data processing has been fitted with an extension of indefinitely scalable computing resources and the integration of end-to-end applications utilizing machine-based technologies. The cloud-based systems will accommodate all data that is processed in the data centers as far as possible and of the highest standard [6–9].

9.7.2 METHOD OF DATA HANDLING AS A GROUP (MDHG) ALGORITHM

This is a class of inductive algorithms, with a strong optimization mechanism that models mathematics and machine programs, and multiparameter data sets.

9.7.3 DRONE OPERATING STEPS

1. Deploy a compute engine instance with the frame work of machine learning images from Drone.
2. Compute engine launch with abstracted image.
3. Enter a deployment name for application emergency services that will be root of virtual machine.
4. Set the entire frame work to PyTorch and choose the required zone.
5. GPU type being chosen.
6. Check the combination for the data center images.
7. Initialize NVIDIA driver that can be used using GPU.
8. Check GPU quota.
9. Enable the deployment stage image caught from drone.
10. After implementation the page will be updated and the data center status will be noted with the amount of utilization.

Table 9.7 provides details about the model of drone used, the layer on which the data is stored, the overall capacity of the drone to catch items and the distance to be flown, and the top acceleration at which the drone can operate.

TABLE 9.7   Usage of Drone

Drone Type	Layer	Units	Max Load (Kms)	Maximum Speed (Km/Hr)
D1	1	13	8.55	35
D2	1	10	2.03	40
D3	1	10	3	50
D4	2	2	5.1	90
D5	2	4	4.8	105

9.8   DEEP LEARNING (DL) USE IN DRONE

Many in-depth learning advances for UAV preparation and situational awareness have been published. Planning activities include creating approaches to specific problems without providing a dynamic system to manage the world model or autonomous abilities or techniques. Planning is important whether the environment or activities of the robot are unstructured, complex or if there is variety. Route, transfer, routing, or coercive scheduling are common activities. Tasks of special awareness enable robots to understand their own state and the state of their environment, e.g., robot state evaluations, self-location, and mapping are examples of these activities:

1. **Planning:** Path planning is presented in Ref. [57] for collaborative research and rescue missions with a deep study. This research, in which a UAV investigates and charts the world in which the robot will move, emphasis on shortened average time of implementation (i.e., course, and exploration).
2. **Situational Awareness:** With the aid of DL [59], a cross-view object localization is achieved. While the study is viewed as the answer Some UAVs for image processing have been used for UAV location and only ground-level photos were used for such experiments. The approach is to search a database of Raw image details in order to identify the nearest visual characteristics (i.e., landmark) that then matches the characteristics derived from the query image (Figure 9.20).

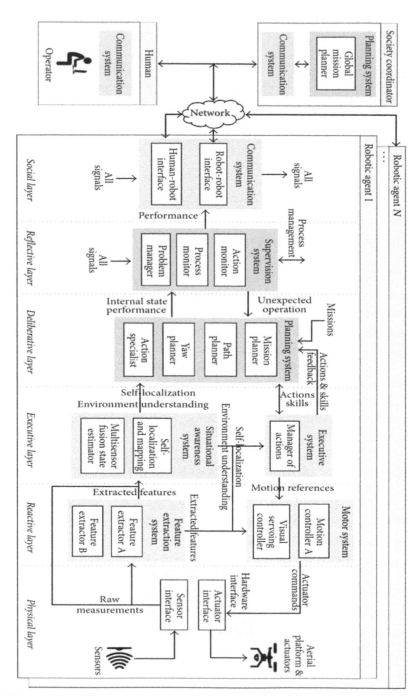

FIGURE 9.20   UAV layered architecture protocol for moving [62].

9.9 PRINCIPLE WORKING OF DEEP LEARNING (DL) IN DRONE

9.9.1 STRUCTURE OF PATH PLANNER

In this segment we continue with a summary of the static model of the evolutionary algorithm (EA) approach for resolving the route planning issue (continue place and time does not changing). For those finding a more comprehensive explanation of any of their behaviors, general EA-based route discovery algorithms [10, 11].

A collection of solutions and random modifications are used by EA-based path-planning algorithms to allow ordered, stochastic work. Of further detail see Figure 9.21. The participants are uniformly planted. Instead, the algorithm is looped. At-point (or generation), by changing the current members (a change) or by introducing new paths (members) to the population first.

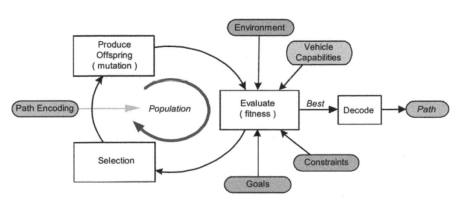

FIGURE 9.21 Structure of drone moving planning.

One or three new (reproduction) participants to combine. The costing function also helps the representatives of the greater community to obtain the health of the group. The population (selection) is limited to its original size depending on certain results.

Our track encoding is a sequence of basic end-to-end clustered splines (termed segments). Two forms of divides exist: straight lines and continuous curves of radius. For further detail, see Figure 9.22. Includes range, radius, and end speed parameters. The distance parameters of speed and change are restricted to maintaining motion inside the automobile.

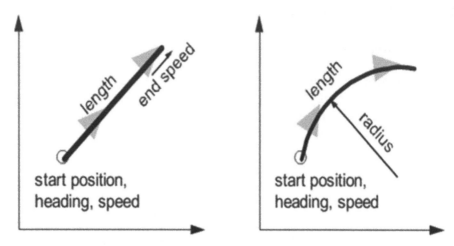

FIGURE 9.22   Drone path segment.

There are four types of mutation:

1. **Propagate and Mutate:** The specifications for one or more parts are automatically modified and then the corresponding segments are moved in order Start enforcing limits at the end stage. A random collection of the Mutation of the first section and the number of repetitions to be modified. See Figure 9.23(a) for further detail.
2. **Crossover:** Use a dot-to-point-joint function to suit the start segments of one track and the end segments of another. See Figure 9.23(b) for more details.
3. **Go to Objectives:** Select a randomized segment close to the end line and the segment is used-top-joint feature to suit the goal direction directly from the starting position of that section. For more information see Figure 9.23(c).
4. **Mutate and Match:** Adjust one segment or more section parameters, determines the corresponding current final point for such segments (like the Mutate and Propagate segments) and connects to a new route section launch. For more detail, see Figure 10.23(d).

A point-to-point-join feature allows many mutation mechanisms. This role specifies the criteria for two parts which are related to the finishing location, thus following the consistency specifications, at the start of the section and heading. The task Continuity will not be applied without this feature.

Figure 9.24 illustrates the existence of the geometric solution required to create such joints.

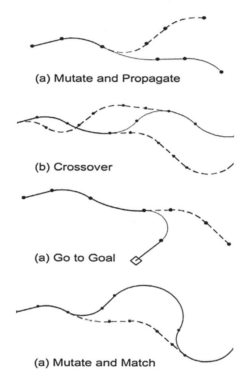

(a) Mutate and Propagate

(b) Crossover

(a) Go to Goal

(a) Mutate and Match

FIGURE 9.23   Path mechanisms of drone.

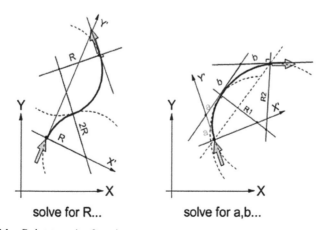

solve for R...   solve for a,b...

FIGURE 9.24   Point to point function.

9.10   DRONE CONTROLLER

Tiny unmanned aerial systems (UAS) or drones may face substantial safety threats given their numerous uses. Tools to track, identify, and combat such vehicles must also be identified. Drone operates to instantly identify the radio frequency (RF) signals it emits from drone controllers. The signal range is tracked by an RF sensor array. A CNN is programed to predict the drone controller's behavior in relation to the sensor. A CNN is used to forecast its efficiency. From this stage, the locations of the controllers can be determined as long as two sensors are installed at least within a suitable distance (Figure 9.25).

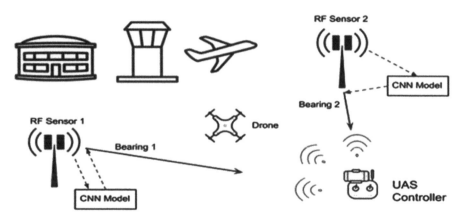

FIGURE 9.25   Structure of drone controller.

9.11   DRONE NAVIGATION

A modern method for autonomous aerial drone navigation in preset directions utilizing an on-board camera just visual feedback and depending not on the Worldwide Positioning Systems (GPS). It is focused on the use of a deep Coevolutionary Neural Network (CNN) and a regression to produce the drone controls. In fact, the device is adaptable to real life situations utilizing several auxiliary navigation paths that create a 'navigation shell' for data increase. Software that shows daily trips or visits to the same locations such as environmental surveillance and desertification, parcel / hello distribution and drone-basis WLAN is a method suitable for automation of drone navigation. It would substitute human operators by a proposed algorithm, improve precision of GPS-oriented map navigation, minimize GPS spoofing problems, and enable navigation to GPS environmentally devoid (Figure 9.26).

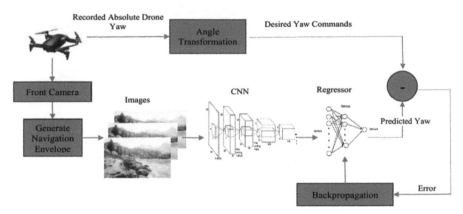

FIGURE 9.26   Structure of drone navigation.

9.12   DEEP LEARNING (DL) DRONE SURVEILLANCE SYSTEM

As video surveillance program is installed, tremendous quantities of data are produced. For the Law enforcement compliance of smart city programs, a clear and appropriate indicator of identification is required. Gait identification is one of the required bio-metric awareness steps. Automatic indexing of human behavior in the monitoring region by way of a gait recognition enables suspicious behaviors to be detected and thus holds the field in sequence, in number. In time. In fact, with electronic computer developments, remote computers that promote the hybrid of static-mobile surveillance systems are able to gait recognition. The in-depth research system to detect suspect human behaviors is indicated by continuous streaming content to understand individual human actions. It consists of a special shot multi-box detector (SSD), a distributed teaching focused on V3, which extracts spatial features by coevolutionary neural networks (CNNs). The spatial functions are consequently built into the behavior processing deep framework of LSTM. The KTH Human Action data collection is educated in six action groups: Running, ride, man-clapping, tossing, and boxing. The network comprises of six action groups (Figure 9.27).

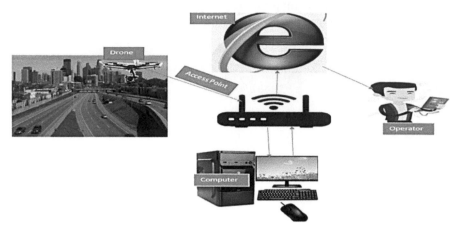

FIGURE 9.27   Structure of drone surveillance system.

9.12.1   IDENTIFICATION AND UNDERSTANDING OF HUMAN ACTS

The goal for the introduction of video surveillance device identification is the capacity to identify and understand local behavior. The current pipeline has been guided by Montes et al.'s recent approach, [10] with an optimized adjustment until practice is understood. It requires local intervention and, above all, individual localization, can give national action awareness in comparison to previous strategies. The proposed human detection and recognition pipeline with a profound learning approach can be seen in Figure 9.28.

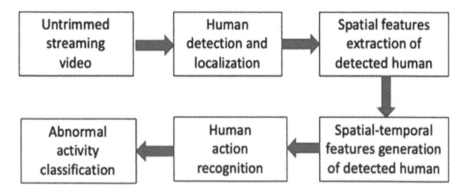

FIGURE 9.28   Diagram of drone detect human activity.

9.12.2 *SINGLE SHOT MULTI-BOX DETECTOR USED BY HUMAN DETECTION*

The CNNs are more reliable and efficient at detection pace have developed over the years. Single Shot Multi-Box Detector is most of the almost accomplished CNNs-standard efficiency while maintaining processing costs small (Figure 9.29).

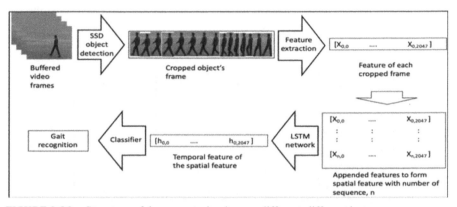

FIGURE 9.29   Structure of drone capturing human different-different image.

9.13   ML AND DL BASED IoD ALGORITHM

Effective image processing methods for different activities are in operation and work as well as ML and DL approaches. One has to wonder, what will my results be and what is accessible information / data? Take into account that every form of ML methodology needs enormous data sets and intensive preparation. So, the best solution could be to have only a limited number of image processing software available. Once large data sets and different activities are managed, ML- or DL-approaches potentially go beyond automated solutions for image processing. In other terms, because image processing software functions and becomes more complicated, the more probable ML / DL methods become, as their data size increases, to be the best means of solving them.

Throughout the area of inspection and repair, several implementations of ML and DL algorithms are currently located. For numerous inspection activities companies such as Sky-Futures and Scopito use separate ML and/ or DL approaches. For example, photographs are automatically used to detect irregularities on power lines insulators and algorithms on metallic surfaces to detect corrosion. Detection rates range between 80% and 90% according to Sky-Futures. Another example is Ardenna, who is actually updating the tech company from Computer Vision Technology to the BNSF railway company's ML train examination, and it identifies over 30 losses in one case. Some AI methods now are being used in the electricity, agriculture, immovable, buildings, and forestry field for data processing. There appears to be an infinite number of AI data analysis applications nowadays and just a tiny fraction of what is actually present on the market are only those examples listed.

9.13.1   ALGORITHM OF ML AND DL BASED IoD ALGORITHM

- **Step 1**: Input: img: Image from the UAV front camera.
- **Step 2:** Initialization: command ← TAKE-OFF.
- **Step 3:** While command! = LAND do.
- **Step 4:** Position ← Trained-Classifier (img).
- **Step 5:** If position = Center then. Actuate UAV in PITCH_FORWARD direction.
- **Step 6:** Else if position = Left then. Actuate UAV in ROLL_RIGHT direction to move towards the center;
- **Step 7:** Else if position = Right then. Actuate UAV in ROLL_LEFT direction to move towards the center;
- **Step 8:** Else if position = End then send LAND command;
- **Step 9:** Img ← Extract the next frame from the UAV front camera.

9.14   PARADIGM OF MLDL BASED IoD

While other businesses are switching from computer vision to mainstream ML methods, it appears like the first moves are being made in the drone industry with fundamental computing algorithms. Recent advances in the technology industry, i.e., GPUs, have allowed DL to be used with its price-to-performance ratio and required hardware resources. Although GPUs provide far more processing capacity, DL algorithms take a reasonable amount of time and millions of images are mostly required to conduct any function with DL in a reliable way. Therefore, if you have access to a large-scale data set of pictures and adequate computing capacity, DL methods could be a better option because conventional ML and CV methods are typically outsourced, particularly in picture recognition.

9.14.1   MOTION PLANNING

Let us all come to Movement Planning now. Sense and Avoid technology, and BVLOS flights are a powerful tool for situational awareness and, in a broader sense. The prerequisite to schedule movements is typically to catch the atmosphere, namely the understanding of the system. In this respect, the drone visualizes the environment, for example with SLAM technology. This helps the drone to not actually recognize, but the distance to, precisely what is in the area. DL is used in the context of motion planning to detect, recognize, and subsequently establish an appropriate flight route for objects such as human, biking or cars (Figure 9.30).

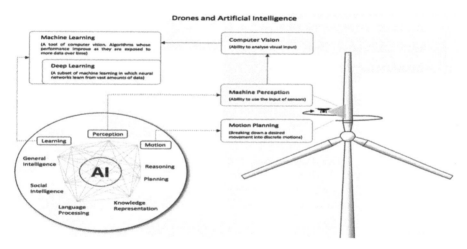

FIGURE 9.30   Paradigm of MLDL based IoD.

9.15   CONCLUSION AND FUTURE WORK

We also implemented a basic route planning method utilizing evolution-based algorithms to take the complexity into account when determining environmental hazard parameters and adjusting during period, certain criteria. In addition, we have introduced a framework for building a static dynamic planner by introducing a state update feature. The planner 's configuration the route produced was so continuously throughout the UAV Restrictions on sphere and velocity and maneuvering. Through implementing the UAV scenarios to adjust its course to avoid a variety of other aircraft operating in its immediate region we have shown the actions of the route planning algorithm. Without extreme movements of evasion or prior experience The UAV was able to clear the single aircraft safely from the probable trajectories of the other aircraft. The manager was in a role to determine the danger of a fuel accident and to implement overly cautious avoidance.

KEYWORDS

- **deep learning**
- **expectative-maximization**
- **machine learning**
- **principal component analysis**
- **radio access network (RAN)**
- **radio resources (RRA)**
- **unmanned air vehicles (UAV)**

REFERENCES

1. Khatib, O., (1986). Real-time obstacle avoidance for manipulator and mobile robots. *International Journal of Robotics Research, 5*(1), 90–98.
2. Mitchell, J. S. B., & Keirsey, D. M., (1984). Planning strategic paths through variable terrain data. *Proc. of the SPIE Conference on Applications of Artificial Intelligence, 485,* 172–179. Arlington, VA.
3. Stentz, A., (1994). Optimal and efficient path planning for partially-known environments. *Proc. of the 1994 International Conference on Robotics and Automation, 4,* 3310–3317. Los Alamitos, CA.
4. Fogel, D. B., & Fogel, L. J., (1990). Optimal routing of multiple autonomous underwater vehicles through evolutionary programming. *Proc. of the 1990 Symposium on Autonomous Underwater Vehicle Technology,* 44–47. Washington, D.C.
5. Capozzi, B. J., & Vagners, J., (2001). Evolving (Semi)-Autonomous Vehicles. *Proc. of the 2001 AIAA Guidance, Navigation and Control Conference.* Montreal, Canada.
6. Szegedy, C., Liu, W., & Jia, Y., (2014). *Going Deeper with Convolutions.* 1409.4842.
7. Hochreiter, S., & Schmidhuber, J., (1997). Long short-term memory. *Neural computation, 9,* 1735–1780.

8. Darrell, T., & Pentland, A., (1993). Space-time gestures. In: *Computer Vision and Pattern Recognition, 1993: Proceedings CVPR'93, IEEE Computer Society Conference* (pp. 335–340). IEEE.

9. Yamato, J., Ohya, J., & Ishii, K., (1992). Recognizing human action in time sequential images using hidden Markov model. In: *Computer Vision and Pattern Recognition, Proceedings CVPR'92., IEEE Computer Society Conference* (pp. 379–385). IEEE.

10. Montes, A., Salvador, A., Pascual, S., & Giro-i-Nieto, X., (2016). *Temporal Activity Detection in Untrimmed Videos with Recurrent Neural Networks.* arXiv preprint rXiv:1608.08128.

11. Li, Z., Gavrilyuk, K., Gavves, E., Jain, M., & Snoek, C. G., (2018). Video LSTM convolves, attends and flows for action recognition. *Computer Vision and Image Understanding, 166*, 41–50.

12. Andre, E., Brett, K., Roberto, A. N., Justin, K., Susan, M. S., Helen, M. B., & Sebastian, T., (2017). Dermatologist-level classification of skin cancer with deep neural networks. *Nature 542*, 115–118.

13. Kai, A., Marc, P. D., Miles, B., & Anil, A. B., (2017). *Deep Reinforcement Learning: A Brief Survey, 34*(6), 26–38. IEEE Signal Processing Magazine.

14. Gheisari, M., Wang, G., & Bhuiyan, M. Z. A., (2017). A survey on deep learning in big data. *IEEE International Conference on Computational Science and Engineering (CSE) and IEEE International Conference on Embedded and Ubiquitous Computing (EUC).*

15. Samira, P., Saad, S., Yilin, Y., Haiman, T., Yudong, T., Maria, P. R., Mei-Ling, S., et al., (2018). A survey on deep learning: Algorithms, techniques, and applications. *ACM Comput. Surv., 51*(5), 1–36.

16. Vargas, R., Mosavi, A., & Ruiz, R. (2017). Deep learning: A review. *Advances in Intelligent Systems and Computing, 5*(2).

17. Buhmann, M. D., & Buhmann, M. D., (2003). *Radial Basis Functions, 270.* Cambridge University Press.

18. Akinduko, A. A., Mirkes, E. M., & Gorban, A. N., (2016). SOM, Stochastic initialization versus principal components. *Information Sciences, 364, 365*, 213–221.

19. Chen, K., (2015). Deep, modular neural networks. In: Kacprzyk, J., & Pedrycz, W., (eds.), *Springer Handbook of Computational Intelligence* (pp. 473–494). Springer Berlin Heidelberg: Berlin, Heidelberg.

20. Ng, A. Y., & Jordan, M. I., (2001). On discriminative vs. generative classifiers: A comparison of logistic regression and naive bayes. *Proceedings of the 14th International Conference on Neural Information Processing Systems: Natural and Synthetic* (pp. 841–848). MIT Press: Ancouver, British Columbia, Canada.

21. Bishop, C. M., & Lasserre, J., (2007). Generative or discriminative? Getting the best of both worlds. *Bayesian Statistics, 8*, 3–24.

22. Zhou, T., Brown, M., Snavely, N., & Lowe, D. G. (2017). "Unsupervised Learning of Depth and Ego-Motion from Video," *2017 IEEE Conference on Computer Vision and Pattern Recognition (CVPR)*, 6612–6619, doi: 10.1109/CVPR.2017.700.

23. Chen, X. W., & Lin, X., (2014). Big Data Deep Learning: Challenges and Perspectives. *IEEE Access, 2*, 514–525.

24. LeCun, Y., Kavukcuoglu, K., & Farabet, C., (2010). Convolutional networks and applications in vision. In: *Proceedings IEEE International Symposium on Circuits and Systems.*

25. Georgios, G., Bogdan, V., Alexander, S., & Andy, Z., (2014). Lean GHTorrent: GitHub data on demand. *Proceedings of the 11th Working Conference on Mining Software Repositories* (pp. 384–387). ACM: Hyderabad, India.

26. AI-Index, (2019). *Top Deep Learning GitHub Repositories*. AI Index; Available from: https://github.com/mbadry1/Top-Deep-Learning (accessed on 19 November 2021).

27. Manuel, F., Eva, C., & Sen'en, B., (2014). Do we need hundreds of classifiers to solve real world classification problems?. *J. Mach. Learn. Res., 15*(1), 3133–3181.

28. Yann, L. Leon, B., Yoshua, B., & Patrick, H., (1998). Gradient-based learning applied to document recognition. *Proceedings of the IEEE* (pp. 2278–2324).

29. LeCun, Y., & Bengio, Y., (1998). Convolutional networks for images, speech, and time series. In: Michael, A. A., (ed.), *The Handbook of Brain Theory and Neural Networks* (pp. 255–258). MIT Press.

30. Graham, W. T., Rob, F., Yann, L., & Christoph, B., (2010). *Convolutional Learning of Spatio-Temporal Features*. Computer Vision-ECCV, Springer Berlin Heidelberg,

31. Ng, A., (2018). *Convolutional Neural Network. Unsupervised Feature Learning and Deep Learning (UFLDL) Tutor*. Available from: http://ufldl.stanford.edu/tutorial/supervised/ConvolutionalNeuralNetwork/ (accessed on 19 November 2021).

32. Christian, J. S., Harold, C. B., Stefan, H., & Bernhard, S., (2013). A machine learning approach for non-blind image deconvolution. *IEEE Conference on Computer Vision and Pattern Recognition.*

33. Radford, A., Metz, L., & Chintala, S., (2015). *Unsupervised Representation Learning with Deep Convolutional Generative Adversarial Networks*. CoRR. abs/1511.06434.

34. Jolliffe, I. T., (2002). *Principal Component Analysis* (2nd edn., xxix, p. 487) Springer series in statistics. New York: Springer.

35. Kuniaki, N., Hiroaki, A., Yuki, S., & Tetsuya, O., (2013). Multimodal integration learning of object manipulation behaviors using deep neural networks. *IEEE/RSJ International Conference on Intelligent Robots and Systems.*

36. Hinton, G. E., & Salakhutdinov, R. R., (2006). Reducing the dimensionality of data with neural networks. *Science, 313*(5786), 504.

37. Wang, M., Li, H. X., Chen, X., & Chen, Y., (2014). Deep learning-based model reduction for distributed parameter systems. *IEEE Transactions on Systems, Man, and Cybernetics: Systems, 46*(12), 1664–1674, Dec. 2016, doi: 10.1109/TSMC.2016.2605159.

38. Ng, A., (2018). *Autoencoders; Unsupervised Feature Learning and Deep Learning (UFLDL) Tutorial*. [cited 20187/21/2018]; Available from: http://ufldl.stanford.edu/tutorial/unsupervised/Autoencoders (accessed on 19 November 2021).

39. Teh, Y. W., & Hinton, G. E., (2001). *Rate-coded Restricted Boltzmann Machines for Face Recognition*, 908–914.

40. Hinton, G. E., (2012). A practical guide to training restricted Boltzmann machines. In: Montavon, G., Orr, G. B., & Müller, K. R., (eds.), *Neural Networks: Tricks of the Trade* (2nd edn., pp. 599–619). Springer Berlin Heidelberg.

41. Hochreiter, S., & Schmidhuber, J., (1997). *Long Short Term Memory, 9*, 1735–80.

42. Metz, C., (2016). *Apple is Bringing the AI Revolution to Your Phone*. In Wired.

43. Gers, F. A., Schmidhuber, J., & Cummins, F. A., (2000). Learning to forget: Continual prediction with LSTM. *Neural Computation. 12*(10), 2451–2471.

44. Junyoung, C., Caglar, G., KyungHyun, C., & Yoshua, B., (2014). *Empirical Evaluation of Gated Recurrent Neural Networks on Sequence Modeling*. eprint arXiv:1412.3555.

45. Kyunghyun, C., Bart, V. M., Caglar, G., Dzmitry, B., Fethi, B., Holger, S., & Yoshua, B., (2014). *Learning Phrase Representations using RNN Encoder-Decoder for Statistical Machine Translation.* eprint arXiv:1406.1078,

46. Naul, B., Bloom, J. S., Pérez, F., & Van, D. W. S., (2018). A recurrent neural network for classification of unevenly sampled variable stars. *Nature Astronomy, 2*(2), 151–155.

47. Maryam, M N., Flavio, V., Taghi, M K., Naeem, S., Randall, W., & Edin, M., (2015). Deep learning applications and challenges in big data analytics. *Journal of Big Data.*

48. Goodfellow, I., Bengio, Y., & Courville, A., (2016). *Deep Learning - Adaptive Computation and Machine Learning* (Vol. xxii, p. 775). Cambridge, Massachusetts: The MIT Press.

49. Gavin, H. P., (2016). The levenberg-marquardt method for nonlinear least squares curve-fitting problems. http://people.duke.edu/~hpgavin/ce281/lm.pdf.

50. Xavier, G., & Yoshua, B., (2010). Understanding the difficulty of training deep feedforward neural networks. *Proceedings of the Thirteenth International Conference on Artificial Intelligence and Statistics, PMLR* (pp. 249–256).

51. Martens, J., (2010). Deep learning via Hessian-free optimization. *Proceedings of the 27th International Conference on International Conference on Machine Learning,* 735–742. Omnipress: Haifa, Israel.

52. Escalante, H. J., Montes, M., & Sucar, L. E., (2009). Particle swarm model selection. *J. Mach. Learn. Res.,* 405–440.

53. Shrestha, A., & Mahmood, A., (2016). Improving genetic algorithm with fine-tuned crossover and scaled architecture. *Journal of Mathematics.*

54. Sastry, K., Goldberg, D., & Kendall, G., (2005). Genetic algorithms. In E. K. Burke & G. Kendall (Eds.), Search methodologies: introductory tutorials in optimization and decision support techniques (pp. 97–125). Berlin: Springer. ISBN: 0387234608.

55. Goldberg, D. E., (2013). *The Design of Innovation: Lessons from and for Competent Genetic Algorithms.* Springer, Boston, MA.

56. Risto, M., Jason, L., Elliot, M., Aditya, R., Dan, F., Olivier, F., Bala, R., et al., (2017). *Evolving Deep Neural Networks.* CoRR, abs/1703.00548.

57. Duchi, J., Hazan, E., & Singer, Y., (2011). Adaptive sub-gradient methods for online learning and stochastic optimization. *J. Mach. Learn. Res., 12,* 2121–2159.

58. Kingma, D. P., & Ba, J., (2014). *Adam: A Method for Stochastic Optimization.* CoRR, abs/1412.6980,

59. Ioffe, S., & Szegedy, C., (2015). *Batch Normalization: Accelerating Deep Network Training by Reducing Internal Covariate Shift.* CoRR, abs/1502.03167.

60. Nitish, S., Geoffrey, H., Alex, K., Ilya, S., & Ruslan, S., (2014). Dropout: A simple way to prevent neural networks from overfitting. *J. Mach. Learn. Res., 15*(1), 1929–1958.

61. Huang, G. B., Zhu, Q. Y., & Siew, C. K., (2006). Extreme learning machine: Theory and applications. *Neurocomputing. 70*(1), 489–501.

62. Tang, J., Deng, C., & Huang, G. B., (2016). Extreme learning machine for multilayer perceptron. *IEEE Transactions on Neural Networks and Learning Systems, 27*(4), 809–821.

63. Liu, J., Gong, M., Miao, Q., Wang, X., & Li, H., (2015). A multi objective sparse feature learning model for deep neural networks. *IEEE Transactions on Neural Networks and Learning Systems, 26*(12), 3263–3277.

64. Siamak, M., Carlos, A., Raghvendra, M., Rocco, L., & Johan, A. K. S., (2015). Multiclass semi-supervised learning based upon kernel spectral clustering. *IEEE Transactions on Neural Networks and Learning Systems, 26*(4), 720–733.

65. Langone, R., (2015). *Kernel Spectral Clustering and Applications*. CoRR, abs/1505.00477.

66. Conneau, A., et al., (2016). *Very Deep Convolutional Networks for Natural Language Processing*. CoRR, 2016. abs/1606.01781.75.

67. Dong, W., & Zhou, M., (2017). A supervised learning and control method to improve particle swarm optimization algorithms. *IEEE Transactions on Systems, Man, and Cybernetics: Systems, 47*(7), 1135–1148.

68. Vapnik, V., & Izmailov, R., (2015). Learning using privileged information: Similarity control and knowledge transfer. *J. Mach. Learn. Res., 16*(1), 2023–2049.

69. Sampson, J. R., (1976). Adaptation in natural and artificial systems (John H. Holland). *SIAM Review, 18*(3), 529–530.

70. Krpan, N., & Jakobovic, D., (2012). Parallel neural network training with OpenCL. *Proceedings of the 35th International Convention MIPRO*.

PART IV

DRONES IN SMART CITIES

CHAPTER 10

Smart Cities and the Internet of Drones

MEHTAB ALAM,[1] AKSHAY CHAMOLI,[1] and NABEELA HASAN[2]

[1]Jamia Hamdard, New Delhi, India, E-mail: mahiealam@gmail.com (M. Alam)

[2]Jamia Millia Islamia, New Delhi, India

ABSTRACT

Unmanned aerial vehicles (UAVs) or drones are certainly going to be the future. They ploy in the air and keep track of the occurrences in their vicinities. They do not need any human to control them or to operate them. They gather data in form of images and sounds with various sensors, camera, and mics inbuilt in them, and further transmit them to the gateway where data can be processed and information can be taken out from it. In the near future they will be working as aerial and portable base stations (BS) for users as well as the government acting like gateways, collecting data from its own sensors as well as various other sensors deployed on roads, bridges, traffic lights, vehicles, buildings, and various other places. They will help us gather big data for analysis and in the same time making smart cities even smarter.

10.1 INTRODUCTION

Today, we are facing challenges regarding the natural resources and sustainability in human growth and development. Governments and cities are willing to invest in smart innovations and research to develop policies for improving quality of living and sustainability. Smart City concept is one of the most buzzed topics now a days. With easy access to wireless technology

The Internet of Drones: AI Applications for Smart Solutions. Arun Solanki, PhD, Sandhya Tarar, PhD, Simar Preet Singh, PhD & Akash Tayal, PhD (Eds.)

and internet the concept of smart city is being accepted worldwide. There a number of countries who have already deployed the concept fully in their cities. New York, London, Paris, Tokyo, Singapore is a few to name [1]. A number of researches have already been carried out regarding internet of things (IoT) in smart cities.

Integration of unmanned aerial vehicles (UAV) (drones) into smart cities would make them more secure and robust. Internet of drones (IoD) is a much raw topic. With IoT devices and sensors installed and working, the drones can connect and collect the data from these devices, which will lead to efficient and optimized way of data collection, aggregation, and offloading of the data. Drones can be remotely controlled and the data can be monitored in real-time. The data can be transmitted to the cloud and be monitored at a later time as well [2]. Programs can be written and implemented for various scenarios, which when met, can alert the person monitoring the area for any abnormalities or risks.

It is believed that technology and IoT is the backbone of all these smart cities [3]. From distributing the traffic and re-routing to ensure easy passage, to managing street lights and traffic lights, to implementation of different types of sensors and actuators to sense the environment and turn the lights on and off to conserve energy, to implementation and maintenance of low emission zones to reduce inner-city traffic and to reduce population and improve the air quality. Everything that happens inside a smart city requires communication system, data analysis and at times artificial intelligence (AI) to conserve resources and at the same time manage and improve the lives of the dwellers of the city.

In this chapter we would discuss the challenges faced by today's cities in fully accepting and deploying the much hyped and waited smart city with drones concept. We would try to deploy the smart city concept with the integration of the UAV (drone) in such a way that would be aiming for sustainable growth and development of the cities since sustainability is the need of the hour. These smart cities tend to invest in research and innovation of information and communication technology (ICT) and developing policies towards improving quality of life and sustainability.

This chapter is divided into seven sections. In Section 10.1, we give a brief introduction to our chapter. In Section 10.2, we have illustrated the smart city concept with a few current real-time implementations of the technology. Section 10.3 talks about the IoD concept, with in-depth explanation for "Drones," implementation of drones in the smart city paradigm and classification of these UAV's. In Section 10.4, we have come up with a few security issues and attacks possible on the IoD network. In Section 10.5, we portray some possible real-time implementations of the IoD paradigm and its benefit to the particular industry. In Section 10.6, some of the most common and basic challenges in IoD implementation have been discussed. The final Section 10.7 includes the conclusion and future scope of the study.

10.2   SMART CITIES

The population of the world is increasing at a rapid pace. More than 45% of human population stays in the urban areas, which is expected to increase to 70% in a decade or so [4]. Providing necessary services, facilities, and infrastructure to people in a sustainable way is very important. It is important to use the natural resources in a beneficial way, not in a way which leads them towards exhaustion. With the increase of population, increased the demand of services and requirements, which in turn lead the way for smart city project. It aims at organizing and simplifying the daily life of the user through things connected to the internet and to each other [5]. A smart city is a self-sustained and a self-contained town which utilizes sensors, smart devices, and wireless communication to gather big data, analyzes the data, and retransmits it to carry out the desired actions. In simple terms, it helps organize daily life and make it easier and faster.

Smart city use case includes smart transportation, smart economy, smart people, smart living, smart management in pursuit of improving the quality of life, and new and smart infrastructure, smart parking, smart water system, smart pollution monitoring system, smart roads, smart lights, smart healthcare, smart waste management system and many more. All these when implemented together will make the city smart. They will help in prudent management of natural resources through the involvement of the public and the government (Figure 10.1) [6].

10.2.1   SMART CITY REAL-TIME IMPLEMENTATIONS

There are a number of smart-city use cases which when implemented together give us a smart-city. We discuss a few in the next section.

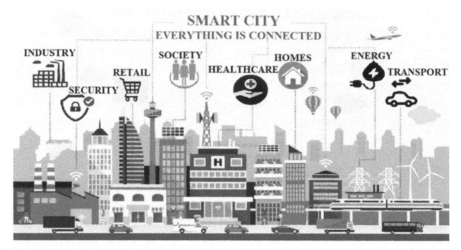

FIGURE 10.1 Smart city use cases.

10.2.1.1   HEALTHCARE

The most talked about smart city use case is in the healthcare sector. Smart wearable devices are the immediate future. Currently patients can interact with doctors or physicians via visits, tele, and text communications. No ways were there for doctors to monitor patients' health continuously and make recommendations correctly. Smart wearable devices will make remote monitoring possible. Patients' data will constantly be monitored which will further help in keeping them safe and healthy, and at the same time empower doctors deliver superlative care. Remote monitoring helps in reducing hospital visits and stays. Health cost will also be reduced significantly. Patients, physicians, hospitals as well as insurance companies will benefit with the implementation (Figure 10.2) [7].

10.2.1.2   SMART TRANSPORTATION

In larger cities, transportation is a huge concern. Everyone needs some kind of transportation system every day. Everyone wants to reach their destination on time. Smart transportation would be the solution. Cameras, sensors, smart traffic lights, smart toll booths, smart vehicles, GPS devices are included in smart transportation. All these devices communicate with each other and the gateway for processing of the data. Vehicle to vehicle communication is also possible. It will help in shifting traffic from roads which are more congested and at the same time it will help avoid roads where accidents have occurred. Smart cameras will capture the details of vehicles jumping red lights of not following road safety rules. Smart sensors will also check the vehicles health and fitness and report prior to breakdown. Smart sensors in public vehicles will report correct location, estimated time arrival, and number of empty seats. All these systems will help to improve efficiency, reduce pollution, time wastage and mis happenings on the road (Figure 10.3) [8].

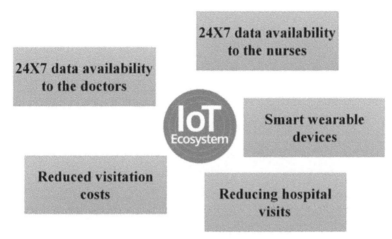

FIGURE 10.2   Smart healthcare.

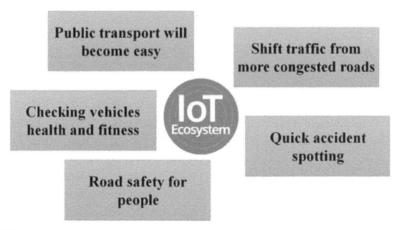

FIGURE 10.3   Smart transportation.

10.2.1.3 SMART PARKING

As urbanization increases, the traffic and the number of vehicles also increases. With this increase, rises the need of a parking system, a system which could save time, fuel, efforts, and most importantly, space. Sensor modules, low power lights are deployed around the space. These devices send data about the free space if available. If a vehicle enters the parking area, the app searches for an empty spot and the lights turn on directing the driver towards the empty space. Once there the driver can park the car and leave. There would be a device to capture the vehicles details including the time of entrance which would be required for billing purpose. The model would definitely save time and fuel for the rider (Figure 10.4) [9].

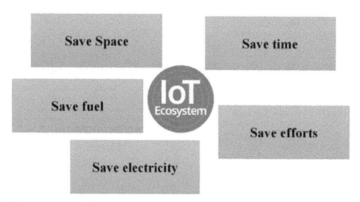

FIGURE 10.4   Smart parking.

10.2.1.4 SMART LIGHTING

Saving energy is the first step towards sustainability. The more energy we conserve, the better. Smart lights have the ability to change color, increase or decrease brightness and when deployed with smart sensors and cameras can automatically switch on and off themselves according to the needs (Figure 10.5).

For example, street lights can automatically switch off themselves in the morning. The lights on a less used street can have motion sensors, which turn of the light when someone is near or passes by, rest of time it is in sleep mode or less brightness, saving energy. Smart light in industries, offices, rail stations, airports will definitely help reduce electricity cost. The data collected by sensors can be used to infer patterns, which will help us identify areas or time when energy usage is maximum [10].

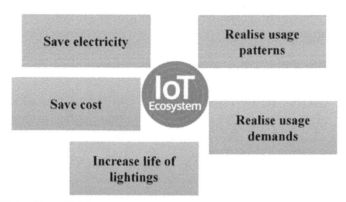

FIGURE 10.5   Smart lighting.

10.2.1.5   *SMART POLLUTION MONITORING SYSTEM*

With growing industrialization and vehicles, increases the pollution in air, water, and soil. Smart sensors are installed in industrial chimneys, vehicle exhausts, roadways, canals, etc., which sense and transmit data to the gateway for further processing. The sensors in the chimneys would sense the nature of the vapors and report dispensaries if any. If the industrial standards are not maintained in the waste matter, necessary actions can be taken. If the vehicles produce more exhaust waste, the owner should get a warning and if it is not corrected, they should be punished. Sensors at the roadways would constantly monitor the air quality and necessary steps should be taken, such as plantation of more green areas near more polluted regions. Sensors would provide enough data which will help to curb and reduce soil and water pollution (Figure 10.6) [11].

10.3   INTERNET OF DRONES (IoD)

Internet of drones (IoD) can easily be stated as the subset of IoT. IoT is its backbone, it provides controlled and coordinated access to supervised airspace for UAV [12], also known as drones. With the continued decrease in the size of the sensors, processors, camera units, digital memory and ubiquitous wireless connectivity, these drones are rapidly finding a number of new and productive uses in enhancing the way of living. Some of its uses can be military surveillance, disaster surveillance, package delivery, traffic, and wildlife surveillance, agriculture, inspection of infrastructures, bridges, tunnels, in cinematography, etc. Drones need proper navigation system and airspace management.

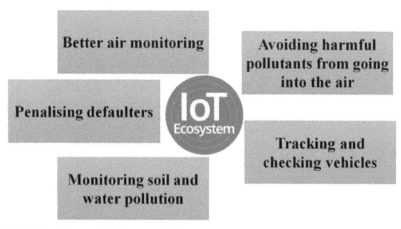

FIGURE 10.6   Smart pollution monitoring.

To make IoD a reality, drone simulations have to be performed which would make sure that they are safe and reliable. For widespread use and acceptance, they should come at a low cost and low maintenance. The software and/or apps controlling them should be certified for safety, following effective communication security protocols, should have an efficient power management system. They should be built with the ability to operate in challenging environments [13].

As per the reports by ResearchandMarkets.com, in 2016, the UAV market had reached $8.02 billion cap which increased to $25.59 billion in 2018. It is estimated to grow at 8.45% during the forecast period, 2019–2029 [14]. According to Business Intelligence, the sale drones was standing at $8 billion in 2016 which is all set to cross $12 billion in 2021 [15]. The growth is estimated to occur in three segments of the industry (Figure 10.7):

- Consumer Drone shipments will surpass 29 million by 2021 [15];
- Commercial Drone will surpass 8,05,000 by 2021 [15]; and
- Government Drone budget was allocated $2.9 billion in 2016 which is forecasted to increase by 10% every year [15].

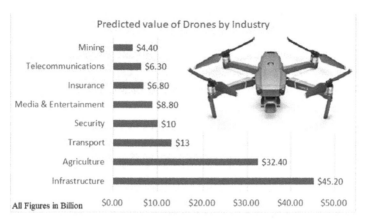

FIGURE 10.7   Growth of drone industry.

Figure 10.7 gives a broad representation of drone adoption in various industries.

In Figure 10.8, we have tried to demonstrate and extend the IoT-smart city use cases explained in Figure 10.1 by adding the drone layer. The drones add an extra layer for data processing and reducing the load on the cloud server [16]. They enable faster transmission of data and at the same time reduce the load on the gateways. They can also work as data producers as well as data consumers.

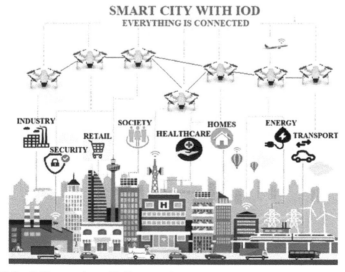

FIGURE 10.8   IoD extension of IoT-smart city.

10.3.1 DRONES

Drones were developed in the 20th century, and since then they have had a lot of attention. They are flexible and have dynamic capabilities to carry out real-time operations so they can be implemented in almost any scenario. Since their first production, huge technological advancements have been made. Some of them are, reduced weight, better battery performance and time of flight, a host of a large number of sensors and receivers, seamless integration of cameras and microphones [17]. The most important and blessed feature recently integrated in the drones is the integration of AI, machine learning (ML) and deep learning (DL) techniques which helps in independent operation [18].

The main components of an UAV are-controller, communication system, sensors, camera, a power source (battery) and mechanical components which help in flying. Figure 10.9 illustrates the components of a basic drone aircraft.

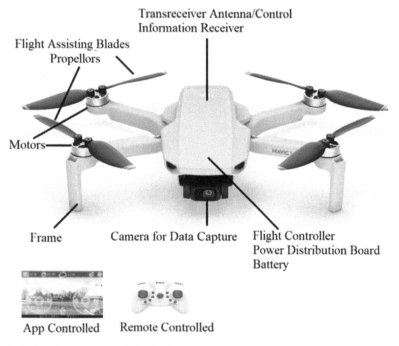

A Basic Drone Components

Transreceiver Antenna/Control Information Receiver

Flight Assisting Blades Propellors

Motors

Frame   Camera for Data Capture   Flight Controller Power Distribution Board Battery

App Controlled   Remote Controlled

FIGURE 10.9   Components of a basic drone.

Next, we discuss four basic steps required in the deployment of IoD:

➢ **Step 1:** This step consists of installation and deployment of the smart devices like sensors, detectors, actuators, camera systems, monitors, and deciding the covering area of a drone. These devices are basically data generators/producers, but can be made to work as data consumers as well.

➢ **Step 2:** Data generated by these producers is in analog forms, which is required to be transformed to digital form, which can then be used for further processing or safe keeping.

➢ **Step 3:** Once the data is converted to digital forms, the data is sent to the next node in the network. Some of these devices would connect to the drones flying over them which would work as access point or gateway for them. These drones would then either process the data on their own or if required send it forward for further processing [19].

➢ **Step 4:** Either the drones would analyze the data, apply some analytics, and would reply to these devices with the inferences or insights for performing some tasks. It will also be able to share data among other drones and other aircrafts nearby [20].

Today, drones are classified as MAV (micro or miniature air vehicles), NAV (nano air vehicles), VTOL (vertical take-off and landing), LASE (low altitude, short endurance), LASE close, LALE (low altitude, long endurance), MALE (medium altitude, long endurance) and HALE (high altitude, long endurance) [21].

10.3.2   RELEVANT NETWORKS

The architecture of IoD involves three different network structures. In this section we will discuss them briefly.

10.3.2.1   AIR TRAFFIC

Since the drones are devices which fly and use the air space, it is very important to keep the airspace managed and clean. Before deploying the drones, it is very important to understand and learn the routes and timings of the currently active aircrafts. Once this is done, we will have to carefully organize the airspace on the basis of number of drones available, active time capability of each drone, the area of surveillance/service falling under each drone and the number of sensors or IoT devices present in the covered area. Each drone has a limited flying time, and a limited range, therefore, if the area is very large or a large number of sensors and devices are present, me might need more number or drones. We will require a larger number of drones to sweep a small area in urban regions and lesser number of drones to sweep the same or even bigger area in rural regions [22].

10.3.2.2   CELLULAR NETWORK

With air space sorted out, next is the cellular networks. There are a number of cellular towers in a town. Cellular networks have hexagonal coverage areas. They use wireless communications standards for sending and receiving data. They use different frequency channels to give service to different users. In an urban area, as stated earlier, there will be a large number of wireless devices requesting access. The number of towers will increase; therefore, they will be closely situated. When drones are introduced in the system, we have to keep a check on the available and restricted frequency spectrum. For the most basic implementation, a single drone might be capable of covering a few hexagons, but in case the number of IoT devices increases we might need to increase the number of service drones. With all this increase in towers, mobile devices, IoT devices and the drones, it is a tough task to decrease interference. We need to deal with interference and see to it that there are sufficient resources at the drones, the towers, and the gateways to sustain the demand [23].

10.3.2.3   INTERNET

As the technology is getting cheaper, the number of smart devices getting connected to internet is increasing. With increase in the number of devices, increases the data usage and data requirements. Devices are connected to the internet via wires or wirelessly. The wired connections are of no interests to us. The wireless ones use gateways, routers, and drones. The drones work as wireless access points for these devices. The IoT devices can directly connect to the drones and get can transmit their data to them, which will be processed at the drones or at the upper levels or the data centers. The areas with a huge number of devices can use the services of the drones to make the system congestion free and to share the load of the gateways [23].

10.3.3 SOME TERMINOLOGIES IN IoD (TABLES 10.1 AND 10.2)

TABLE 10.1   Classifications of UAVs as per Department of Defense USA [24]

Category	Type of Size	Max Take-Off Weight	Altitude of Operating-ft	Airspeed (Knots)
Category 1	Small	0–20	<12,000	<100
Category 2	Medium	21–55	<3,500	<250
Category 3	Large	<1,320	<18,000	<250
Category 4	Larger	>1,320	<18,000	Variable
Category 5	Largest	>1,320	<18,000	Variable

TABLE 10.2   Classification of UAV's as per Technical Characteristics [24]

UAV Type	Technical Attributes
Very small	Size: 30–50 cm, Rotary Wings-Insect size, Lightweight, Range: 5 km, Flight time: 20–45 mins
Small	Size: 50 cm to 2 m, Fixed wings, Range: 20 to 400 m, Flight-10 to 50 m/s
Medium	Wingspan: 5 to 10 m, Payload: 100 to 200 kg, Altitude 10,000 to 30,000 ft, Flight time: 24–48 hours. Range: 100–700 Nautical Miles
Large	Wingspan: 35–55 ft, Payload: 1,000 kg, Range: 1,000–1,500 km.

10.4   SECURITY ISSUES

Security is main concern in wireless network. The IoD gathers data from a large number of actuators and sensors. These data can be sensitive sometimes. This data can be intercepted by intruders before it reaches the drones hovering above. The keys can be compromised and this could lead to privacy breach. If any how the intruders are able to alter the access control information, then the configuration of the IoD can be altered with ease. This alteration, even if nominal, can produce a lot of abnormalities in IoD. In extreme situations, a node can be compromised and the services with that particular node can be can be altered or turned off, which might lead to multiple task failures [25].

10.4.1   ATTACKS ON IoD

The attacks on IoD can be classified in five broad areas. We have displayed them in pictorial form.

10.4.1.1   PRIVACY

Privacy is one of the most important concern on internet today. The data is accumulated by sensors and sent to the IoD for processing. The attackers can target IoD and steal the data [26]. In traffic analysis, the attacker examines the traffic/ packets. These packets contain various information like location, type of sensor, mac address and the data captured by the sensors. In Interception, the attacker intercepts the data going and coming from the IoD [27]. Only via traffic analysis, large amount of sensitive data can be collected. Encrypted data can also be collected and if decrypted, the whole network can be compromised. Location of the sensor transmitting the data can also the found.

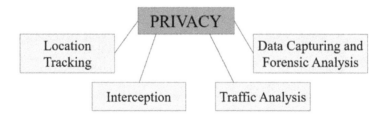

10.4.1.2   INTEGRITY

Integrity means that the data should be in accurate state and be consistent. The transfer of data should not be altered between the path from the sender to receiver [28]. Substitution/ altercation of information means adding wrong/ false/ or new information in the original message. Access control modification is the modification of ground rules of the IoD network. Man-in-the-middle means an attacker is sitting in between the IoT sensor and the IoD. All the data coming from the sensor can be viewed/altered by the attacker and all the data coming from the IoD to the sensor can also be seen and altered. Message forgery means the login request messages, and other sensitive data are forged from the previous attempts to connect to the IoD.

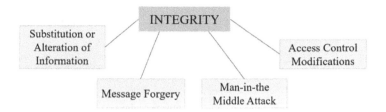

10.4.1.3 CONFIDENTIALITY

Confidentiality means the data will not be leaked to non-legitimate users [29]. The attacker masquerades as a legitimate user and gains access to the IoD network in identity spoofing. Unauthorized access happens when some unauthorized user gains access to the IoD network with some other users' credentials. In replay attacks, the attacker sniffs some data from the IoD network and tries to bypass the security by sending requests again and again to the IoD server. Eavesdropping means unauthorized interception of the current IoD communication.

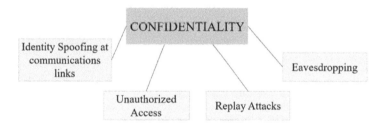

10.4.1.4 AVAILABILITY

Availability means the services of IoD is available as and when need, or always available [30]. Physical attacks are on the hardware components of the network. DoS and DDoS attack is denial of the legitimate users from using the services of the network. The attacker floods the server to create such attacks. In GPS spoofing means the attacker manipulate the GPS data of the device. GPS data is one of the key components of the drones. In channel jamming the attacker disrupts the communication channel. Routing attacks consist of flooding of the server, node isolation, etc.

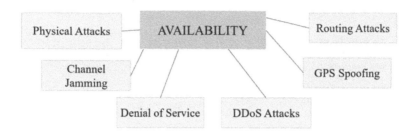

10.4.1.5   TRUST

Trust means the confidence of the user on the service provider that their data is safe and private. In IoD trust is between the IoD developing and deploying organizations [31]. Using of false IoD means replacing the legitimate IoD components with fabricated or forged components. Key loggers are used to steal the keyboard data that the user types or sends over the network. Third party violations include false IoD certification. In firmware replacement attack the attacker exploits the device when the device is sent for repair or maintenance purpose and then ships back to the user in tempered condition with the idea to steal data.

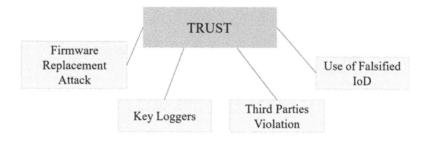

10.5   IoD REALTIME IMPLEMENTATIONS

In this section we discuss some of the real-time IoD implementations.

10.5.1   CONSTRUCTION AND INFRASTRUCTURE

One of the most widely use of IoD can be in construction planning and management. Software's have been developed which analyze construction progress, quality of the materials used, elapsed time, etc. And when dispensaries occur, or the timeline is missed, an alert with the details is sent to the authorities.

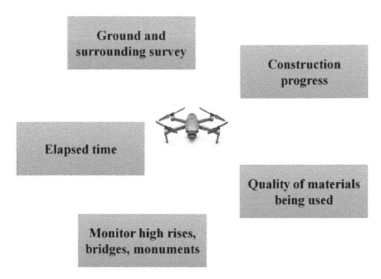

Ground and surrounding surveying is also a critical part which is performed by the drones. Flying cameras are used monitor buildings, high rises, bridges, monuments for any failures. They can alert the authorities for potential failures well in advance which can help in averting disasters and saving lives and money [32].

10.5.2 EMERGENCY/DISASTER RELIEF

Drones can be very helpful in case of disasters. Floods, hurricanes, landslides, earthquakes, etc., disconnect the towns from the rest of the world. Drones can be used to access the intent of damage caused, locate victims and most of all deliver essentials to them. Food, clothing, medicines, etc., can be delivered to the victims or people forced to be in adverse environment conditions. Now a days, due to COVID 19 breakdown, movement of people is very difficult. Drones can be very helpful in delivering essentials such as food items and medicines. They can be used to monitor areas and check whether proper lockdown and quarantine is being observed. They can also be used to spray disinfectants city wide [33].

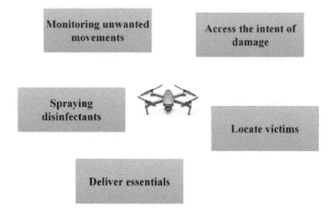

10.5.3 SMART CITY MONITORING

Transformation of legacy cities to smart cities was due to the widespread acceptance of IoT. But now, it is time for IoD. Drones are the future of smart cities. Drones will make the smart cities more secure and safe with crime detection. Search and rescue, tracking or vehicles, people, or other objects as and when required. Stationed cameras may not be able to capture the full view and details, so the flying drones will be more robust and secure. With drones flying above, every corner of the city will be covered, which will make monitoring very easy. It will also help in monitoring the traffic on the streets, report accidents. Drones can help in reading electric meters, water meters, gas meters without any human involvement. They can also be used to monitor the air quality of the city. They can check for unauthorized construction, gatherings, or other activities [34].

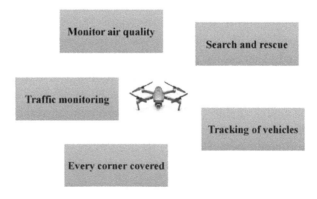

10.5.4   SMART AGRICULTURE

Use of IoD in agriculture is still a challenge due to limited or no connectivity in rural areas. Drones will help the farmers to gather big data from their farms, automate some processes, and improve efficiency by good decision making. Drones can be used to monitor the farm; animal intrusion can be detected well in advance. Cattle tracking can be done efficiently. Cattle health can behavior can be managed efficiently. Sensors deployed in the soil and around the filed can send data to the drones for faster and real-time monitoring and decision making. Drones can check the health of the crops and tress on regular basis. Drones can be used in aerial dispersal of seeds and spraying insecticides and pesticides [35].

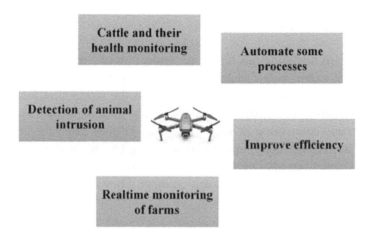

10.5.5   WILDLIFE CONSERVATION

The changing climate and poachers are the two main reasons that impact the health of wildlife throughout the world. As per World Wildlife Fund, on an estimate, thousands of species go extinct each year. To help overcome this problem, conservationists are looking for different and innovative methods to better study and protect different ecosystems, wildlife, and their habitats. Drones can easily be used to monitor and track animals in their natural habitats and environments without creating any disturbance and alarm to them. Drones can be used to follow endangered species and transmit all the information about them to the caretakers and researchers. Drones will also help combat poachers and recognize and punish them if needed [36].

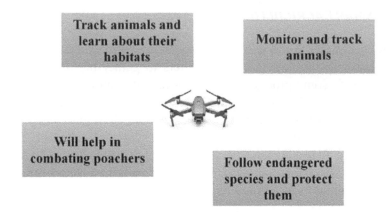

10.5.6   HEALTHCARE

Today's modern medicine has had huge impact on preventing diseases and increasing life expectancy of humans, and raising the living standards. But there are a large number of rural areas around the world that lack access to the high quality of healthcare, mainly due to lack of infrastructure like hospitals, roads connecting them to the cities. Drones can very easily and efficiently deliver medical supplies to such regions without any delay or problems. Drones can also be used to deliver blood, small to medium sized equipment's and medicines in a timely manner. Further they can be used to carry samples of the patients to the hospitals for performing tests [37].

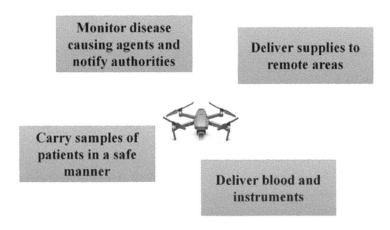

10.5.7   WEATHER FORECASTING

At present the task of data collection for weather forecasting is carried through stationery structures or the geospatial imaging satellites. Scientists are regularly working on developing better hardware and software solutions for data collection and prediction of the global weather. Drones can prove to be handy and useful for the purpose. They offer diverse options for climate data collection. They can travel to remote regions, or in deep sea regions and collect data useful for weather prediction. They can also be programed to physically follow different weather patterns as they develop and move. Drones can also be used to collect oceanic and atmospheric data from above the surface of the ocean [38].

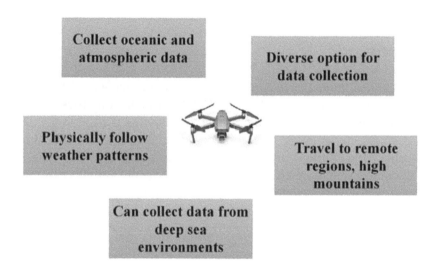

10.5.8   ENERGY

Alternate energy resources are becoming more and more popular, but the fossil fuels still remain a key energy source for the world. Regular examination and observation of the infrastructure used to extract, refine and transport the oil and gas is a very important part of the industry and is also needed to ensure compliance with regulations and standards.

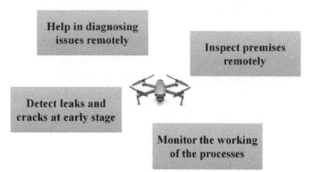

Drones can be easily used to inspect and monitor the premises remotely. When equipped with different types of sensors they can detect various leaks and crack faster. With high resolution camera some issues will can be diagnosed remotely and efficiently [39].

10.5.9   MINING

Mining is a department which requires regular measurements and assessment of the physical materials around the site. Substances like ore, rock, minerals are stored in the form of stockpiles which needs to be measured for radiations, quantity, and amount. With high quality and unique cameras, drones are capable of capturing volumetric data of the stockpiles. They can also serve in mining operations from the air itself. This exponentially reduces the risk connected having people to survey the stockpiles, on the ground. They can be used to boost the security of the area as well [40].

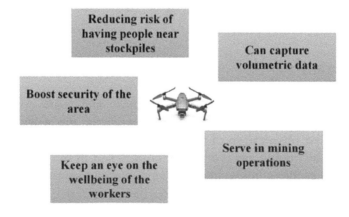

10.5.10 INSURANCE

Inspection is a major part of an insurance claim. Drones can used extensively for the purpose. Insurance companies can use drones to scale and observe structures and vehicles insured by them. Drones can provide detailed assessments with high quality cameras, which can be used to accept or reject the claim without involvement of people on road.

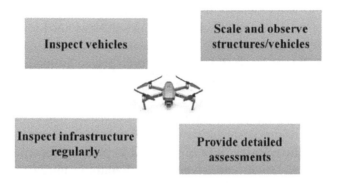

Drones can also be used to inspect infrastructure on a regular basis for defects and damages. Same in case of vehicles, the drones can be used to inspect vehicles once in a while to understand the patterns of damages and risks which would help insurance companies decide whether to provide insurance to users or not [41].

10.6 CHALLENGES OF IoD

IoD is currently counted into an emerging technology. One of the biggest challenges is 24×7 connectivity to the internet for real-time monitoring. There are a number of security threats are discussed earlier. Next, we discuss some of the challenges:

1. **Low Overhead and Dynamic Load Balancing:** The memory, storage, energy, and delays are still falling short when compared to the amount of data being generated and transmitted which leads to resource degradation and low efficiency. Therefore, supporting tasks which consist of low overheads is still a challenge to work upon. Dynamic load balancing, if implemented in the right way can be considered as an effective solution.

2. **Survivability and Lifetime:** Drones are battery operated, so they should be back to the station before the battery runs out. Survivability is overdone by unwanted signals and starvation conditions which lead to exhaustion of resources. Improving the resources without increasing the weight and size of the drones requires attention from the research communities [42].

3. **Low Cost of Functioning and High Throughput:** Cost is also one of the biggest factors in deployment and acceptance of IoD. Drones are sold at a relatively higher price range then the acceptance rate. Therefore, low, and effective costing solutions need to be found for enhancing and increasing the deployment of these drones.

4. **Performance and Reliability:** Performance is measured on the basis of efficiency with which the job is done and the efficiency of resource utilization. We need to discover more efficient and reliable methods to deploy and run the IoD in more crowded environments. Optimal resource utilization will lead to better performance [43].

5. **Low Failure Rates:** IoD is designed to perform specific tasks. But the current failure rate is very high. Failure in internet connection, battery, cameras, exhaustion of resources are a few to mention. Failure degrades the overall performance of the IoD. Steps need to be found to overcome these types of failure and make the system more robust [44].

6. **IoD Needs Better Security Options:** Recent advancements in IoD are on data rate and security of data. New methods need to implemented to deal with various attacks possible on the drones and the network [45].

10.7 CONCLUSION AND FUTURE SCOPE

In the foreseen future, IoT is going to transform everything with drones. In this chapter we tried to put forward the concept and the new terminology of IoD. IoD is the combination of IoT, which include, the cloud, big data, smart sensors, actuators, etc. IoD has capabilities to perform extreme tasks if and when deployed correctly. We have a complete network of a number of drones connected to different sensors which are data producers, drones are connected to one another as well for data sharing and transfer. Still a lot of work has to be done in the field, as there are a number of vulnerabilities and issues in the system. Cost, reliability, and maintenance being at the top of the list.

KEYWORDS

- **big data**
- **drones**
- **information and communication technology**
- **sensors**
- **smart city**
- **sustainability**
- **unmanned aerial vehicles**

REFERENCES

1. Lee, R. S. T., (2020). Smart city. In: *Artificial Intelligence in Daily Life* (pp. 321–345). Springer Nature.
2. Caillouet, C., Giroire, F., & Razafindralambo, T., (2019). Efficient data collection and tracking with flying drones. *Ad Hoc Networks, 89,* 35–46.
3. Mohanty, S. P., Choppali, U., & Kougianos, E., (2016). *Everything You Wanted to Know About Smart Cities: The Internet of Things is the Backbone, 5*(3), 60–70. IEEE Consumer Electronics Magazine.
4. Alam, M., Khan, I. R., & Tanweer, S., (2020). *IoT in Smart Cities: A Survey, 89–101. Juni Khyat.*
5. Al-Hader, M., Rodzi, A., Sharif, A. R., & Ahmad, N., (2009). SOA of Smart City Geospatial Management. In: *Computer Modeling and Simulation, EMS '09. Third UKSim European Symposium* (pp. 25–27).
6. Schaffers, H., Komninos, N., Pallot, M., Trousse, B., Nilsson, M., & Oliveira, A., (2011). Smart cities and the future internet: Towards cooperation frameworks for open innovation. *The Future Internet,* pp. 431–446.
7. Hui, K., (2016). A Secure IoT-based healthcare system with body sensor networks. *IEEE Access,* 10288–10299.
8. Rathore, M. M., Ahmad, A., Paul, A., & Jeon, G., (2015). Efficient graph-oriented smart transportation using internet of things generated big data. In: *11th International Conference on Signal-Image Technology & Internet-Based Systems (SITIS)*. Bankok.
9. Khanna, A., & Anand, R., (2016). IoT based smart parking system. In: *International Conference on Internet of Things and Applications (IOTA)*. Pune.
10. Martirano, L., (2011). A smart lighting control to save energy. In: *Proceedings of the 6th IEEE International Conference on Intelligent Data Acquisition and Advanced Computing Systems*. Prahue.

11. Parmar, G., Lakhani, S., & Chattopadhyay, M. K., (2017). An IoT based low cost air pollution monitoring system. In: *International Conference on Recent Innovations in Signal processing and Embedded Systems (RISE)*. Bhopal.

12. Cavoukian, A., (2012). Privacy and drones: Unmanned aerial vehicles. *Information and Privacy Commissioner of Ontario*, 1–30. Ontario, Canada.

13. Crowe, W., Davis, K. D., Cour-Harbo, A. L., Vihma, T., Lesenkov, S., Eppi, R., Weatherhead, E. C., et al., (2012). Enabling science use of unmanned aircraft systems for arctic environmental monitoring. *Arctic Monitoring and Assessment Program*, 1–30.

14. *Global Unmanned Aerial Vehicle (UAV) Market: Focus on VLOS and BVLOS UAVs using Satellite Communications – Analysis and Forecast, 2019–2029*. Published on 2020. [Online]. Available: https://www.researchandmarkets.com/reports/4773748/global-unmanned-aerial-vehicle-uav-market (accessed on 19 November 2021).

15. Insider Intelligence. (2020). *Commercial Unmanned Aerial Vehicle (UAV) Market Analysis-Industry trends, Forecasts and Companies*. [Online]. Available: https://www.businessinsider.com/commercial-uav-market-analysis?IR=T (accessed on 19 November 2021).

16. Khan, S., Ali, S. A., Hasan, N., Shakil, K. A., & Alam, M., (2019). Big data scientific workflows in the cloud: Challenges and future prospects. In: *Cloud Computing for Geospatial Big Data Analytics. Studies in Big Data* (pp. 1–28). Springer International Publishing.

17. *What is a Drone: Main Features & Applications of Today's Drones*. [Online]. Available: https://www.mydronelab.com/blog/what-is-a-drone.html (accessed on 19 November 2021).

18. Colomina, L., & Molina, P., (2014). Unmanned aerial systems for photogrammetry and remote sensing: A review. *ISPRS Journal of Photogrammetry and Remote Sensing*, 79–97.

19. Guillen-Perez, A., Sanchez-Iborra, R., Cano, M. C., Sanchez-Aarnoutse, J. C., & Garcia-Haro, J., (2016). WiFi networks on drones. *ITU Kaleidoscope: ICTs for a Sustainable World (ITU WT)*.

20. Shan, L., Miura, R., Kagawa, T., Ono, F., Li, H. B., & Kojima, F., (2019). Machine learning-based field data analysis and modeling for drone communications. *IEEE Access*, 79127–79135.

21. Watts, A. C., Ambrosia, V. G., & Hinkley, E. A., (2012). Unmanned aircraft systems in remote sensing and scientific research: Classification and considerations of use. *Remote Sensing*, 1671–1692.

22. Foina, A. G., Sengupta, R., Lerchi, P., Liu, Z., & Krainer, C., (2015). Drones in smart cities: Overcoming barriers through air traffic control research. In: *Workshop on Research, Education and Development of Unmanned Aerial Systems (RED-UAS)*. Cancun.

23. Sekander, S., Tabassun, H., & Hossain, E., (2018). *Multi-Tier Drone Architecture for 5G/B5G Cellular Networks: Challenges, Trends, and Prospects*, 96–103. IEEE Communications Magazine.

24. Nayyar, A., Nguyen, B. L., & Nguyen, N. G., (2019). The internet of drone things (IoDT): Future envision of smart drones. In: *First International Conference on Sustainable Technologies for Computational Intelligence*.

25. Choudhary, G., Sharma, V., Gupta, T., Kim, J., & You, I., (2018). Internet of drones (IoD): Threats, vulnerability, and security perspectives. In: *The 3rd International Symposium on Mobile Internet Security*. Cebu, Phillippines.

26. Pauner, C., Kamara, V., & Viguri, J., (2015). Drones. current challenges and standardization solutions in the field of privacy and data protection. In: *ITU Kaleidoscope: Trust in the Information Society*. Barcelona, Spain.

27. Nabeela, H., Chamoli, A., & Alam, M., (2020). Privacy challenges and their solutions in IoT. In: *Internet of Things (IoT)*. Springer, Cham.

28. Hartmann, K., & Steup, C., (2013). The vulnerability of UAVs to cyber attacks-an approach to the risk assessment. In: *5th IEEE International Conference Cyber Conflict (CyCon'13)*. Tallinn, Estonia.

29. Akram, R. N., Markantonakis, K., Mayes, K., Habachi, O., Sauveron, D., Steyven, A., & Chaumette, S., (2017). Security, privacy and safety evaluation of dynamic and static fleets of drones. In: *36th IEEE Digital Avionics Systems Conference (DASC'17)*. St. Petersburg.

30. Yampolskiy, M., Horvath, P., Koutsoukos, X. D., Xue, Y., & Sztipanovits, J., (2013). Taxonomy for description of cross-domain attacks on cps. In: *2nd ACM International Conference on High Confidence Networked Systems*. Philadelphia.

31. Oleson, K. E., Hancock, P. A., Billings, D. R., & Schesser, C. D., (2011). Trust in unmanned aerial systems: A synthetic, distributed trust model. In: *International Symposium on Aviation Psyshology*. Dayton.

32. Li, Y., & Liu, C., (2018). Applications of multirotor drone technologies in construction management. *International Journal of Construction Management, 401–412*.

33. Alam, M., Parveen, R., & Khan, I. R., (2020). Role of information technology in COVID-19 prevention. *International Journal of Business Education and Management Studies, 65–75*.

34. Won, J., Seo, S. H., & Bertino, E., (2017). Certificateless cryptographic protocols for efficient drone-based smart city applications. *IEEE Access, 3721–3749*.

35. Puri, V., Nayyar, A., & Raja, L., (2017). Agriculture drones: A modern breakthrough in precision agriculture. *Journal of Statistics and Management Systems, 507–518*.

36. Jiménez, L. J., & Mulero-Pázmány, M., (2019). Drones for conservation in protected areas: Present and Future. *Drones, 3*(1), 10.

37. Amukele, T., (2019). Current state of drones in healthcare: Challenges and opportunities. *The Journal of Applied Laboratory Medicine, 4*(2), 296–298.

38. Leuenberger, D., Haefele, A., Omanovic, N., Fengler, M., Martucci, G., Calpini, B., Fuhrer, O., & Rossa, A., (2020). Improving high-impact numerical weather prediction with lidar and drone observations. *Bulletin of the American Meteorological Society, 101*(7), E1036–E1051.

39. *38 Ways Drones Will Impact Society: From Fighting War to Forecasting Weather, UAVs Change Everything*. C B Insights [Online]. Available: https://www.cbinsights.com/research/drone-impact-society-uav/ (accessed on 19 November 2021).

40. Shahmoradi, J., Talebi, E., Roghanchi, P., & Hassanalian, M., (2020). A comprehensive review of applications of drone technology in the mining industry. *Drones, 4*(3), 34.

41. Luciani, T. C., Distasio, B. A., Bungert, J., Sumner, M., & Bozzo, T. L., (2016). *Use of Drones to Assist with Insurance, Financial and Underwriting Related Activities*. USA Patent 14/843,455.

42. Long, T., Ozger, M., Cetinkaya, O., & Akan, O. B., (2018). *Energy Neutral Internet of Drones,* 22–28. IEEE Communications Magazine.
43. Li, J., & Han, Y., (2017). Optimal resource allocation for packet delay minimization in multi-layer UAV networks. *IEEE Communications Letters,* 580–583.
44. Sharma, V., Kumar, R., & Rana, P. S., (2015). Self-healing neural model for stabilization against failures over networked UAVs. *IEEE Communications Letters,* 2013–2016.
45. Kamthan, S., Singh, H., & Meitzler, T., (2017). UAVs: On development of fuzzy model for categorization of counter-measures during threat assessment. In: *Unmanned Systems Technology XIX.* Anaheim, California.

CHAPTER 11

The Internet of Drones for Enhancing Service Quality in Smart Cities

ADITI SAKALLE,[1] PRADEEP TOMAR,[1] HARSHIT BHARDWAJ,[1]
UTTAM SHARMA,[1] and ARPIT BHARDWAJ[2]

[1]Department of Computer Science and Engineering, University School of Information and Communication Technology, Gautam Buddha University, Greater Noida – 201310, Gautam Budh Nagar, Uttar Pradesh, India

[2]Department of Computer Science and Engineering, BML Munjal University, Gurugram, Haryana.

ABSTRACT

The concept of Smart cities is basically, the cities that contain a lot of intelligent things to enhance the quality of life. The technologies such as drones, robotics, artificial intelligence (AI) and IoT, are the need for a balanced and quick responding ecosystem in smart cities, by increasing their connectivity, energy efficiency and service quality (SQ). A Smart city requires automated intelligent systems which keeps every part of the city connected. Unmanned arial vehicles popularly known as drone along with the Internet can alter the traditional way of doing things. This internet of drone (IoD) can play an essential role in a variety of smart urban applications, for example; weather forecasting, healthcare, communications, transport, agriculture, safety, and protection, environmental reduction, service delivery, and e-disposal, etc. Internet of drones (IoD) can alter the idea of the surveillance because of its mobility and fundamentally change the method and viewpoint of data collection and input to governments, businesses, and people. This chapter

The Internet of Drones: AI Applications for Smart Solutions. Arun Solanki, PhD, Sandhya Tarar, PhD, Simar Preet Singh, PhD & Akash Tayal, PhD (Eds.)
© 2023 Apple Academic Press, Inc. Co-published with CRC Press (Taylor & Francis)

will discuss the application of the Internet of Drone (IoD) to make a Smart City.

11.1   INTRODUCTION

With the pressures of rapid growth and development faced by today's cities, more metropolitan centers continue to invest in information technology and development of methods for science and creativity to enhance the quality of life and sustainability. Smart cities have advanced technologies that can increase the quality of life, save lives and function as a balanced network of energy intelligently and collaboratively. New connectivity technology, such as drones, robots, artificial intelligence (AI), and the internet of things (IoT), is required for increased mobility, energy efficiency and quality of service (QoS) in an intelligent community [1].

The delivery of efficient infrastructure and a reduction in services costs while making such services and facilities ubiquitous is an important aspect of smart cities. Several recent studies have improved the characteristics of the smart cities [2, 3] and various aspects [4, 5] and focused on sustainability [6, 7], and were carried out in various areas and different aspects. The fact that smart services and ICT solutions are integrated effectively is a common feature of all these studies. In recent years, drone technology has made the city more intelligent. Today, a smart city without drone services is difficult to envisage [8].

In the past few year drones have been proving a very beneficial technology for building and sustaining a smart city. In recent years, drones have been used in many vertical applications such as agriculture, manufacturing, factory inspections and surveys. The improved sensing and communication capability coupled with declining prices fuels this huge popularity of drones. Modern drones are now available with a wide range of capabilities for wireless communication, 4k camera including Wi-Fi, Bluetooth, 4 G, RFID, etc., as well as the sensors for temperature, humidity, and contaminated air, and GPS. Lastly, drones are available in several form factors, which have also increased mechanical performance, such as airspeed, level of autonomy, stability, and maneuverability. Such innovations made drones a very appealing forum for launching numerous IoT applications.

IoT adoption has helped to bring in important improvements and implementation such as intelligent homes, intelligent motorways, intelligent parking, and intelligent grids, and so on, thanks to IoT. The key premise of

IoT is that anything, for example, computers, homes, home appliances, cars, and even natural objects, should be linked and shared through the internet to create an intelligent environment and a massive global network for enhancing service quality (SQ) in smart cities. Our lives, communities, and the climate rely on the creation of these applications.

To read values for air, water, and power meters from the sky in an unparalleled advantage, IoT operators may, for example, use the most sophisticated IoT Gateway techniques like LoRa [9] and 5G [10] Network Connectivity. A drone with the right sensors will capture data from hard-to-reach locations in a smart city from different air quality and move this data in real-time via LoRa [9] or even 4G to an IoT server. The researchers have already initiated ground-breaking ventures using drones as the primary tool for IoT sensing and data transmission. A big part of the IoT is drones (UAVs).

11.2 PROFESSIONAL DRONE

There are IoD systems which are already working as a professional drone worker for enhancing SQ in Smart Cities. Working professional IoD systems can sometimes be misunderstood with Drones for hobbies. Professional drones perform important tasks. Some of the applications with drones are as follows:

- Trade drones may carry out missions, ranging from entertainment, security, agriculture to the distribution of parcels.
- They have a vast variety of extra tools and functions to support their planned applications. These may cover:
  o Place information and data transmitting radios and GPS;
  o Video and still cameras to film objects and individuals, speeding criminals or action sports;
  o Sensors for the location of living organisms, environmental systems tracking and more for light, heat and temperature;
  o Capacity to convey and distribute payloads, including freight, emergency equipment and packets;
  o Fertilizer in agricultural use to spread seeds or fertilizer;
  o Recharge solar panels to increase battery life.
- For the above reasons, industrial design requirements for working drones are required:
  o They have a variety of environmental factors to deal with;
  o They also have more propellers, depending on their weight;

○   They are usually fitted with light, sensors, and other functions of industrial standard.
- They can be called to move items from one place to another.
- They will provide Wi-Fi to otherwise inaccessible areas, anywhere internet access is accessible on request.
- They will show us what goes on in situations that we may have to speculate or where we must control conditions in distant environments that are difficult to control. They are fitted with video cameras.
- They will help us monitor objects or individuals and alert us to changing atmospheric environments, fitted with RFID antennas or sensors.

Unlike other IoT gadgets on the Internet, drones have huge lithium batteries and are mounted in smaller ships, and they transmit a vast volume of data within a limited period and called the internet of drones (IoD). Drone systems require a secure high-bandwidth connection to provide real-time navigation, telemetry, and payload control. Also, drones are needed on-board to make a variety of crucial decisions. Researchers have also developed a tailored protocol to satisfy these unique drone specifications. This protocol is used to connect drones effectively to cloud-based business apps, i.e., internet accessing drones, in real-time.

11.3   DRONE-API

It is necessary to understand the actual effect of a command specified in the API of a drone [11]. The drone must be flown independently using computer algorithms in many of the IoT applications so that it can be maneuvered quickly and efficiently at the right place in the proper time. At any stage of the complicated flight, any small maneuverability errors can spread rapidly throughout the remainder of the flight and cause major performance problems for the application target. Because of its three-dimensional (3D) usability and reliability, multirotor drones have recently been used in a number of smart city applications and have been extensively employed as IoT-devices. Drones may be used together with IoT to perform multiple short and long-term tasks involving low altitude travel. The IoD era is expected to fuel more development in the sector. Drone technology has been rising.

The Drones Internet is an interface that provides organized access for unmanned aerial vehicles (UAVs), also referred to as drones, to protect air space, which is popularly known as IoD.

Features required in the API of Drone so that the IoD system can work properly are as follows:

- Access to real-time telemetry and payload data;
- Real-time control;
- Comprehensive RESTful APIs for quick connectivity and fast third-party integration;
- Contact with drones via a safe link connection;
- Authentication based on the token for exchanging information from drone to the third party;
- Suitable for dealing with flexible bandwidth, flexible duration, intermittent drone-cloud connections, like 4G/LTE, Wi-Fi, etc.

11.4   IoD APPLICATION INTO SMART CITY SOLUTIONS

11.4.1   INTELLIGENT TRAFFIC CONTROL IN SMART CITIES

Through a look in the sky, the people at the ground can help them handle traffic accurately and lead the source of the congestion. When a single UAV is deployed for several hours, it consistently looks at traffic nodes in greater detail. Intelligent transport administration is the gateway to intelligent cities.

The density of traffic is seen as one of the big challenges facing intelligent cities. That is when the number of cars rises abruptly for many reasons, including peak hours, renovation, major incidents, or injuries. It can happen anywhere in a smart city at any moment. To alleviate such conditions, effective, and efficient strategies and technology are therefore required. Static street cameras can provide some details, but not a comprehensive congestion reports [12]. The drone is a vital tool to capture and transmit congestion data in real-time. Several technologies have been used to interact and analyze data gathered by drones efficiently and safely, such as the following versions are available in [13, 14]. Traffic jams can also occur due to efforts to locate parking spots where parking spaces are small. This leads to cars entering the road at low-speed many times. D'Aloia et al. [15] have suggested drones to monitor the situation in certain scenarios like finding a Parking lot and parking spaces free. Collaborative IoD and IoT gadgets in cars also allow the user to easily locate vacant car parks. Figure 11.1 shows the applications of IoD into smart cities.

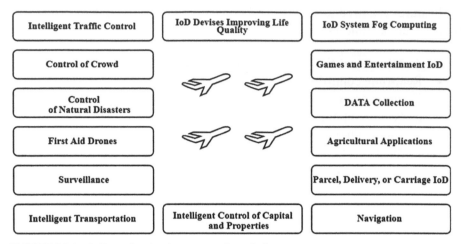

Intelligent Traffic Control	IoD Devises Improving Life Quality	IoD System Fog Computing
Control of Crowd		Games and Entertainment IoD
Control of Natural Disasters		DATA Collection
First Aid Drones		Agricultural Applications
Surveillance		Parcel, Delivery, or Carriage IoD
Intelligent Transportation	Intelligent Control of Capital and Properties	Navigation

FIGURE 11.1   IoD application into smart city solutions.

11.4.2 CONTROL OF CROWD IN SMART CITIES

Smart policing is made possible by using Internet drones to track crowds. IoD devise will aid police officers in security initiatives in smart cities and the protection of the law. With the quick deployment of drones, data gathering and support for police activities can be quickly deployed in diverse situations such as theft or road hits [16]. IoD strategies like human behavior identification, [17] identification of movements [18], etc., will also allow communities to become smarter and more effective, as opposed to the human workforce, by working with IoD devices.

11.4.3 CONTROL OF NATURAL DISASTERS

IoD can effectively assess life risks by actively tracking and enforcing emergency management initiatives. In addition to evaluating various circumstances to determine rescue needs especially in areas not reached by humans. In China, speaker drones are used to alert people that they are wearing masks to protect them from coronaviruses.

In smart cities, emergency systems play a very significant role in saving lives. Some emergencies exist in locations that are difficult to access, and delays in bringing them physically can impair response efficiency. IoD will also play a significant role in enhancing the response to emergencies in smart cities. Sensor-equipped drones will serve as null respondents in smart cities for victims in a very crowded environment. In this scenario, the drone will take the victim to the point, find a clear road, report to an emergencies team, and monitor the victim's status [19]. Terrorist attacks and disasters, both natural and manmade, are examples of disasters which affect intelligent cities and are extremely difficult to cope with. It is incredibly necessary to monitor, handle, and plan certain disasters. The earlier the solution enters the impacted region, the more cost savings will be minimized for the government, and the greater will be the possibility of losing more lives.

The air network was proposed as one technology that can provide broadband communications services like a high-altitude network and balloon tattered [20], while drones were lately proposed as an even more powerful and efficient system for disaster detection and prevention broadband communication [21]. Also, drones can be used for delivering instruments and medical supplies to the needy, because of their agility.

11.4.4   FIRST AID DRONES IN SMART CITIES

An IEEE Spectrum post recently covered the battle between an IoT based drone fitted with first aid and an ambulance. The robot was triumphant. The technology is designed to provide patients with a swift life-saving medical supply, just for example, if two people walk down a street and one collapse, a scenario will come into play. In cases like these, sudden cardiac arrest (SCA) is a potentially life-threatening cause. It relies on the availability of an appropriate AED or automatic defibrillator equipped to revive the heart. The capacity to preserve the victim life. Although this device is frequently kept in office buildings in first aid cabinets, it is not available elsewhere.

The article states, "Arriving 90–120 seconds faster than the ambulance in the busy city in Northern Iraq, the Drone Device was identified in the Sensor newsroom. Of course, it is also important to provide an ambulance and first responders; however, the arrival of life-saving equipment will allow passers-by during the emergency response to react rapidly.

11.4.5   SURVEILLANCE IN SMART CITIES

The surveillance data are important and vulnerable, and the drones can be useful, inter alia, to gather such data from inaccessible locations as sites, war grounds and mountain tops, etc. However, this data and the drones may be exploited with malicious reasons due to the underlying open net.

In Ref. [22] a new authentication scheme to shield unauthorized users from drone access has been investigated. For tracking and drone security, [22] an upgraded framework has been implemented using only light-weight hash and symmetrical encryption/decoding operations. Formal, informal, and automatic approaches illustrate the reliability of the proposed system. The model introduced is better suited for protecting the tracking data transmitted by drones, as many of the security functions are given and immune to certain established attacks. This requires less calculation time than other systems.

11.4.5.1 *FLYING POLICE EYE IN SMART CITIES*

Travel police are being more modern trained to allow safe road traffic. The most used tools for the enforcement of traffic laws remain speed and CCTV cameras. If the speed limit of a vehicle is surpassed, a static or handheld speed camera may detect it, and if the vehicle has a stop signal, the driver is detected by a nearby CCTV camera. The static cameras become less effective as drivers become acquainted with time.

It inspired the introduction of mobile cameras, which can be mounted at various locations (in some cases undisclosed places) to ambush drivers and change their driving actions when entering their area under the angle of these cameras. State-of-the-art technologies allow the versatility of speed cameras and the movement of them. This state-of-the-art handheld speed camera is typically installed in police cars on the way to identify cars that break the laws on traffic in the area around them. The same technologies can be built into a UAV, make a moving speed camera or other applications.

Completely integrated IoD systems which may handle any or duties of traffic police officers. For example, an IoD device will travel over a road to avoid a certain identification vehicle to normal car driving. A traffic light can be turned to red in front of the car by the IoD panel. The same IoD device will travel on a road and collect any speed or breach traffic regulations for vehicles. The problem here may be the restriction of a drone's top speed against a fast road car. The drone can overcome this restriction by allowing it to fly high above sea level to get an overview to compensate for the speed limit.

11.4.6 INTELLIGENT TRANSPORTATION IN SMART CITIES

Automation of vehicles is one step towards fully integrated transport networks, but some transport network members such as public service officers, public management and support services will also need to be integrated. The last miles to fully controlled end-to-end transport, IoD units may be used to allow systems. IoD can be used to draw up traffic routes that provide consistency and to allow surveyors to effectively map long corridors while projects are underway, and to gather detailed data to help earlier decisions. IoD, particularly for the last mile deliveries in congested areas, may also be used for transport and distribution. Application of IoD units for intelligent transportation system:

- Traffic monitoring;
- Highway maintenance;
- On-road accidental help;
- Path clearance and navigation;
- Speed monitoring on highway;
- License checking.

11.4.7 INTELLIGENT CONTROL OF CAPITAL AND PROPERTIES

IoD can help in strategic planning and maintenance in infrastructure along with monitoring of smart city technologies networks and facilities.

11.4.8 NAVIGATION IN SMART CITIES

IoD devices have been proved very helpful in building a successful navigation system for a smart city. But there are several challenges in developing a smart navigation system to increase the city's smartness. The IoD system may break barrier systems in city-wide performance of its duties, such as high buildings and trees. The deployment of drone fleets in specific geographical regions must also be effectively controlled.

The small battery power of these drones is one of the main obstacles in the management of drones. Indeed, to perform longer flights, drones would need to refuel their batteries again and again. Another difficulty is to ensure that fleet members travel safely and accident-free particularly as they fly low altitude in urban environments, for example in the field of site management applications [23]. Finally, if multi-tasks are to be conducted at the same time, an operation plan should be given for each IoD device outlining the precise timetable to be implemented to ensure maximum energy-efficient mission coverage and for collision-free navigation, and service. Various routing and trajectory modeling methods for maximizing drone operations have been developed. Neither has considered the three problems that have previously been discussed. A statistical approach was applied to Ref. [24] in public safety communications systems to find a quick way to optimize coverage of the drones deployed. This chapter does not answer the energy dilemma.

In Ref. [25] a problem for a fixed-wing IoD system used for safe photography applications was developed for routing and trajectory design. Few researched the issue of scheduling while operating a fleet of IoD, but it is always adapted to different realms or requirements because of the difficulty of the task. The sophistication of the problem ensures that the problems related to the individual program are generally discussed. In Ref. [26] the author has developed an iterative low-complexity algorithm that specifies the urban drone fleet's collide-free navigation and scheduling solution. The energy limit takes account of the IoD system and its capacity to charge their batteries at a docking station in the smart downtown. Optimal schedules and shortest routes for drones are suggested by algorithms. They are combined iteratively. Collide between drones is prevented by static hovering. Results demonstrate how many drones abandon their journeys for the gains of the whole fleet and their effect on device efficiency.

11.4.9   PARCEL, DELIVERY, OR CARRIAGE IoD SYSTEM IN SMART CITIES

For package distribution, Amazon plans to use IoD units. A partnership between CVS and UPS was also mentioned in a recent article to allow drones to supply drugs. A broad variety of cases of distribution use are being investigated and subject to regulatory approvals on business entry. Possibly, the distribution of pizzas and other fast foods will not be far away until drone delivery for deliveries is omnipresent. Of course, drones differ widely depending on their purpose and the payload they aim to carry. However, the "mean carrying power for technical drones is 20 to 220 kg," or about 44 to 485 libraries, is stated in an article in Drone Tech World.

11.4.10   AGRICULTURAL APPLICATIONS OF IoD UNITS

Imagine sending a drone to execute a field mission instead of a massive gas-eating rig with multiples. Not all agricultural activities can be done with a UAV, but reductions in time and cost are important in situations where IoD devices can do the work. Crop spraying will be a simple IoD devices method for planting, but there are more: For fertilization and pesticide management:

1.   **Crop Seeds:** Seeds are expected to become a commonplace procedure with the aid of nutrients in the long run as the drones can do this function efficiently and with excellent accuracy and can reduce the cost of planting crop planting by 85%.
2.   **Soil Analysis:** Drones will offer an early glimpse into the crop cycle, according to a technology review report, by creating detailed 3-D maps enabling farmers to schedule planting patterns for the crops, as well as fertilization and irrigation.

11.4.11   DATA COLLECTION IN SMART CITIES

Effective IoD devices may also be used in real-time to capture and distribute data. Traditional approaches typically require the exchange of data gathered by multi-hops and relays, which leads to delays and mistakes. In gathering and transmitting data from IoD devices to the main station, IoD devices have an important role to play [27]. The energy available for IoT systems can be reduced by collective drones by reducing the distance between the battery data to be transmitted (IoT system for the local drone, not IoT, to a remote base station (BS)) [28]. Several active data collection and cooperation techniques have been established between drones and IoT devices.

11.4.12   GAMES AND ENTERTAINMENT IoD UNITS

Sports and entertainment activities are perhaps one of the most popular applications for commercial drones today since they provide excellent aerial and video imagery [29, 30].

11.4.13 IoD SYSTEM FOG COMPUTING

Interestingly, the latest paradigm recently implemented in drone-based fog computing is one. The benefits of drone versatility based IoD fog computing [31], durability, and overall ease include weather prediction that makes smart cities happy. Either fixed fog device fee, or restores damage or damaged unit for service and recovery (SAR), teams, may benefit from the advantages of IoD systems.

11.4.14 IoD DEVICES IMPROVING LIFE QUALITY IN SMART CITIES

The detection and regulation of air emissions in the Western developed world recently became important challenges. Sensors are typically used to capture pollutant data used for surveillance purposes [32, 33, 35]. The capacity to transmit these small sensors are therefore minimal, therefore not appropriate for the transmitting of information in real-time [35, 36]. To prevent this, sensors can be moved via drones and then the data phase and forward transmission can be divided into two parts: the on-board sensor is capable of sensing / polluting data, and the drone's contact systems can effectively transmit information to a standalone sensor over much larger distances than possible. Four gas sensors and a particle number (PNC) detector onboard a hexacopter have been mounted for the very best mounting points. Their research focused on the airflow behavior and the performs assessment of CO_2, CO, NO_2 and NO sensors for the measurement of hazardous gaseous emissions in a specific geographical area. Also, they developed drone systems for assessed point source emissions [38, 39]. Several other future Drone [40] technologies use has also been investigated to communicate with soil-moisture sensor sensors, remote crop tracking, water quality monitoring, and remote sensor deployment.

11.5 TYPES OF DRONES

1. **Multi-Rotor:** If you want to put a small camera in the air for a short period, you cannot comply with the multi-rotor. These are the fastest and most accessible way to get an eye in the sky and are suitable for aerial photography because they give you such a clear command of the location and picture. The detrimental effects of multi-rotors are reduced endurance and speed, which render it unsuitable for

aerial mapping on a large scale, long-lasting tracking and long-haul tracking of rivers, roads, and power lines [41–43].

2. **Fixed-Wing:** Fixed drones use an aircraft as a standard wing to provide the lifting rotor rather than the vertical lifting rotor. As a result, they only have to use energy to step forward, not stay in the cold, because they are much heavier. Therefore, they can travel longer distances, chart even wider areas and track their point of interest for long periods. Gas engines can now be used as their energy supply as well as improving reliability, and many UAVs can stay aloft for 16 hours or more with the improvement in fuel energy densities (Table 11.1) [44–46].

TABLE 11.1   Pros and Cons

	Pros	Cons	Typical Uses
Multi-rotor	• Accessibility • Ease of use • VTOL and hover flight • Good camera control • Can operate in a confined area.	• Brief bursts of travel • Room for limited loads	Audio and aerial inspection Aerial imagery
Fixed-wing	• Long stamina • Broad coverage of the region • Pace of fast flight	• Start and rehabilitation require a lot of space. • VTOL / hover is not available. • Further preparation is required, tougher to fly.	Inspection of aerial mapping, pipeline, and power line
Single-rotor	• Flight VTOL and flover • Long-lasting (gas energy) • Greater power for payload	• More hazardous • Further preparation is required, tougher to fly. • Costs	Laser screening Air LIDAR
Fixed-wing hybrid	• Flight VTOL and long-lasting	• Not suitable for any flight or flight forward • Progress underway	Drone supply

3. **Single-Rotor Helicopter:** There is just one rotor and a tail rotor governing the heading of a multi-rotor, although it has a large

spectrum of rotors. Hubs are popular for manned aviation, but there actually is just a small void in the drone world. Much more powerful than a multi-rotor, a single-rotor helicopter can be powered for longer by a gas engine. The longer the blade of the rotor and the slower it gets, the more potent it is, the more aerodynamics it is. This is why a quad-copter functions more effectively than an octocopter and the large prop diameter of unique long-term quads. A Heli mono-rotor achieves impressive results with very long blades, almost more like a spinning wing than a propeller [47–49].

4. **Fixed-Wing Hybrid VTOL:** The benefits of UAVs with fixed wing and their floating capability are new hybrids. In essence, both of them are only flanged models with vertical lift motors on. In manufacturing there are various types. Another is a "tail sitter," which resembles a traditional aircraft, but is situated on their earth dots, points directly to the departure before the aircraft reaches its normal course or a type of "tilt-rotor" which rotates from point upwards to the horizontal location of the rotor, and even entire wing with an attached propeller [37, 50].

11.6 CONCLUSION AND FUTURE SCOPE

Internet of Drone is enhancing SQ in Smart Cities. IoD can be one of the key elements of intelligent Cities, they are also faced with many obstacles in science and deployment, including energy limits and drone rules for flights. We expect research and development efforts in academic and industrial areas to pave the road to successful incorporation of drones into smart cities. Smart cities consist of insightful, intelligent things and work continuously to increase the quality of life, save lives, and preserve capital. In this chapter different application of IoD, devices are studied which enhance the SQ in a smart city. Recently, a significant addition in developing the many real-time implementations of smart cities has been the benefit of drone technologies. IoD machinery is designed in various ways to enhance civic life and at the same time protecting our atmosphere. In order to improve quality of life, minimize use and attain high levels of productivity and many other benefits, IoD innovations will dramatically change our lives. Since Drone's Internet is a modern technology, many possibilities and uses remain unclear and can be seen as the potential scope of this segment.

KEYWORDS

- **drone**
- **internet of drone (IoD)**
- **internet of things**
- **service quality**
- **smart city**
- **unmanned arial vehicles**

REFERENCES

1. Alsamhi, S. H., Ou, M., Mohammad, S. A., & Faris, A. A., (2019). Survey on collaborative smart drones and the internet of things for improving the smartness of smart cities. *IEEE Access, 7,* 128125–128152.
2. Barrionuevo, J. M., Berrone, P., & Ricart, J. E., (2012). Smart cities, sustainable progress. *IESE Insight, 14,* 50–57.
3. Lytras, M. D., & Anna, V., (2018). Who uses smart city services and what to make of them: Toward interdisciplinary smart cities research. *Sustainability.*
4. Bibri, S. E., & John, K., (2017*). Smart Sustainable Cities of the Future: An Extensive Interdisciplinary Literature Review, 31,* 183–212. Sustainable cities and society.
5. Hollands, R. G., (2008). Will the real smart city please stand up? Intelligent, progressive, or entrepreneurial? *City, 12*(3), 303–320.
6. Azevedo, G., André, L., Jeferson, C. A., Maurício, D. S. S. G., Martius, V. R. Y R., & Carlos, A. P. S., (2018). Smart cities: The main drivers for increasing the intelligence of cities. *Sustainability, 10*(9), 3121.
7. Errichiello, L., & Roberto, M., (2018). Leveraging smart open innovation for achieving cultural sustainability: Learning from a new city museum project. *Sustainability, 10*(6), 1964.
8. Mohammed, F., Ahmed, I., Nader, M., Al-Jaroodi, J., & Imad, J., (2014). UAVs for smart cities: Opportunities and challenges. *International Conference on Unmanned Aircraft Systems (ICUAS)* (pp. 267–273). IEEE.
9. *LoRa Alliance: Wide Area network for Internet of Things.* https://www.lora-alliance.org/ (accessed on 19 November 2021).
10. López-Pérez, D., Ming, D., Holger, C., & Amir, H. J., (2015). Towards 1 Gbps/ UE in cellular systems: Understanding ultra-dense small cell deployments. *IEEE Communications Surveys & Tutorials, 17*(4), 2078–2101.
11. Fotouhi, A., Ming, D., & Mahbub, H., (2017). Understanding autonomous drone maneuverability for internet of things applications. *IEEE 18th International Symposium on A World of Wireless, Mobile and Multimedia Networks (WoWMoM)* (pp. 1–6). IEEE.
12. Barmpounakis, E. N., Eleni, I. V., & John, C. G., (2016). Unmanned Aerial Aircraft Systems for transportation engineering: Current practice and future challenges. *International Journal of Transportation Science and Technology, 5*(3), 111–122.

13. Chen, Y. M., Liang, D., & Jun-Seok, O., (2007). Real-time video relay for UAV traffic surveillance systems through available communication networks. *IEEE Wireless Communications and Networking Conference*, 2608–2612. IEEE.

14. Alvarenga, J., Nikolaos, I. V., Kimon, P. V., & Matthew, J. R., (2015). Survey of unmanned helicopter model-based navigation and control techniques. *Journal of Intelligent & Robotic Systems, 80*(1), 87–138.

15. D'Aloia, M., Maria, R., Ruggero, R., Marianna, N., & Leonardo, P., (2015). A marker-based image processing method for detecting available parking slots from UAVs. In: *International Conference on Image Analysis and Processing* (pp. 275–281). Springer, Cham.

16. Hull, L., (2016). *Drone Makes First UK 'Arrest' as Police Catch Car Thief Hiding Under Bushes*. Mail Online.

17. Popoola, O. P., & Kejun, W., (2012). Video-based abnormal human behavior recognition: A review. *IEEE Transactions on Systems, Man, and Cybernetics, Part C (Applications and Reviews), 42*(6), 865–878.

18. Kiryati, N., Raviv, T. R., Ivanchenko, Y., & Rochel, S., (2008). Real-time abnormal motion detection in surveillance video. In: *19th International Conference on Pattern Recognition* (pp. 1–4). IEEE.

19. Cochez, M., Jacques, P., Vagan, T., Kyryl, K., & Tero, T., (2014). *Evolutionary Cloud for Cooperative UAV Coordination*. Reports of the Department of Mathematical Information Technology. Series C, Software engineering and computational intelligence.

20. Alsamhi, S. H., Mohd, S. A., Ou, M., Faris, A., & Sachin, K. G., (2018). *Tethered Balloon Technology in Design Solutions for Rescue and Relief Team Emergency Communication Services* (pp. 1–8). Disaster Med. Public Health Prep.

21. Tuna, G., Bilel, N., & Gianpaolo, C., (2014). Unmanned aerial vehicle-aided communications system for disaster recovery. *Journal of Network and Computer Applications, 41*, 27–36.

22. Ali, Z., Shehzad, A. C., Muhammad, S. R., & Al-Turjman, F., (2020). Securing smart city surveillance: A lightweight authentication mechanism for unmanned vehicles. *IEEE Access, 8*, 43711–43724.

23. Ashour, R., Tarek, T., Fahad, M., Eman, H., Yasmeen, A. K., Malak, E., Maha, K., et al., (2016). Site inspection drone: A solution for inspecting and regulating construction sites. In: *2016 IEEE 59th International Midwest Symposium on Circuits and Systems (MWSCAS)* (pp. 1–4). IEEE.

24. Košmerl, J., & Andrej, V., (2014). Base stations placement optimization in wireless networks for emergency communications. *IEEE International Conference on Communications Workshops (ICC), 200*–205. IEEE.

25. Kim, J., Daewoo, L., Kyeumrae, C., Jeongho, K., & Dongin, H., (2014). Two-stage trajectory planning for stable image acquisition of a fixed wing UAV. *IEEE Transactions on Aerospace and Electronic Systems, 50*(3), 2405–2415.

26. Bahabry, A., Xiangpeng, W., Hakim, G., Gregg, V., & Yehia, M., (2019). Collision-free navigation and efficient scheduling for fleet of multi-rotor drones in smart city. *IEEE 62nd International Midwest Symposium on Circuits and Systems (MWSCAS)*, 552–555.

27. Jawhar, I., Nader, M., Al-Jaroodi, J., & Sheng, Z., (2013). Data communication in linear wireless sensor networks using unmanned aerial vehicles. *International Conference on Unmanned Aircraft Systems (ICUAS)*, 492–499. IEEE.

28. Jawhar, I., Nader, M., Al-Jaroodi, J., & Sheng, Z., (2014). A framework for using unmanned aerial vehicles for data collection in linear wireless sensor networks. *Journal of Intelligent & Robotic Systems, 74*(1, 2), 437–453.
29. Kim, S. J., Yunhwan, J., Sujin, P., Kihyun, R., & Gyuhwan, O., (2018). A survey of drone use for entertainment and AVR (augmented and virtual reality). In: *Augmented Reality and Virtual Reality* (pp. 339–352). Springer, Cham.
30. Andersen, C. E., (2014). *Games of Drones: The Uneasy Future of the Soldier-Hero in Call of Duty: Black Ops II, 12*(3), 360–376. Surveillance & Society.
31. Raj, P., & Anupama, R., (2018). *Handbook of Research on Cloud and Fog Computing Infrastructures for Data Science.* IGI Global.
32. Alsamhi, S. H., Ou, M., Mohammad, S. A., & Faris, A. A., (2019). Survey on collaborative smart drones and internet of things for improving smartness of smart cities. *IEEE Access, 7*, 128125–128152.
33. Klimkowska, A., Lee, I., & Choi, K., (2016). Possibilities of UAS for maritime monitoring. *The International Archives of Photogrammetry, Remote Sensing and Spatial Information Sciences, 41*, 885.
34. Wang, J., Erik, S., Brian, O., & Travis, D., (2015). *A New Vision for Smart Objects and the Internet of Things: Mobile Robots and Long-Range UHF RFID Sensor Tags.* arXiv preprint arXiv:1507.02373.
35. Wall, T., & Torin, M., (2011). Surveillance and violence from afar: The politics of drones and liminal security-scapes. *Theoretical Criminology, 15*(3), 239–254.
36. McNeal, G. S., (2016). Drones and the future of aerial surveillance. *Geo. Wash. L. Rev., 84*, 354.
37. Gu, H., Ximin, L., Zexiang, L., Shaojie, S., & Fu, Z., (2017). Development and experimental verification of a hybrid vertical take-off and landing (VTOL) unmanned aerial vehicle (UAV). In: *2017 International Conference on Unmanned Aircraft Systems (ICUAS)* (pp. 160–169). IEEE.
38. Jensen, O. B., (2016). Drone city-power, design and aerial mobility in the age of smart cities. *Geographica Helvetica, 71*(2), 67.
39. Berntzen, L., Adrian, F., Cristian, M., & Noureddine, B., (2019). A strategy for drone traffic planning dynamic flight-paths for drones in smart cities. In: *SMART 2019, The Eighth International Conference on Smart Cities, Systems, Devices and Technologies.*
40. Alsamhi, S. H., Ou, M., Samar, A. M., & Sachin, K. G., (2019). Collaboration of drone and internet of public safety things in smart cities: An overview of QOS and network performance optimization. *Drones, 3*(1), 13.
41. Li, Y., & Chunlu, L., (2019). Applications of multirotor drone technologies in construction management. *International Journal of Construction Management, 19*(5), 401–412.
42. Shukla, D., & Narayanan, K., (2018). Multirotor drone aerodynamic interaction investigation. *Drones, 2*(4), 43.
43. Hsiao, K., & Yufeng, H., (2015). *Multi-Rotor Aerial Vehicle.* U.S. Patent Application 29/502,242, filed.
44. Kang, Y., & Hedrick, J. K., (2009). Linear tracking for a fixed-wing UAV using nonlinear model predictive control. *IEEE Transactions on Control Systems Technology, 17*(5), 1202–1210.

45. Euston, M., Paul, C., Robert, M., Jonghyuk, K., & Tarek, H., (2008). A complementary filter for attitude estimation of a fixed-wing UAV. In: *2008 IEEE/RSJ International Conference on Intelligent Robots and Systems* (pp. 340–345).
46. Quigley, M., Michael, A. G., Stephen, G., Andrew, E., & Randal, W. B. N., (2005). Target acquisition, localization, and surveillance using a fixed-wing mini-UAV and gimbaled camera. In: *Proceedings of the 2005 IEEE International Conference on Robotics and Automation* (pp. 2600–2605).
47. Qi, X., Juntong, Q., Didier, T., Youmin, Z., Jianda, H., Dalei, S., & ChunSheng, H., (2014). A review on fault diagnosis and fault tolerant control methods for single-rotor aerial vehicles. *Journal of Intelligent & Robotic Systems, 73*(14), 535–555.
48. Songchao, Z., Xue, X., Sun, Z., Zhou, L., & Jin, Y., (2017). Downwash distribution of single-rotor unmanned agricultural helicopter on hovering state. *International Journal of Agricultural and Biological Engineering, 10*(5), 14–24.
49. Qi, X., Juntong, Q., Didier, T., Dalei, S., Youmin, Z., & Jianda, H., (2016). Self-healing control design under actuator fault occurrence on single-rotor unmanned helicopters. *Journal of Intelligent & Robotic Systems, 84*(14), 21–35.
50. Tian, Y., (2018). *Hybrid VTOL Fixed-Wing Drone.* U.S. Patent Application 29/609,928, filed.

PART V

APPLICATIONS OF DRONES IN BUSINESS AND
DISASTER RELIEF MANAGEMENT

CHAPTER 12

Internet of Drones Applications in Aviation MRO Business Services

BHARATI SINGH and SAURABH TIWARI

Department of Transportation, School of Business, UPES, Dehradun, Uttarakhand – 248007, India, E-mails: bharti.singh012@gmail.com (B. Singh), stiwari@ddn.upes.ac.in (S. Tiwari)

ABSTRACT

The aim of this research is to identify the optimized outcomes to the aviation maintenance repair and overhaul (MRO) business service provider with the effective use of "DRONES" in their daily operation. Many engineering companies are using the drone services to create a pavement survey for airport pavement management. High-definition images, which were captured from the low altitude by the drone, can be used for airport maintenance to establish a repair program. The airport planner who keeps track on the pavement condition use drone to decide whether they need any replacement. In the airport sector, the drone application is mainly used for engineering and construction projects. Drones help in collecting the land survey data before they begin the project. It helps to evaluated, monitor the project stages, and document the progress of the project during the entire phase of process construction. Many airlines and other aviation stakeholder like MRO service provider have identified that drone can assist them to reduce cost and the time required on Maintenance repair overhaul activities on the components of the aircrafts. The damages, which occurred due to lighting, can be done through visual inspection. If the visual inspection is done manually, it will take nearly four to six hours. But unmanned aerial vehicle (UAVs) can finish this visual inspection within half an hour. These drones are used to find the problematic areas in the fuselage and other parts for the scheduled maintenance. It helps us to plan the repairs that are needed. Drone will increase the efficiency of the repair work by delivering the spare parts to the MRO personnel very quickly. Micro drones are used to inspect inside the aircraft engines. The automation of defect detection helps engineers to find the potential damage in the parts, which can be inspected visually at the later stage. This study implies that MRO Business Services require research and investment in technologies like drones, machine learning (ML) and cloud applications to support the maintenance and repair process. This will help in more effective maintenance and component defect identification, optimized technician schedule for each work plan as well as improved turnaround times for aircrafts.

12.1 INTRODUCTION

Aviation industry is one of the fastest growing sectors around the world with a quick progression both in terms of generating revenue and providing services. This industry can be considered as a part of logistics and supply chain management. Because of its versatile service nature, it can be treated as a service industry in supply chain management, in which the services offered are not only permitted to handle passengers they have been to cargo movement also. For this reason, it has been treated as a service industry with all its operations being performed from different parts of the world. Though there are some serious incidents and accidents occurred in the past, Aviation in a short period of time has created its own mark in the area of transportation by handling the passengers efficiently and effectively to reach their destination without any major difficulties faced at the time of voyage. From the accidents that have taken place, safety is the most important factor in the industry to make passengers believe that they are at safest and fastest mode of transportation. Any firm which is interlinked with aviation industry have to give priority to safety parameter in the first place, as the mode of transportation is by air any action which will leads to disastrous endings has to be monitored and mitigate the reasonable factors which leads to damage may be in terms of death of passengers.

Lack of monitoring safety parameters at the time of maintenance work will bring loss to airlines in terms of destruction of aircraft, loss of man power (cabin crew) and also in terms of good will from their customers (passengers) and also the interested parties who had stake in their organization. The MRO sector which will looks after the maintenance work will attend the aircraft with all its efforts to make the aircraft "Fit to fly," engineering department will work for this maintenance purpose and will follow the guidelines of regulatory authorities without missing any checks prescribed by them. There are some airlines which have their own MRO facilities from where they will get this maintenance works and make the aircraft free from all the minor parameters which will help the aircraft to airborne safely and reach their destination without any disturbances. Maintaining aircraft is not that easy as we see other maintenance, repair, and overhauling firms in other industries. Aviation industry in India has grown in a fastidious and will be proliferated in coming years, attention towards optimization of expenditure on aviation related service sectors is of utmost important. Based on NOVONOUS report for 2016–2020, airlines in India spent around 13–15% of their revenue in this service which makes up second highest cost after ATF (air turbine fuel). India's MRO was fragile with over 90% being expended outside India to countries such as Sri Lanka, UAE, Singapore, etc.

India, with its growing aircraft fleet, strategic location, pool of engineering expertise and lower labor costs, appears to have great potential to become a global MRO hub. India's current MRO market is estimated to be around $700–900 million, with estimates from local authorities in India at the higher end of this range. Boeing forecasts this market to grow at a 7% compound annual growth rate and, to reach as much as $1.2 billion by 2020. With the increase in the number of civil and military aircrafts, more, and more global MRO companies are planning to offer engineering services by forming joint ventures with Indian firms. For instance, GMR Hyderabad International Airport and MAS Aerospace Engineering, a wholly-owned subsidiary of Malaysia Airlines, have set up a 50:50 joint venture airframe MRO company in Hyderabad [69].

12.2 LITERATURE REVIEW

Sun et al. [47], says that 3D modeling is one of the important for aircraft MRO industry. It is a very accurate surface model of aircraft, very labor intensive and time consuming. The author proposed a solution of automated 3D scanning of an aircraft by using robotics, AI, and other sensing technologies. Sun et al. [47] says that Regional connectivity scheme (RCS) consists of UAV, Unmanned Ground Vehicle (UGV) Unmanned aircraft general and a manipulator. It has a two-phase scanning method. The author says that these 3D model can be used to analyze to identify cracks, and other defects which normally detected on the surface of the aircraft structures like fuselage, wings, and other components. This technology further detects boring, tedious, etc. It is clearly evident that the solution provided by this process will be more accurate, efficient, and high quality when compared to the human labor.

The author Ismail [4] mentions the human error associated with the military MRO management program. Author discussed about the rising human error in the military aviation maintenance. Author collected quantitative data from various military aviation technicians at MRO. The data collection consists of questionnaires subdivided into two sections. The results of the study indicated that aviation fuel cost and labor cost challenges contribute to human error. From the data he analyzed and suggested that UAV systems should be implemented at the maintenance program to reduce the human error.

Manda and Chaitanya, [41] says that aircraft maintenance consists of many small complex activities knows as MRO. The main problem lies with the attention of the oversight authorities. It was done through close monitoring with outsourcing agencies known as 'Service level agreements.' The author says that Indian aviation industry expends more of their financial resources on MRO than manufacturing, design, and R&D activities. Israel Aerospace Industries is providing design for aerospace solutions to make indigenous aircraft design in this competitive industry. The global delivery and outsourcing have become the key strategies for aerospace industry with cost savings to gain competitive advantage.

The Australian Aircraft Industry [48] a defense point speaks about the aircraft industry of Australia from the defense point of view. They discussed about the relationship between the national security policies and the industry requirement in their report. The report says that the Australian aircraft industry provides three services including design facility, technical expertise, and engineering support. The service dependence falls within engineering support, maintenance support and supply support. It clearly indicates that aircraft industry follows two principle higher workload and great stability in that workload.

Jennifer [36] analyzed the long- and short-range fleet readiness by providing aviation MRO industry to the current potential expansion. He discussed the strategy development upon a complete industrial analysis. This analysis includes various factors such as definition of industry, external forces acting upon the industry, key success factors and industry structures.

Lester [39] identifies emerging trends and forecasts of the military aviation MRO industry. He also explores the opportunities of the UAV in the MRO market. Lester [39] indicate that the opportunity is immense and as a result in future, there is a scope to conduct a market study of UAV in MRO market.

Koslosky [37] discussed the transformation of different industry according to the new technologies such as Augmented Reality (AR), Artificial Intelligence (AI), Big Data, and Internet of Things (IoT) Industry 4.0. Author suggested that the aviation industry is investing heavily on the latest technology, technologies which supports the commercial airlines. The investment in this industry is primarily driven to the manufacturing of the aircraft. The technology must support to the back office of commercial airlines. He discussed about the digital transformation of the aviation industry and their benefits. The use of drones and how the drones operations will be useful in the inspection and the associated cost benefits for the MRO industry. He clearly analyzes how the aviation industries are moving towards digitization and this helps the airlines in reducing their cost by implementing the new technologies which comes in the market.

Hussein deals with the major causes of delay in delivery time in the MRO industry. He specifies certain challenges faced in the MRO industry, which are competitive services, predictive maintenance, effective collaboration, technology utilization and supply chain with the primary focus on the causes for the delivery delay. He identified few causes such as presence of poor supply chain, poor process design and scarcity of skilled man power. The usage of the drone in the MRO industry will increase not only increase the efficiency but also helps in increasing work process design and efficiency [35].

Kostopoulos et al deals with the innovative approach for inspection of aircraft composite materials. He emphasized on the autonomous inspection of the aircraft parts using infrared thermography (IRT) and phased array. These inspections are non-destructive investigations, which do not damage any part during the inspection. The use of this new design robotics for this inspection and the technology will be very useful to make the aircraft safer during the inspection with lowest possible cost. These types of inspection are need because of the rapid expansion of the industry and use of composite material in the aviation industry [38].

Robbins and Keller discussed about the implementation of advanced robots in the inspection of the fuselage will reduce the cost, time, and improve the quality. There are some technologies which are being evaluated imaging and laser ultrasonic to inspect the aircraft without even touching the aircraft.

The author did simulation on different inspection technique for different components of aircraft such as landing gear, fuselage, etc., The inspection detects cracks, holes, etc. He emphasized on the use of non-contact laser ultrasonic technologies to inspect the exterior of the aircraft by the MRO industry [45].

Miranda et al proposes a new approach to inspect and detect the exterior screw on the aircraft services. The uses of UAV with LIDAR (light Detection and Ranging) technology to locate the exterior part on the aircraft automatically with a precise image. He uses a neural network method to characterize zone of interest to extract only the image of screws. The uses of computer vision algorithms to evaluate the visible screw and detect the missing one and loose ones [42].

Bouarfa et al. used a deep learning to automate the visual inspection for the aircraft maintenance. The author says this technology will be more accurate in finding the damage and reduce aircraft downtime. This technology also helps to reduce accidents during inspections. The objective of the chapter is to demonstrate the automatic detection of aircraft dents. The author says small dataset size has been used for training and it gave a promising result. The author also says this method can be used to detect other type of damages such as lightning strike, paint damage, cracks, and holes. It will be helpful for the aircraft engineers to take decision on support system [29].

Blokhinov et al. discussed the visual inspection for the aircraft surface using programmable UAV. In this chapter the author faced a problem of drone's indoor navigation due to the satellite signal is weak. He also focused on the algorithm program and software to detect the damage on the aircraft surface based on the video analysis. He conducted this testing in a closed hanger where the condition was almost expected for the experiment [27].

Almadhoun et al. [26] propose a planning algorithm for aircraft inspecting using UAV. This algorithm can be used to inspect various structures like bridges, buildings, ships, and any large structures. These inspections are hard task for human to perform and of critical nature, sometimes they may miss some damage. The author says this algorithm is a time saving and resource intensive task, it can perform as efficiently and accurately. In this chapter author has introduced a "Search space coverage path planner." The proposed method follows admissible waypoint to fully cover the complex structure. The author says this algorithm can also predicts the coverage percentage.

Malandrakis et al focus on the NDT (non-destructive testing) inspection on the wing panels. The authors work addresses the challenges and the improvement in the defect detection by the UAV. The author fitted field of view camera and an ultraviolet torch on the UAV for implementing imaging inspection. The programed UAV to perform the complete mission and stream video, in the real time and send it to the ground station. The people in the ground station will detects the defects. The author used MATLAB in order to assess the behavior of the system. The author tested the UAV in the lab with the six-meter wing panel [40].

UAS drone for MRO inspections [51], intelligent aerospace, (Orient Aviation-Orient Aviation, discussed about the integration of Donecle drone in the MRO operations. AAR Miami MRO facility is the first to use the fully automated drones in the MRO operation [6]. It is more cost efficient and drive operational where it will be more precise, drone is fitted with laser position so it can safely perform end-to-end visual inspections. This drone is programed to detect any structural damage such as paint quality, markings, and lighting strikes. One completed scan covered by drove is equivalent to several maintenance task performed by personnel.

The Innovative Use of Drones in Aircraft Maintenance and Repair [50], discussed about the use of drones in the aviation MRO industry. The author says the innovation of drones in the robotics age where useful for MRO and 3D printing of non-critical parts. The author also mentioned the value of drone in aviation MRO. The author mentioned that EasyJet is said to experimenting drone to inspect the fuselage of their airplanes. They are using 3D scanners for hail damage in the fuselage. The also mentioned that drones will be more useful in the scheduled maintenance task.

Drone MRO Taking Shape-Aviation Maintenance Magazine [33], speaks about the MRO industry for drones. Air taxis are almost about to launch once the regulations are set for them. The author speaks about the growth of the drone in various sectors like surveillance, agriculture, oil, and gas, etc. The author says that like aircraft, the drones also need schedule maintenance program. The author mentioned that regulation for drones and air taxis are not yet fully defined. They are under type certification or continuing airworthiness. The author says that UAS needs a unique skilled labor because it is not just airframe, it consists of command-and-control technology and payload handling.

The future of MRO emerging technologies in aircraft maintenance elaborates Uniting Aviation [49], speaks about the emerging technologies in the aviation MRO industry. The author mentions maintenance is the major contributor to the airlines in operating costs, flight delays and cancellation. The author says that each and every airlines spend more money on maintenance than on fuel or crew. To make the process in the MRO industry feasible they are about to launch augmented and mixed reality and Big data. These technologies will increase the efficiency of the workers and delivery the aircrafts on time.

This chapter Airbus using drones to visually inspect aircraft [23] unify, speaks about the recent demonstration of drone-based aircraft maintenance tool by airbus. The airbus has developed advanced inspection drone to operate inside a hangar. The airbus (Airbus launches advanced indoor inspection drone to reduce aircraft inspection times and enhance report quality-Commercial Aircraft Airbus says the drones in equipped with visual camera and obstacle detection sensor, flight planner and inspection analysis tool. The airbus made initial demonstration to several airlines who showed interest. The airbus says the inspection takes only three hours including 30 minutes of image capture by the drone.

The chapter Boeing Invests [28] in Global Drone MRO Network with an Eye Toward UAM Repair-Aviation Today, speaks about the developing MRO station for unmanned aircraft. Already two companies are working together with Boeing to provide MRO services and supply chain management for commercial and civil drones. 170 plus repair shops will be open across 40 countries by robotic skies network, each certified by civil aviation authority or FAA.

Digital twins [32], AI, mobile apps, and drones, speaks about the innovation in maintenance techniques and recent research that has been conducted in the maintenance industry. Nowadays airlines spend more money on maintenance rather than on fuel and crew. So, airlines are adopting new techniques and innovation to reduce the cost and time and increase the efficiency in the MRO industry. A SITA has claimed that airlines will use AI in the next three years. This will help to airlines to reduce cost by correctly predict the right moment to repair certain parts. Airlines are also trying to use drones for autonomous visual inspection. These inspection can identify cracks, hole, loosen screws, and paint quality on the surface of the aircraft.

Aircraft Inspection Drones [25] Size, share, Market 2018 Key Manufacturers, Demands, Industry Share, Size, Status and Forecasts to 2028, (Unmanned Aerial Vehicle Market, UAV Size, Share, system and Industry Analysis and Market Forecast to 2024 [52] MarketsandMarkets™, analyzed the market for the aircraft inspection drone. Normally MRO services are essential part of the aircraft life. To reduce the human error during the inspection process, the MRO industry brought drone to increase the efficiency of the inspection. These drones a detect hail, lightning strike on the surface and other damage easily.

The drone inspection will take only 3 hours to inspect a complete aircraft when compare to the human inspection. The author also speaks about the different market segmentation of the drone and regional outlook. The author speaks about the market participants and the competitors.

Airbus innovation [21] for military aircraft inspection and maintenance-Defense-Airbus, discussed about using new technologies like drones, augmented reality, and artificial intelligence in the MRO operations. The airbus is jointly developing drones and augmented reality inspection with military service to inspect A400M heavy lift aircraft. The airbus mentioned that these drones are equipped with high-definition cameras and 3D augmented reality LIDAR along with AI to accurately inspect the aircraft. The Airbus [22] says "The use of Drone and Augmented Reality inspections will significantly reduce inspection time, while also lowering the risk of inadvertent damage caused by traditional inspection methods that require a build-up of scaffolding and the deployment of mobile equipment around the aircraft." The Airbus uses mathematical logarithms and deep learning to increase the precision and to create more robust and autonomous system.

Aircraft Inspection Drones [24] Technology Paves the Way to Future, speaks about the how UAV is taking shape in MRO industry and how the technological advancement support drone. Technology advancement such as machine learning equipped with laser widen the scope of application of drones [25]. The drones have proven efficiency in detecting damages caused by external factors. Those the application of drone is widen the stringent regulation affecting the growth of the inspection growth in their market. The drone market segmentation classified into two types; automated and manually operated drones. Further the drones are classified into two types' single rotor and multi rotor aircraft inspection drone.

The research work 5 New MRO [55] Technologies that will change the aviation industry, discusses about the adopting AI and machine learning into the maintenance industry. The new generation complex aircraft have increased volumes of data for airline maintenance teams to manage. The chapter warns us to stay alert to drone technology because soon it will be part of daily MRO operations. (AI, drones, and sustainability in commercial aviation [3]. BlueSky News Barriers to the adoption of drone technology in inspections include certification and ensuring devices are stable enough to capture precise images. Laser technology will however help yield accurate images onboard sensors can sense the drone's environment and position the device with an accuracy down to centimeters- and if drones gain full certification for use in commercial aviation" [19].

In this research work flying, clinging, and crawling [34]-using robots in MRO, the author speaks about the difficulty faced in the MRO when aircraft face damage due to bird strike, accidental damage, FOD strike and collision with service vehicles. To speed up the service in the MRO, they are looking at alternative ways other than human inspection. The author suggested to use robotics system to inspect the aircraft. The author uses different kinds of robots such as drones and crawling robots. These robots will able to perform visual inspection from the outer surface of the aircraft to the engine fan blade inspection. The robots are equipped with LIDAR and obstacle detector and programed by mathematical algorithm.

In this chapter ST Engineering and Air NZ trial drones for aircraft inspections [34] Flight Global, Air New Zealand uses drones for aircraft inspections-Airline Ratings, Air New Zealand and ST engineering did a trial of using Drone in their MRO operation. This experiment was conducted at the MRO's provider's facility near Changi Airport [20]. The Air New Zealand plane undergone heavy maintenance with UAV. The engineers used drone to inspect the outer surface with high-definition image. The drone took planned routes to capture the image. The result of the experiment shows that inspection time has reduced from six hours to two hours. The author also says that time for inspection varies according to the aircraft type.

In this chapter [30] Delta wants to use drones on the airfield for MRO, Delta Wants to Use Drones on the Airfield for MRO, speaks about Delta airlines through its Delta TechOps developing an UAV to inspect an aircraft outside the hangar. They conduct this experiment at Hartsfield-Jackson Atlanta International Airport. The author says that it will take few years to get acceptance from the regulators. The author mentioned that Donecle already developing drone to inspect aircraft inside the hangar.

In this chapter MRO Drone and Ubisense [43] join forces with new Smart Hangar Solution-Geospatial World, n.d.) MRO and Ubisense are pairing to launch new innovative smart hangar for aircraft inspection. They call it as "Hangar of the future" The chapter says this smart hangar use MRO drones with RAPID aircraft inspection. Ubisense says that "RAPID drone speeds up and streamlines the inspection process while Smart Space optimizes the repair readiness of the Smart Hangar, reducing the time to get the plane back into service" (Table 12.1).

TABLE 12.1 Literature Review

SL. No.	Journal Name/ Conference Proceedings	Title	Author	Key Finding/Results	Limitations, Research Gaps and Future Scope
1.	*Sensors* (ISSN 1424–8220; CODEN: SENSC9)	Piezoelectric Transducer-Based Structural Health Monitoring for Aircraft Applications	Xinlin Qing , Wenzhuo Li, Yishou Wang and Hu Sun	The author says that the solution of the process will be more accurate, efficient, and high quality when compared to the human labor	Need study on system integration, airworthiness compliance in aircraft applications,
2.	The British University in Dubai	A Literature Based Examination of Human Error in Military Aviation Maintenance Management Programs	ISMAIL ABDULRAHMAN AL ARIF	The author concludes in this research that unmanned ariel device (UAV) systems should be implemented at the maintenance program to reduce the human error.	The work is limited to aviation technicians.
3.	International Journal of Marketing and Technology (IJMT) ISSN: 2249–1058	Inventory Management Best Practices for Aircraft Servicing INDUSTRY IN INDIA	Manda and Chaitanya	Indian aviation industry expends more of their financial resources on maintenance, repair, and overhaul (MRO) than manufacturing, design, and R&D activities.	Key strategies for aerospace industry with cost savings to gain competitive advantage
4.	KPMG.com.au	Research and Economic Modeling of the Aircraft Manufacturing and Repair Services Industry in Australia	KPMG	The author says that the Australian aircraft industry provides three services including design facility, technical expertise, and engineering support.	Innovation activity in the field defense MRO

TABLE 12.1 *(Continued)*

SL. No.	Journal Name/ Conference Proceedings	Title	Author	Key Finding/Results	Limitations, Research Gaps and Future Scope
5.	Journal of Quality in Maintenance Engineering	Airline maintenance strategies-In-house vs. outsourced-An optimization approach	Jennifer M. Johnson [37]	The author also explores the opportunities of the unmanned ariel device (UAV) in the maintenance, repair, and overhaul (MRO) market.	Limitation of the UAV devices.
6.	Aeronautics and Aerospace Open Access Journal	Commercial aviation in a digital world: a cyber-physical systems approach for innovative maintenance	Leonardo Borges Koslosky	Digital transformation of the aviation industry and their benefits in MRO business services	A cost reduction is possible for those who embrace such changes
7.	academia.edu	Delivery time delay, its cause, and remedies in MRO service in Dejen aviation engineering industry	Kamil Hussen	The author says that usage of the drone in the maintenance, repair, and overhaul (MRO) industry will increase the efficiency	Poor supply chain, poor process design and scarcity of skilled man power in MRO.
8.	IFAC conference paper (Elsevier)	Autonomous Inspection and Repair of Aircraft Composite Structures	Vassilis Kostopoulos	Uses of new design robotics for the inspection of aircraft defect	Technological impact on a direct impact on reducing turnaround time
9.	catsr.vse.gmu. edu/	Design of a System for Aircraft Fuselage Inspection	Rui Filipe Fernandez, Kevin Keller, Jeffery Robbins	The author recommended that maintenance, repair and overhaul (MRO) used your non-contact laser ultrasonic technologies to inspect the exterior of the aircraft	Limitation of the technologies in repair process

TABLE 12.1 *(Continued)*

SL. No.	Journal Name/ Conference Proceedings	Title	Author	Key Finding/Results	Limitations, Research Gaps and Future Scope
10.	Springer chapter	On the Optimization of Aircraft Maintenance Management	Duarte DinisAna Paula Barbosa-Póvoa	Challenges faces by the MRO services provider in task scheduling and resource allocation.	Lack of Research opportunity
11.	Emerald insight Vol 27 issue 1	Procedure structuring for programming aircraft maintenance activities	Viviane Souza Vilela Junqueira, Marcelo Seido Nagano, Hugo Hissashi Miyata	Author also uses computer vision algorithms to evaluate the visible screw and detect the missing one and loose ones.	Technological acceptability by maintenance center
12.	International Journal of Production Research	Flexibility in Service Parts Supply Chain: A Study on Emergency Resupply in Aviation MRO	Aghil Rezaei SomarinSobhan Sean AsianSobhan Sean AsianFariborz JolaiFariborz JolaiSonglin ChenSonglin Chen	To maintain optimized inventory policy for resupply emergency components	Complexity in contractual agreement with the stakeholders.
13.	Transportation Research Procedia 25 (2017) 136–148	Developing a Conceptual Model of Organizational Safety Risk: Case Studies of Aircraft Maintenance Organizations in Indonesia	Bouarfa et al	Importance of Human factors at workplace to reduce fatal loss.	Challenges to develop safety outcomes

TABLE 12.1 *(Continued)*

SL. No.	Journal Name/ Conference Proceedings	Title	Author	Key Finding/Results	Limitations, Research Gaps and Future Scope
14.	Elsevier (Automation in construction)	LiDAR-equipped UAV path planning considering potential locations of defects for bridge inspection	Neshat Bolourian	The author says this algorithm is a time saving and resource intensive task, it can perform as efficiently and accurately	Identifying the criticality of higher risk level job level
15.	IFAC conference paper (Elsevier)	Design and Development of a Novel Spherical UAV	K. Malandrakis	The author used MATLAB in order to assess the behavior of the system. The author tested the unmanned ariel device (UAV) in the lab with the six-meter wing panel.	Challenges by UAV to achieve trim state
16.	Intelligent Aerospace (GATN)	Article: AAR launches Donecle drone technology integration for MRO aircraft inspections	MIAMI, Fla.,	Drone is programed to detect any structural damage such as paint quality, markings, and lighting strikes. One completed scan covered by drove is equivalent to several maintenance task performed by personnel.	Agreement related to technological agreements
17.	Mainblades	Easyjet makes drone inspection a reality in aviation MRO	Viktoriya Zoriy	Author says that UAS needs a unique skilled labors because it is not just airframe, it consists of command-and-control technology and payload handling.	Challenges in MRO business in component inspections and work scheduling.

TABLE 12.1 (*Continued*)

SL. No.	Journal Name/ Conference Proceedings	Title	Author	Key Finding/Results	Limitations, Research Gaps and Future Scope
18.	Aerospace, Drones, Tech	Airbus using drones to visually inspect aircraft	Aerospace, Drones, Tech	The airbus has developed advanced inspection drone to operate inside a hangar. The airbus made initial demonstration to several airlines who showed interest. The airbus says the inspection takes only three hours including 30 minutes of image capture by the drone	Complexity in contractual agreement with the stakeholders
19.	Elsevier	Some approaches to comparative assessment and selection of unmanned aerial systems	Kulyk, Mykola Silkov, Valeriy Samkov, Alexei	The results of the research allow us to consider the influence of characteristics on a global assessment of UAS and to make correct decisions about the selection of UASs	Study limited to the use of Unmanned Aerial Vehicle.
20.	Elsevier	Justification of thrust vector deflection of twin-engine unmanned aerial vehicle power plants	Kulyk, Mykola Kharchenko, Volodymir Matiychyk, Mykhailo	Changing the PPs thrust vectors and respective change of the horizontal components value of their thrust forces is an effective way of eliminating-adverse diving and nose up moments of the UAV.	Technological impact on direct impact on reducing turnaround time

TABLE 12.1 (Continued)

SL. No.	Journal Name/ Conference Proceedings	Title	Author	Key Finding/Results	Limitations, Research Gaps and Future Scope
21.	Journal: IEEE Access	Unmanned Aerial Vehicles (UAVs): A Survey on Civil Applications and Key Research Challenges	Shakhatreh, Hazim Sawalmeh, Ahmad H. Al-Fuqaha, Ala Dou, Zuochao Almaita, Eyad Khalil, Issa Othman, Noor Shamsiah Khreishah, Abdallah Guizani, Mohsen	Identify the key challenges for UAV civil applications such as charging challenges, collision avoidance and swarming challenges, and networking and security related challenges	One of the important challenges is to preserve the privacy of sensitive information (e.g., location) from other vehicles and drones. Since usually there is no encryption on UAVs on-board chips, they can be hijacked and subjected to man-in-middle attacks-originating up to two kilometers away
22.	Journal : International Journal of Vehicle Design	Future Developments in Military Aviation Propulsion.	Lewis, G.	1) Turboramjets probably best for high Mach cruise aircraft 2) Field still wide-open for space launchers 3) Turboramjets most fuel economical .	The research work focusses on development on military MRO for Engine Component

TABLE 12.1 *(Continued)*

SL. No.	Journal Name/ Conference Proceedings	Title	Author	Key Finding/Results	Limitations, Research Gaps and Future Scope
23.	Journal : International Journal of Transportation Science and Technology	Unmanned Aerial Aircraft Systems for transportation engineering: Current practice and future challenges	Barmpounakis, Emmanouil N. Vlahogianni, Eleni I. Golias, John C.	The future of the use of UAV to transportation will be dominated by advanced algorithms and tools to ensure: 1) the safe and effective navigation of UAV above transportation infrastructure ? their energy efficient use 2) the mining of critical information based on the predictive analytics	Study focusses only on Military MRO business operation.
24.	Journal of Transportation Security	Civil unmanned aircraft systems and security: The European approach	Huttunen, Mikko	It would appear that airports are smart and prudent enough to protect aviation without any legal obligation to do so. Still, given the looming threat of mass delays and mid-air collision, it might be appropriate to make anti-drone equipment a mandatory feature at the busiest airports within the EU	Technological impact on a direct impact on the reducing turnaround times not covered in this chapter.

TABLE 12.1 *(Continued)*

SL. No.	Journal Name/ Conference Proceedings	Title	Author	Key Finding/Results	Limitations, Research Gaps and Future Scope
25.	International Journal of Mining, Reclamation and Environment	On the application of drones: a progress report in mining operations	Said, Khadija Omar Onifade, Moshood Githiria, Joseph Muchiri Abdulsalam, Jibril Bodunrin, Michael Oluwatosin Genc, Bekir Johnson,	Complete adoption and integration of drone technology into mining operation will increase optimization, reduce the cycle of operations, and limit the risks of exposing workers to hazardous environments thus adding value to the mining industry. This	Complexity in contractual agreement with the stakeholders
26.	Journal :Ad Hoc Networks	Impact of drone route geometry on information collection in wireless sensor networks	Skiadopoulos, Konstantinos Giannakis, Konstantinos Tsipis, Athanasios Oikonomou, Konstantinos Stavrakakis, Ioannis	The main outcome of the study is that the circular trajectory yields the lowest number of messages, with the square route being close. Since every route has the same distance, it is conjectured that the order of symmetry for each shape is responsible for the discrepancy among each case.	The study focusses on only on wireless sensor technology.

TABLE 12.1 *(Continued)*

SL. No.	Journal Name/ Conference Proceedings	Title	Author	Key Finding/Results	Limitations, Research Gaps and Future Scope
27.	Journal : Transportation Research Part B: Methodological	Drone routing with energy function: Formulation and exact algorithm	Cheng, Chun Adulyasak, Yossiri Rousseau, Louis Martin	Generated benchmark instances for the drone routing problem and conduct extensive numerical experiments to evaluate the effects of valid inequalities and user cuts	There are multiple directions for future work. For instance, the visual process is essentially based on color detection. Higher robustness in the visual processing maybe reached by employing machine learning methods in computer vision
28.	Journal : Ad Hoc Networks	Drone assisted Flying Ad-Hoc Networks: Mobility and Service oriented modeling using Neuro-fuzzy	Kumar, Kirshna Kumar, Sushil Kaiwartya, Omprakash Kashyap, Pankaj Kumar Lloret, Jaime Song, Houbing	The proposed communication framework is tested to comparatively evaluate the performance with the state-of-the-art protocols considering metrics related to flying ad-hoc networks environments. The simulation results show that D-IoT outperforms the state-of-the-arts protocols.	Reliability centric development needs to be added before using it for security-oriented drone monitoring applications. In the future research, the authors will focus on consideration of energy utilization as performance metric including the design modification. The authors will also explore the work in diverse scenarios, and applications. Declaration

TABLE 12.1 *(Continued)*

SL. No.	Journal Name/ Conference Proceedings	Title	Author	Key Finding/Results	Limitations, Research Gaps and Future Scope
29.	Robotics and Autonomous Systems	Autonomous drone race: A computationally efficient vision-based navigation and control strategy	Li, Shuo Ozo, Michaël M.O.I. De Wagter, Christophe de Croon, Guido C.H.E.	Developed the highly efficient snake gate detection algorithm for visual navigation, which can detect the gate at 20 HZ on a Parrot Bebop drone. Then, with the gate detection result, we developed a robust pose estimation algorithm which has better tolerance to detection noise than a state-of-the-art perspective-n-point method.	There are multiple directions for future work. For instance, the visual process is essentially based on color detection. Higher robustness in the visual processing maybe reached by employing machine learning methods in computer vision.
30.	Journal of Air Transport Management	Managing the drone revolution: A systematic literature review into the current use of airborne drones and future strategic directions for their effective control	Merkert, Rico Bushell, James	Results suggest that security, privacy and acceptance concerns, whilst significant and relevant, are not as dominant as they have been in previous periods-with the use of drones in various ecosystems providing an opportunity for researchers to examine their introduction and impact on those with whom they interact.	Further work is needed to understand potential impacts of drone usage (e.g., fatalities due to accidents), subsequent potential risk trade-offs and adjustment/formulation of new regulation [47]. The safety/cost trade-off will be an important one to contribute to the setting of appropriate safety rules that facilitate the industry without constraining it unnecessarily, including the development of low altitude airspace management systems to support the increased deployment. Acknowledgements

TABLE 12.1 *(Continued)*

SL. No.	Journal Name/ Conference Proceedings	Title	Author	Key Finding/Results	Limitations, Research Gaps and Future Scope
31.	Applied Mathematical Modeling	An improved stochastic programming model for supply chain planning of MRO spare parts	Li, Ling Liu, Min Shen, Weiming Cheng, Guoqing	This study suggests two avenues for future research. 1) Future researchers can attempt to use robust optimization to handle the uncertain parameters that cannot be observed repeatedly and to extend the stochastic programming model to a robust optimization model in which the decision makers only need to obtain the upper and lower bounds of the uncertain parameters	Future researchers can model users' benefits as another optimization objective of the model. It is envisioned that combining multi-objective programming with multi-choice programming to formulate a new optimization problem is possible. Acknowledgements
32.	Procedia Manufacturing	Improving MRO order processing by means of advanced technological diagnostics and data mining approaches	Seitz, Melissa Lucht, Torben Keller, Christian Ludwig, Christian Strobelt, Rainer Nyhuis, Peter	The results presented show that wherever an implementation is possible, technological diagnosis based on known cause and effect relations should be preferred to improve capacity, throughput time and material planning	Future research activities will focus on the evaluation of data from the entire product life cycle, following the hypothesis that the quality of predicting MRO event dates and expenditures can thus be further improved

TABLE 12.1 *(Continued)*

SL. No.	Journal Name/ Conference Proceedings	Title	Author	Key Finding/Results	Limitations, Research Gaps and Future Scope
33.	Materials Today: Proceedings	Indian MRO industry: Business retention and development opportunities pre COVID-19	Karunakaran, C.S. Ashok Babu, J. Khaja Sheriff, J	Increased MRO investment opportunity will provide a healthy competition among the MRO investors which will increase the quality of maintenance. This focused development approach will retain the existing $900 Million USD MRO expenditure inside India which is 10% of the expenditure now. Retaining 100% MRO business will contribute 0.03% of India's $2.72 Trillion USD economy.	There are multiple directions for future work. For instance, the visual process is essentially based on color detection. Higher robustness in the visual processing maybe reached by employing machine learning methods in computer vision
34.	Journal of Civil and Environmental Engineering	Drones and Possibilities of Their Using	Kardasz, Piotr Doskocz, Jacek	Due to the much more favorable parameters of energy density the attempts are carried out to replace previously used cells (e.g., of lithium polymer) with fuel cells [6]. Due to the fact that in unmanned flying apparatuses lengthening of flight time is a critical factor in many cases, there are made the attempts to use fuel cells	Potential risks associated with the widespread use of drones require the use of complex solutions and the introduction of deliberate regulation aiming at effective protection of citizens' privacy

TABLE 12.1 (*Continued*)

SL. No.	Journal Name/ Conference Proceedings	Title	Author	Key Finding/Results	Limitations, Research Gaps and Future Scope
35.	Proceedings Annual Reliability and Maintainability Symposium	Maintenance of a Drone Fleet	Segal, Amir Bot, Yizhak	Fleet operators can reduce operation costs without jeopardizing performance by using analytic tools such as the APM optimizer. The modeling and optimization methods which were used in the example are applicable to any fleet, and are therefore relevant to many industries: defense, rolling stock, aviation, and mining. Furthermore,	Potential risk associated with the widespread use of drones require the use of complex solutions and the introduction of deliberate regulations mining at effective at effective protection of citizen privacy.
36.	Journal of Geographical Sciences	Iterative construction of low-altitude UAV air route network in urban areas: Case planning and assessment	Xu, Chenchen Liao, Xiaohan Ye, Huping Yue, Huanyin	The proposed methods and initial air routes network in typical demonstration areas can be used as a normative reference for the planning of national UAV low-altitude air routes in urban areas and further promoting the development of the construction technology of air route networks in low-altitude three-dimensional traffic network system.	Future research activities will focus on the evaluations of the data from the entire product cycle, following the hypothesis that the quality of predicting MRO events date and expenditure can thus be further improved

TABLE 12.1 *(Continued)*

SL. No.	Journal Name/ Conference Proceedings	Title	Author	Key Finding/Results	Limitations, Research Gaps and Future Scope
37.	Procedia Social and Behavioral Sciences	Innovating General Aviation MROs through IT: The Sky Aircraft Management System SAMS	Ucler, Caglar Gok, Orhan	It has been shown that GA-MRO organizations can be innovated radically towards sustainability using SAMS, which is an integrated and stand-alone tool suitable for the management and support of MRO organizations in GA by incorporation of all stakeholders.	There are multiple directions for future work. For instance, the visual process is essentially based on color detection. Higher robustness in the visual processing maybe reached by employing machine learning methods in computer vision
38.	Transportation Research Interdisciplinary Perspectives	Drones for parcel and passenger transportation: A literature review	Kellermann, Robin Biehle, Tobias Fischer, Liliann	The vertical move into low level airspace appears to be an intuitive, simple, and, as shown, historically long-aspired move. However, such claims will have to prove themselves against the disillusions of former transportation innovations.	Future research should make sure to clarify the relatedness of findings to the respective use case. This in turn would help professionalize the discourse on drones as a transportation medium

TABLE 12.1 *(Continued)*

SL. No.	Journal Name/ Conference Proceedings	Title	Author	Key Finding/Results	Limitations, Research Gaps and Future Scope
39.	Internet of Things	Security analysis of drones systems: Attacks, limitations, and recommendations	Yaacoub, Jean-Paul Noura, Hassan Salman, Ola Chehab, Ali	Presented a holistic view of the drones/UAVs domains and provided detailed explanation and classification of their use in various domains and for different purposes, in addition to the different lethal/ non-lethal security solutions as part of drones/UAVs countermeasures.	Due to the alarmingly increase in the use of drones by terrorists, further studies, and experiments on how to prevent and counter the UAV threats, imposed by terrorists, will be performed, and conducted as part of future work
40.	Reliability Engineering and System Safety	Software in military aviation and drone mishaps: Analysis and recommendations for the investigation process	Foreman, Veronica L. Favaró, Francesca M. Saleh, Joseph H. Johnson, Christopher W.	There is a widening safety gap between the software-intensive capabilities we create and our understanding of the ways they can fail or contribute to accidents, and hence our ability to prevent accidents	Technological acceptability by maintenance center

TABLE 12.1 *(Continued)*

SL. No.	Journal Name/ Conference Proceedings	Title	Author	Key Finding/Results	Limitations, Research Gaps and Future Scope
41.	International Journal of Remote Sensing	UAVs surpassing satellites and aircraft in remote sensing over China	Liao, Xiaohan Zhang, Yu Su, Fenzhen Yue, Huanyin Ding, Zhi Liu, Jianli	Although incomplete airspace-related regulations have some limitations for the development of UAVs, UAV RS will replace space-borne remote sensing and aircraft remote sensing in many cases. In addition, the trend of a nationwide UAV RS network on top of the existing ground observation network will be invaluable, which is also an important challenge. Disclosure	Due to the alarmingly increase in the use of drones by terrorists, further studies, and experiments on how to prevent and counter the UAV threats, imposed by terrorists, will be performed, and conducted as part of future work
42.	The current aviation MRO IT landscape	Leveraging Information Technology for Optimal Aircraft Maintenance, Repair and Overhaul (MRO)	Sahay, Anant	As the stock market analysts will say, it is trending 'sideways,' Hence the buyers find it difficult to decide on how to place their bets. The IT industry still does not seem to have fully committed to the aviation MRO industry.	Study does not cover strategies for aerospace industry with cost saving to gain competitive advantages.

12.2.1 ANALYSIS

The study mentioned above concludes the below mentioned points:

- The automated 3D scanning of an aircraft by using robotics, AI, and sensing technologies can help in reduce time of inspection and increase accuracy of the defect report.
- UAV systems integration will support the maintenance program to reduce the human error.
- The global delivery and outsourcing have become the key strategies for aerospace industry with cost savings to gain competitive advantage.
- Over all emerging trends in military aviation has gain more attention in better MRO services, there is potential opportunities of the UAV in the MRO market and provide reliable maintenance services.
- Aviation industry has been involved continuously in new technology like AR, AI, big data and IoT, which supports the commercial airlines.
- Application of drones in the MRO industry will increase the efficiency of the overall operations.
- MRO industry are competitive services, predictive maintenance, effective collaboration, technology utilization and supply chain.
- Integration of custom design robotics technology for aircraft and component inspection will be very useful to make the aircraft safer during the inspection with lowest possible cost.
- Implementation of advanced robots in the inspection of the fuselage will reduce the cost, time, and improve the quality.
- Application of drone technology will be more accurate in finding the damage and reduce aircraft downtime.
- Visual inspection of aircraft inspection takes between four and six hours while UAVs can get the job done in half an hour and reduce the turnout time.
- Automation and use of other smart technologies will enhance work processes and quality.
- Commercial Airlines spend more money on maintenance rather than on fuel and crew. Therefore, airlines are adopting new techniques and innovation to reduce the cost and time and increase the efficiency in the MRO industry.

12.2.2   IDENTIFICATION OF CRITICAL FACTORS IN MRO SERVICES

The factors, which will make the MRO business in a sustainable manner for any region has been identified with the help of consulting the experts in the respective sector and the academicians from various universities.

From their recommendations, the following factors are found to be some of the contributors to successful running of MRO in any region across the world. The following listed are the factors for MRO services development:

- Marketing and communication;
- Collaborations among aircraft manufacturers and the carriers;
- Infrastructure and resources availabilities;
- Technological innovation and new technologies adoption;
- Security and safety;
- Economic incentives and R&D investments;
- Education and training;
- Government policies and support.

12.3   PROBLEM DEFINITION

The serious problem that all the Indian and global MROs are facing is that they are not able to integrate the technological resources available for them in the market due to so many factors. There are so many unaddressed problems, Delayed in scheduled maintenance, automation of defect detection, Inspection accuracy, etc. These small problems are creating a huge differential layer with the global MRO's. In addition, the airlines after spending more percent to fuel the next area where they are spending more is on maintenance of aircraft.

This issue has to be taken into serious note where the MRO business in Indian aviation sector has to gain a potential to share a major part in the global aviation MRO sector. To done this; critical success factors are identified for MRO service development in India in Literature review.

12.4   METHODOLOGY AND ANALYSIS

This chapter has opted the qualitative descriptive theory methodology to explore and explain the contextually relevant factors that can improve the MRO service efficiency by integrating Application of drone, this will help to reduce cost and the time spent on maintenance, repair, and overhaul of the aircraft. The research is done under the theoretical framework, which is achieved by expensive literature review.

12.4.1 ANALYSIS

The analysis is articulating the theoretical assumption of the research study which allow us to understand the importance of Internet of drones in the MRO Business.

The key variable factors listed in Section 12.2.2 are further explained in the context why they are selected among the influential factors which will make a sustainable MRO business service provider.

12.4.1.1 MARKETING AND COMMUNICATION

For attracting the airlines, there should be a perfect strategy for occupying space in the market and to compete with their par companies in the sector. Lack of marketing for any service will drag the company into losses even though the MRO is capable of providing the same service which its competitor is providing. As there is healthy competition in this sector, any MRO company has to operate more efficiently to meet the needs of airlines [1].

As the personnel working under maintenance are the crux of business they have to communicate properly about the technical requirements, relay project completions or delays and status updates to the client companies and to the stakeholders of the company. Lack of communication in Indian scenario with the foreign carriers is one of the reasons for low market share in MRO market [17].

12.4.1.2 COLLABORATIONS AMONG AIRCRAFT MANUFACTURERS AND THE CARRIERS

For any MRO sector to be successful there should be a strong relationship with the airline operators and the aircraft manufacturers to help them suggesting the certified MROs based on the type of aircraft they are operating and the maintenance that has to be attended on the particular type of aircraft. In India by 2020 there will be increase of fleet size to around 1,200 says Boeing [28].

To meet the requirement the aircraft manufacturers should make a collaboration with the airlines to facilitate the fleet with sufficient maintenance infrastructure either in case of line maintenance or base maintenance. The Indian Flag carrier Air India along with the global leading aircraft manufacturers Boeing and Airbus made a joint venture of MRO, to conduct some part of MRO to some extent not to look after abroad MRO [54].

12.4.1.3   INFRASTRUCTURE AND RESOURCES AVAILABILITIES

For generating more revenue, the MRO companies need more fleet to attend and for this they need more space for facilitating the fleets. The MROs have to look after the availability of sufficient infrastructure to accommodate the fleets. In Indian scenario if 20 wide body aircrafts are delivered by airline manufacturers one out of them will have the magnanimity of getting service in Indian MRO Company.

But 90% of the Indian MRO work is getting its maintenance work done outside India. The increase in fleet size will obviously needs sufficient number of skilled labors to attend the maintenance work. The total labor requirement by airlines has been estimated and rose from 62,000 (2011 data) to 117,000 by 2017. So, the increase in fleet is going to increase more there is a need to improve the infrastructure facilities and also the man power.

12.4.1.4   TECHNOLOGICAL INNOVATION AND NEW TECHNOLOGIES ADOPTION

Maintenance and inspection of Aircraft is a very complex system, in which employees will perform different tasks in an environment with time pressures and sometimes it is difficult to work in difficult environment surrounding them, which will result in occurrence of error. For easy and error free work MRO companies has to focus on technology and they have to update themselves for adopting new technologies.

In India the employees are using little technology compared to outsiders because of money constraint. As the airworthiness has to be maintained and the time that will take more for attending the fleet, intervening of technology into MRO will be more helpful for the sector to complete the maintenance works very effectively and precisely without any difficulties in handling them.

12.4.1.5   SECURITY AND SAFETY

Aviation industry will always look for the safety and security aspects of passengers. Any incident happened in the middle of the sky will result in loss of lives and also a big burden to airlines, for this reason the aircraft before it has to be airborne should be checked effectively for any defects which is prescribed by the respective aviation regulatory bodies of a country.

Also, the employees who are working in MRO should feel responsible and should follow the safety requirements provided by regulatory bodies. In case of India these MROs has to strictly follow the safety parameters prescribed by DGCA. There were 90 accidents recorded due to technical issues and human factor is the reason for one third of these accidents in 2013 alone. The players in the MRO sector have to strictly adhere to the regulatory bodies and has to audit whether they are reaching the standards or not.

12.4.1.6   ECONOMIC INCENTIVES AND R&D INVESTMENTS

No business wants to end up in losses, so there should be some encouragement from government bodies for establishing MROs and also provide resources to conduct research and development of the operations that are handling. Research will consider the patterns over time to capture a deeper understanding of the breadth of contribution to the knowledge bases underpinning aviation.

By having a R&D, Indian MRO Company can work on the areas such as optimization of resources, control the spare parts inventory, and reduce wastage of both man power and also electricity. R&D will help the MROs to cope up with losses that will occur if they are using more resources than required. Also, the interested ones will invest into the sector by confirming only after the company's place in the market. So, government has to encourage the new entrants into the sector by providing some soft loans.

12.4.1.7   EDUCATION AND TRAINING

As safety starts from ground in aviation industry the personnel who are at maintenance work are to be well trained in handling the operations and also, they ought to be skilled in their specialized areas. Also, there should be initiative from government for setting up training centers for those who are interested. In most cases it will take around 5 to 8 years for a new technical specialist to acquire required practical experience and receive a skilled license [17].

Although India has an advantage of getting personnel at low cost and it has a shortage of qualified MRO personnel who are very much skilled to carry out complicated maintenance works on the latest aircrafts and their related components. This will support the establishment of MRO training centers and educating the interested people [18].

12.4.1.8 GOVERNMENT POLICIES AND SUPPORT

Without any support from government in case of payment of taxes for the maintenance as well as importing of spare parts and also the support in other forms such as provision of other ancillaries for successful running of the business. In case of India earlier till 2016 there are so many taxes such as Value added tax, 12% service taxes, 19% import duties and 13% royalty fee, but there are some exemptions in these taxes and the airline operators are getting some reimbursement for what they are paying.

After the introduction of GST there is a tax rate of about 18% levied on the aircraft maintenance which makes the MRO service more costly as compared with other outside MROs. Therefore, MRO companies need the support from the government in terms of subsidiaries and relaxation from tax payments.

12.5 RESULTS AND DISCUSSION

12.5.1 INSPECTION OF THE AIRCRAFT

Automated drone technology: The automation of defect detection that would enable engineers to pinpoint potential damage that could be inspected visually at a later stage. If this was possible, UAV) images could be compared with existing digital images of defects that have previously been found. This will expertise the time to inspect the aircrafts defects. Aircraft inspection drones can be broadly classified into two types; automated drones and manually operated drones. Two types of drones; Single Rotor Aircraft Inspection Drones and Multi-Rotor Aircraft Inspection Drones [25].

Visual inspection of aircraft that have, or may have been damaged by lightning or other similar occurrences. Manual inspection takes between four and six hours while UAVs can get the job done in half an hour. The new inspection process will take only three hours, including 30 minutes of image capture by the drone, and will improve operator maintenance floor plan.

Application of DRONE in MRO business have proven to reduce the inspection time on a scheduled maintenance by 80% to 90% [27].

12.5.2   SAFETY

In the above study it has been shown that Automated drone Technology increases the workplace safety whilst maintaining a competitive edge. Traditional survey methods are labor intensive, time consuming and often needs dedicated facilities and infrastructure to be conducted. Automated drone technology uses no heavy or cumbersome equipment that may further damage the aircraft, or puts engineers into potentially hazardous situations.

12.5.3   COST REDUCTION

Application of drone integration with the flight control system, automation, and sensor technologies provide innovative 3D visualization and damage localization tools that offer aircraft operators and MRO significant cost saving to their MRO operation. The automated drone captures all the required images with its on-board camera. High quality pictures are transferred to the respective database for detailed analysis using a software application Remote Automated Plane Inspection and Dissemination System (RAPID). This allows the operator to localize and measure visual damage on the aircraft's surface by comparing it with the aircraft's digital mock-up. The software automatically generates an inspection report thus resulting in saving time and AOG which will result in saving cost of the career and Operator. Effective utilization of application of drones helps in finding the problem areas and plan repairs(Maintenance floor plan) that are needed in saving time where else in the case of manually planning of maintenance activity will require more time and this it lead to delayed in deliver resulting in higher cost for the operator.

12.5.4   REDUCE DELIVERY TIME

It has also been discussed in the above analysis that Delivery of spare parts which could help cut down on time spent by MRO personnel in the hangar, ultimately resulting in much more efficient repair work. The integration of automation in defect detection would enable engineers to pinpoint potential damage that could be inspected visually at a later stage. This could save time for the engineers and they can use the UAV images to compare with the existing digital images of defects that have previously been found.

This automation process would save the time in the hanger and optimize effective delivery time of the aircraft. In research done by Robbins and Keller [45] speaks about the implementation of advanced robots in the inspection of the fuselage will reduce the cost, time, and improve the quality. Since this automation is designed for the use inside maintenance hangars, the drone is equipped with a laser-based sensor capable of detecting obstacles and halting the inspection if necessary. This laser-based technology allows the vehicle to fly automatically without the need for remote piloting, with laser positioning, the drone can safely perform end-to-end visual inspections.

12.5.5 DATA SECURITY

The research also sums up to provide integrated database. The maintenance system integration with drone application has significant potential in terms of effectiveness, cost reduction, and traceability. The data is traceable, because all data harvested by the drone is stored in a secure Cloud environment that is accessible at all times to every stakeholder. Aircraft inspection drones has simplified the documentation process associated with each inspection activities done on the component or the aircraft. Database can be designed for each MRO business specifications; the drone is programed to detect any aircraft structural data and store effectively and reduce error. An inspection report after every inspection is automatically generated, providing reliable and traceable data to improve efficiency across the board.

12.6 CONCLUSION AND FUTURE SCOPE

The study concludes that MRO Business Services require research and investment in technologies like drones, machine learning (ML) and cloud applications to support the maintenance process. This will lead to operations that are more effective, optimized technician schedule for each work plan as well as improved turnaround times, which could be achieved through accurate and faster inspections delivery plan and maintenance design plan and execution actions. This will achieve in higher customer satisfaction, which will altogether build richer data pattern for the operator to analyzes that will lead to reliable maintenance plan and execution capabilities in future operations. Besides improving, the accuracy in inspection next generation of drone's automation will still require a human interference to set up more instructions as per the individual business needs. With the help of visual processing algorithms integrated with IT application will enable the drone application to send maintenance work orders straight to the respective technician as soon as a fault has identified.

The future Study will include extensive research and study on Aircraft inspection drones and classification of drones; automated drones and manually operated drones in the MRO Process. Fast-paced error-free inspection process for the growth of aircraft inspection drones in the aviation industry. Extensive study on flight control, automation, and sensor technologies with innovative 3D visualization and damage localization tools that offer aircraft operators and MRO significant cost saving to their operation. The next generation of drones will still require a human controller, but visual processing algorithms combined with IT-systems will enable the drone to send work orders straight to the maintenance crew as soon as a fault has been identified. Effective regulatory policy for the MRO business service provider to improve the scope of growth and alliance with the aviation stakeholders.

KEYWORDS

- **aircraft maintenance program**
- **cloud services**
- **internet of drones**
- **MRO business services**
- **regional connectivity scheme**
- **unmanned aerial vehicle**

REFERENCES

1. Anil, S., Yadava, G. S, & Deshmukh, G., (2011). A literature review and future perspectives on maintenance optimization. *Journal of Quality in Maintenance. Engineering, 17*(1), 5–25. doi: 10.1108/13552511111116222.
2. Lia, S., (2020). *Autonomous Drone Race: A Computationally Efficient Vision-Based Navigation and Control Strategy.* Elsevier. doi.org/10,.1016/j.robot.2020.103621.

3. *New Maintenance, Repair and Overhaul (MRO) Technologies That Will Change the Aviation Industry.* June 23, 2020, from https://blog.satair.com/mro-europe-2021-brighter-times-ahead-aviation (accessed on 30 November 2021).

4. Ismail, A. A., (2018). *A Literature Based Examination of Human Error in Military Aviation Maintenance Management Programs.* 2018 Decemberhttps://bspace.buid.ac.ae/bitstream/handle/1234/1309/2015103173.pdf?sequence=2&isAllowed=y (accessed on 30 November 2021).

5. Kulyk, M., Kharchenko, V., & Matiychyk, M., (2011). *Justification of Thrust Vector Deflection of Twin-Engine Unmanned Aerial Vehicle Power Plants.* doi: 10.3846/16487788.2011.566319.

6. Everaerts, J. (2008). The use of unmanned aerial vehicles (UAVs) for remote sensing and mapping. The International Archives of the Photogrammetry, *Remote Sensing and Spatial Information Sciences 37,* 1187–1192. DOI: 10.1201/9780203888445.ch9.

7. Ayeni, P., Baines, T., Lightfoot, H., & Ball, P., (2011) State-of-the-art of 'Lean' in the aviation maintenance, repair, and overhaul industry. *Proceedings of the Institution of Mechanical Engineers, Part B: Journal of Engineering Manufacture, 225*(11), 2108–2123.

8. Block, J., Ahmadi, A., Tyrberg, T., & Söderholm, P., (2014). Part-out-based spares provisioning management: A military aviation maintenance case study. *Journal of Quality in Maintenance Engineering, 20*(1), 76–95.

9. Dekker, S., (2014). *The Field Guide to Understanding Human Error.* Ashgate Publishing, Ltd.

10. Doyle, E. K., (2004). On the application of stochastic models in nuclear power plant maintenance. *European Journal of Operational Research, 154,* 673–690.

11. Hooper, B. J., & O'Hare, D., (2013). Exploring human error in military aviation flight safety events using post-incident classification systems. *Aviation, Space, And Environmental Medicine, 84*(8), 803–813.

12. Hopp, W. J., & Kuo, Y., (2010). Heuristics for multicomponent joint replacement: Application to aircraft engine maintenance. *Naval Research Logistics, 45,* 435–458.

13. Jeong, K., Choi, B., Moon, J., Hyun, D., Lee, J., Kim, I., & Kang, S., (2016). Risk assessment on abnormal accidents from human errors during decommissioning of nuclear facilities. *Annals of Nuclear Energy, 87,* 1–6.

14. Panagiotidou, S., & Tagaras, G., (2007). Optimal preventive maintenance for equipment with two quality states and general failure time distributions. *European Journal of Operational Research, 180,* 329–353.

15. Rashid, H. S. J., Place, C. S., & Braithwaite, G. R., (2014). Eradicating root causes of aviation maintenance errors: Introducing the AMMP. *Cognition. Technology & Work, 16*(1), 71–90.

16. Shafiee, M., & Chukova, S., (2013). Maintenance models in warranty: A literature review. European. *Journal of Operational Research, 229*(3), 561–572.

17. Simões, J. M., Gomes, C. F., & Yasin, M. M., (2011). A literature review of maintenance performance measurement: A conceptual framework and directions for future research. *Journal of Quality in Maintenance Engineering, 17*(2), 116–137.

18. Lian Ding, Dannie Davies, & Christopher A. McMahon, (2009). The integration of lightweight representation and annotation for collaborative design representation, *Research in Engineering Design 19*(4), 223–238. DOI: 10.1007/s00163-008-0052-3 (accessed on 19 November 2021).

19. *AI, Drones and Sustainability in Commercial Aviation?* | BlueSky News. Retrieved from: https://www.blueskynews.aero/contributors/AI-drones-and-sustainability-in-commercial-aviation-Graham-Grose-IFS.html (accessed on 19 November 2021).
20. *Air New Zealand uses drones for aircraft inspections - Airline Ratings.* Retrieved from https://www.airlineratings.com/news/air-new-zealand-uses-drones-aircraft-inspections/ (accessed on 19 November 2021).
21. *Airbus Innovation for Military Aircraft Inspection and Maintenance - Defense - Airbus.* Retrieved from https://www.airbus.com/newsroom/news/en/2019/05/airbus-innovation-for-military-aircraft-inspection-and-maintenance.html (accessed on 19 November 2021).
22. *Airbus Launches Advanced Indoor Inspection Drone to Reduce Aircraft Inspection Times and Enhance Report Quality - Commercial Aircraft - Airbus.* Retrieved from https://www.airbus.com/newsroom/press-releases/en/2018/04/airbus-launches-advanced-indoor-inspection-drone-to-reduce-aircr.html (accessed on 19 November 2021).
23. *Airbus Using Drones to Visually Inspect Aircraft* | Unifly. Retrieved from https://www.unifly.aero/news/airbus-using-drones-to-visually-inspect-aircraft (accessed on 19 November 2021).
24. Lian Ding, Alex Ball, Jason Matthews, Chris McMahon, & Manjula Patel, (2007). Product representation in lightweight formats for product lifecycle management (PLM), in: *4th International Conference on Digital Enterprise Technology (DET 2007),* DOI: 10.1504/ijplm.2011.038100.
25. *Aircraft Inspection Drones Size, share, Market 2018 Key Manufacturers, Demands, Industry Share, Size, Status and Forecasts to 2028.* Retrieved from: https://www.persistencemarketresearch.com/market-research/aircraft-inspection-drones-market.asp (accessed on 19 November 2021).
26. Almadhoun, R., Taha, T., Seneviratne, L., Dias, J., & Cai, G., (2016). Aircraft inspection using unmanned aerial vehicles. *International Micro Air Vehicle Competition and Conference,* 43–49.
27. Blokhinov, Y. B., Gorbachev, V. A., Nikitin, A. D., & Skryabin, S. V., (2019). Technology for the visual inspection of aircraft surfaces using programmable unmanned aerial vehicles. *Journal of Computer and Systems Sciences International, 58*(6), 960–968. https://doi.org/10.1134/S1064230719060042.
28. Liu, Jian-Hua, Ning, Ru-Xin, Wan Bi-Le, & Xiong Zhen-Qi, (2007). Research of complex product assembly path planning in virtual assembly, Journal of System Simulation 19(9), 2003.
29. Bouarfa, S., Doğru, A., Arizar, R., Aydoğan, R., & Serafico, J., (2020). *Towards Automated Aircraft Maintenance Inspection. A Use Case of Detecting Aircraft Dents Using Mask R-CNN.* https://doi.org/10.2514/6.2020-0389.
30. *Delta wants to use drones on the airfield for Maintenance, Repair and Overhaul (MRO).* https://www.plantservices.com/industrynews/2019/delta-wants-to-use-drones-on-the-airfield-for-Maintenace,repairandoverhaul(MRO)/ (accessed on 19 November 2021).
31. Aditya P., Kumar, U., Galar, D., & Stenström, C., (2015) "Performance measurement and management for maintenance: a literature review," *Journal of Quality in Maintenance Engineering, 21*(1), 2–33, doi: 10.1108/JQME-10–2013- 0067.
32. Wohl, R. (2015). Up in the air: new approaches to the history of American aviation. *Reviews in American History, 43*(4), 687–696.

33. Drone Maintenance, Repair and Overhaul (MRO) Taking Shape – Aviation Maintenance Magazine. (n.d.). June 22, 2020, from https://www.avm-mag.com/drone-mro-taking-shape/ (accessed on 30 November 2021).

34. Flying, clinging and crawling – using robots in Maintenance, Repair and Overhaul (MRO). (n.d.). June 22, 2020, from https://www.aerosociety.com/news/flying-clinging-and-crawling-using-robots-in-mro/ (accessed on 30 November 2021).

35. Hussein, K. (2018). "Delivery Time Delay, Its Cause and Remedies in Maintenance, repair and overhaul (MRO) Service in Dejen Aviation Engineering Industry."

36. Jennifer M. Johnson. (2017). Naval Postgraduate. Security, June, 1–55.

37. Koslosky, L. B, "Commercial aviation in a digital world: a cyberphysical systems approach for innovative maintenance" 2019. *Aeronautics and Aerospace Open Access Journal*, *3*(2), 49–64. https://doi.org/10.15406/aaoaj.2019.03.0081 (accessed on 30 November 2021).

38. Kostopoulos, V., Psarras, S., Loutas, T., Sotiriadis, G., Gray, I., Padiyar, M. J., Petrunin, I., Raposo, J., Fragonara, L. Z., Tzitzilonis, V., Dassios, K., Exarchos, D., Andrikopoulos, G., & Nikolakopoulos, G. (2018). Autonomous Inspection and Repair of Aircraft Composite Structures. *IFAC-PapersOnLine, 51*(30), 554–557. https://doi.org/10.1016/j.ifacol.2018.11.267 (accessed on 30 November 2021).

39. Lester, O, "Maintenance, Repair and Overhaul (MRO)," 2009, Industry Readiness Center Southwest, Naval Postgraduate.

40. Malandrakis, K., Savvaris, A., Domingo, J. A. G., Avdelidis, N., Tsilivis, P., Plumacker, F., Fragonara, L. Z., & Tsourdos, A. "Inspection of aircraft wing panels using unmanned aerial vehicles" (2018). 5th IEEE International Workshop on Metrology for AeroSpace, MetroAeroSpace 2018 – Proceedings, 56–61. https://doi.org/10.1109/MetroAeroSpace.2018.8453598 (accessed on 30 November 2021).

41. Manda, V., & Chaitanya, M. "Aircraft Servicing, Maintenance, Repair & Overhaul- the Changed." (2017). International Journal of Research in Engineering and Applied Sciences (IJREAS), 249–270. https://www.researchgate.net/publication/323880295_aircraft_servicing_maintenance_repair_overhaul-the_changed_scenarios_through_outsourcing (accessed on 30 November 2021).

42. Miranda, J., Larnier, S., Herbulot, A., & Devy, M. "UAV-based inspection of airplane exterior screws with computer vision." (2019). VISIGRAPP 2019 – Proceedings of the 14th International Joint Conference on Computer Vision, Imaging and Computer Graphics Theory and Applications, 4, 421–427.

43. Maintenance, Repair and Overhaul (MRO) Drone and Ubisense join forces with new Smart Hangar Solution – Geospatial World. June 23, 2020, from https://www.geospatialworld.net/news/Maintenace, repair and overhaul (MRO)-drone-and-ubisense-join-forces-with-new-smart- hangar-solution/ (accessed on 30 November 2021).

44. Orient Aviation – Orient Aviation. June 23, 2020, from http://www.orientaviation.com/articles/4726/aar-trials-Maintenace, repair and overhaul (MRO)-drone-inspections-at-miami-shop (accessed on 30 November 2021).

45. Robbins, J., Fernandes, R. F., & Keller, K. (2015). Design of a System for Aircraft Fuselage Inspection. https://catsr.vse.gmu.edu/SYST490/490_2015_AircraftInspection/AircraftInspectionTechnicalReport.pdf (accessed on 30 November 2021).

46. ST Engineering and Air NZ trial drones for aircraft inspections | News | Flight Global. June 23, 2020, from https://www.flightglobal.com/Maintenace,repairandoverhaul(M

RO)/st-engineering-and-air-nz-trial-drones-for-aircraft-   inspections/133128.article (accessed on 30 November 2021).

47. Sun, Y., Zhang, L., & Ma, O, "Robotics-Assisted 3D Scanning of Aircraft" (2020). 1–11. https://doi.org/10.2514/6.2020–3224 (accessed on 30 November 2021).

48. The Australian Aircraft Industry: A Defence Point. (2020). KPMG Edu. https://www.industry.gov.au/sites/default/files/2019–12/australias-aerospace-industry-capability-report.pdf (accessed on 30 November 2021).

49. The future of Maintenance, Repair and Overhaul (MRO): "emerging technologies in aircraft maintenance unit Aviation, June 22 2020 https://www.researchgate.net/publication/295256240_Maintenance_Repair_and_Overhaul_MROFundamentals_and_Strategies_An_Aeronautical_Industry_Overview (accessed on 30 November 2021).

50. "The Innovative Use of Drones in Aircraft Maintenance and Repair." June 22, 2020, from https://primeindustriesusa.com/drones- in-aircraft- maintenance/ (accessed on 30 November 2021).

51. "UAS drone for Maintenance, Repair and Overhaul (MRO) inspections | Intelligent Aerospace." June 22, 2020. https://www.intellige nt-aerospace.com/commercial/article/14068378/uas-drone-for-Maintenance,repairandoverhaul(MRO)-inspections (accessed on 30 November 2021).

52. Unmanned Aerial Vehicle Market, UAV Size, Share, system and Industry Analysis and Market Forecast to 2024 | Markets and Markets TM. (n.d.). June 23, 2020, from https://www.marketsandmarkets.com/Market-Reports/unmanned-aerial-vehicles-UAV-market-662.html (accessed on 30 November 2021).

53. Airbus News Press (2018). https://www.airbus.com/newsroom/press-releases/en/2018/04/airbus-launches-advanced- indoor- inspection-drone-to-reduce-aircr.html (accessed on 30 November 2021).

54. Graham Grose (2020). Bluesky Business Aviation news: https://www.blueskynews.aero/contributors/AI-drones-and-sustainability-in-commercial-aviation-Graham-Grose-IFS.html (accessed on 30 November 2021).

55. Satair Blog (2018). https://blog.satair.com/four-Maintenace, repair and overhaul (MRO)-trends-to-consider Aviation, Transport & Logistics | Staff Reporter, Singapore. https://blog.satair.com/four-mro-trends-to-consider (accessed on 30 November 2021).

56. Lee, G., Ma, Y.-S., Thimm, G. L., & Verstraeten, J. (2008). Product lifecycle management in aviation maintenance, repair and overhaul, *Computers in Industry 59*, 296–303. DOI: 10.1016/j.compind.2007.06.022.

57. Liu Ming, Zuo Hongfu, Geng Duanyang, & Cai Jing, (2006). Expert system of maintenance review board report based on CBR and RBR, *Journal of Beijing University of Aeronautics and Astronautics 5*, 521–525.

58. Yu Fengjie, Ke Yinglin, & Ying Zheng, (2009). Decision on failure maintenance for aircraft automatic join-assembly system, *Computer Integrated Manufacturing Systems 15*(9), 1823–1830.

59. Johannes Christian, Horst Krieger, Andreas Holzinger, & Reinhold Behringer, (2007). Virtual and Mixed Reality Interfaces for e-Training: Examples of Applications in Light Aircraft Maintenance, in: Proceedings of the 4th international conference on Universal access in human-computer interaction: applications and services (UAHCI 2007), p.520. DOI: 10.1007/978–3-540–73283–9_58.

60. Sajay Sadasivan, Deepak Vembar, Carl Washburn, & Anand K. Gramopadhye, (2007). Evaluation of Interaction Devices for Projector Based Virtual Reality Aircraft Inspection Training Environments, in: Proceedings of the 2nd international conference on Virtual reality (ICVR 2007), p.533. DOI: 10.1007/978–3-540–73335–5_58.

61. Lu Zhong, & Sun Yongchao, (2010). Disassembly sequence planning of civil aircraft products for maintainability design, *Acta Aeronautica Et Astronautica Sinica 31*(1), 143–150.

62. Shu Li, Liu Yi, Liu Jia, & Nandong Wang, (2009). The application research on the virtual maintenance in aircraft design, in: Reliability, The 8th International Conference on Reliability, Maintainability & Safety (ICRMS 2009), pp. 678–683. DOI: 10.1109/icrms.2009.5270103.

63. Christiand, Jungwon Yoon, (2007). Optimal assembly path planning algorithm for aircraft part maintenance, in: International Conference on Control, Automation and Systems (ICCAS 2007), 2190–2194. DOI: 10.1109/iccas.2007.4406696.

64. Hao Xu, & Pu Wan, (2011). An improved genetic algorithm for solving simulation optimization problems, *International Journal of the Physical Sciences 6*(10), 2399–2404.

65. Liu Jian-Hua, Ning Ru-Xin, Wan Bi-Le, & Xiong Zhen-Qi, (2007). Research of complex product assembly path planning in virtual assembly, *Journal of System Simulation 19*(9), 2003-(2007).

66. Lian Ding, Alex Ball, Jason Matthews, Chris McMahon, & Manjula Patel, (2007). Product representation in lightweight formats for product lifecycle management (PLM), in: 4th International Conference on Digital Enterprise Technology (DET 2007), DOI: 10.1504/ijplm.2011.038100.

67. Lian Ding, Dannie Davies, & Christopher A. McMahon, (2009). The integration of lightweight representation and annotation for collaborative design representation, *Research in Engineering Design 19*(4) 223–238. DOI: 10.1007/s00163–008–0052–3.

68. Everaerts, J. (2007). The use of unmanned aerial vehicles (UAVs) for remote sensing and mapping. *The International Archives of the Photogrammetry, Remote Sensing and Spatial Information Sciences, 37,* 1187–1192.DOI: 10.1201/9780203888445.ch9.

69. Zaloga, S. A. (2007). UAV in the hand can be worth a lot. Unmanned Systems. Jul/Aug. pp. 25–27.

70. World Wide Aviation, "Indian MRO Industry on Verge of Expanding?," www.mro-network.com, Oct 24, 2018.

CHAPTER 13

Deploying Unmanned Aerial Vehicle (UAV) for Disaster Relief Management

SOMYA GOYAL

Manipal University Jaipur, Jaipur – 303007, Rajasthan, India,
E-mail: somyagoyal1988@gmail.com

ABSTRACT

A drone is an auto-aircraft which is operated by humans but without a human pilot inside the aircraft. It is also called unmanned aerial vehicle (UAV). UAVs originated and deployed for undertakings which are considered dangerous for humans. Drones have found initially, military applications, later on applications from the commercial domains. In this chapter, the use of UAVs for relief-task in case of disasters is proposed. Because when natural disasters strike, the initial striking period is the most critical time. Life-saving task force work to save lives and minimize damage. For this, accurate location information is desirable to carry out rescue operations effectively. The more we know affected areas, the better relief-task can be carried out. In such scenario, drones find applications to scan the affected areas for generating the radiation intensity in case of nuclear disaster, to gather location of hotspots, to look for the victims and to assess the intensity of damage incurred. The drones can play an essential role in disaster relief management. Drones allows the awareness about the situation quickly by utilizing the image capturing techniques and then developing the maps for that. The captured images are useful for strong broadcastings like news and for locating the survivors and to communicate with them. The information

The Internet of Drones: AI Applications for Smart Solutions. Arun Solanki, PhD, Sandhya Tarar, PhD, Simar Preet Singh, PhD & Akash Tayal, PhD (Eds.)

communicated by the UAV or drones, is effective to mitigate the impacts of disasters and it also support better decision-making. I propose to deploy drone for the quick damage-assessment and also help recovery. This chapter includes three case studies where the drones are deployed to face disasters including nuclear accidents, floods, and wildfires. It can be concluded that the application of drones in disaster management provides effective and reliable support to emergency response team (ERT) members and has potential to save lives in situation of natural calamities.

13.1   INTRODUCTION

Drones are unmanned aerial vehicles (UAVs); these are aircrafts which do not require human pilot sitting inside the drone shown in Figure 13.1. These are equipped with a control device which works automatically or remotely controlled by human. They perform flight without man sitting inside. Another name is aerial robot. They find applications in military related fields [1].

FIGURE 13.1   A drone (UAV).

Disaster is a compromising condition that requires pressing activity, a compelling reaction and inside a disaster situation there might be dangers for responders, just as for those influenced. Reaction time is pivotal for influenced people and conditions to be tended to on their necessities. In this specific situation, the objective of this work is to help the operators engaged with the disaster reaction, through an application-upheld community-oriented arrangement utilizing drones. This arrangement expects to gather data from the worked disaster situation, so that, through the joint effort of masters, there is a more noteworthy help for the dynamic made by the mindful operators inside this situation, making it happen in a shorter time, consequently accelerating the reaction to the disaster [1–7].

In todays life, everything is associated by means of web and data is looked for by the internet. At the point when a crisis circumstance happens, it greatly affects society, so it is fundamental to decipher the signs that are created and the reaction groups, in crisis circumstances, ought to have the option to act promptly and successfully following being called. Reaction time, hence, is basic for controlling the antagonistic circumstance. In any case, reaction activities can get befuddling due to the absence of solid and coordinated data about the situation of crisis [8–10]. Subsequently, the coordination of activities loses effectiveness as the absence of data and its discontinuity add to the wasteful portion of the assets included. There is regularly an absence of data with respect to the locale around a crisis; The absence of flying vision of the site and its environmental factors carries confinements to the specialists liable for the reaction. Frequently the situation experienced in these circumstances presents a few hazard factors, both for the group liable for the reaction and for individuals who need some help in this condition [11, 12]. Also, reaction times related with viable activities are fundamentally significant in lessening or in any event, disposing of existing dangers. In this situation, the chance to apply an answer including drones was imagined, planning to help choice making in crisis conditions by getting and imparting helpful data to the reaction groups. Drones are machines that are picking up prominence and are as of now broadly utilized. They can be independent, distantly controlled or furnished with cameras; permitting constant account and survey [13, 14].

The working of ERT is depicted in Figure 13.2. Anticipation is the primary stage and covers activities planned for forestalling debacles or diminishing the effect of the outcomes. Readiness is the second period of catastrophe the board, that incorporates activities to expand the responsiveness of people and associations with the goal that they can act all the more successfully. The appropriate response is the third stage in crisis and catastrophe the board. It covers the arrangement of activities taken to help and help the influenced parts (individuals, creatures, condition, properties, and so forth.), lessening harm and misfortunes, to guarantee the working of the principle frameworks that make up the network foundation. Remaking is the fourth and last stage and incorporates activities for the recuperation of the influenced parts, for example, reconstructing a network, permitting the rebuilding of the ordinary state.

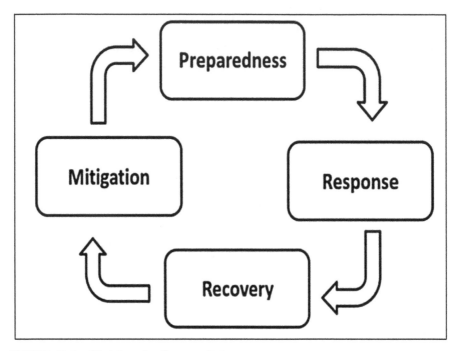

FIGURE 13.2   Workflow for disaster relief team.

The drones can play an essential role in disaster relief management [15]. This work is dedicated to exploring the deployment of drones to gain insight into the situation using technology for mapping and imagery. Drones cover the imagery to look for survivors and hence to rescue them. Another major task is to provide communication for ERT. Based on the information supplied by drone, the effect of the disaster is mitigated successfully [17–20]. Further, it also supports better decision-making. This chapter proposes to deploy drone for the quick damage-assessment and to help recovery.

This chapter is organized as follows: Section 13.2 includes a case study where the drones are deployed to face disasters like nuclear accidents. Section 13.3 covers the application of drones to floods. Section 13.4 describes a case study on utilization of drones in wildfire disaster. Section 13.5 concludes that for special rescue teams, the drone application helps much in a rapid and effective disaster relief management.

13.2   DRONES IN NUCLEAR DISASTER: CASE STUDY

In nuclear disaster, radioactive substances are spread in very large amounts over huge geographical area [21]. The environmental recovery is essential, which in turn requires to identify the areas which are having strong radiations called hotspots. The reflection of such distribution of radioactive rays or substances over a geographical map is then generated. Such maps are very useful to recover from nuclear disaster [22]. Sensors are effectively deployed in literature to monitor the real-life situations [23–28]. Generally, the cracks in the road, and the water sewage from residential areas are hotspot which are necessary to be removed. The reason of these hotspots is majorly due to emission of cesium during the nuclear accident [29]. If the manpower is deployed to recover from accidental spot using survey meters, then, it may result in the high risk of mishap to the workers. The reason is that the reading-meters are not effective to estimate the degree of concentration of gamma radiations along a particular direction. Perhaps the situation may be like workers approach hotspots unintentionally. Therefore, some automatic detection is desirable without human pilot. Such a device (UAV) automatically detects radioactive affected areas. A drone like UAV is utilized with an assembled camera. It flights over huge geographic area and takes the measurements quickly.

To describe in detail, an example of the nuclear accident happened at Fukushima Daiichi Nuclear Power Station (FDNPS), due to tsunami on 11 March 2011 [30]. For recovery, a radiation distribution map is desirable to plan the decontamination in the affected areas. A remote imaging system mounted with Compton camera on a drone is developed to compute the concentration levels of radioactive radiations or substances over the area. This drone can visualize the radioactive distribution as a three-dimensional plot using the Compton camera with the help of light detection and ranging system in three dimensions. It is used to monitor the radioactive hotspots in the surrounding area of disaster spot.

The device is multi-copter drone mounted with a camera, satellite sensor and a sensor for measuring the inertial units. Using this device, a 3D visualization of the radioactive substance is made with respect to the topographic information.

The complete device is shown in Figure 13.3. The device takes two flights. One to measure radiation concentration level with the help of Compton camera. Second, to measure topographical information in 3-Dimensions with the help of 3-Dimensional Light Detection and Ranging (3D-LiDAR).

By the mapping of the 3D topographical model, image is reconstructed of radioactive compounds. The highest concentration areas having the radioactive activity can easily be identified.

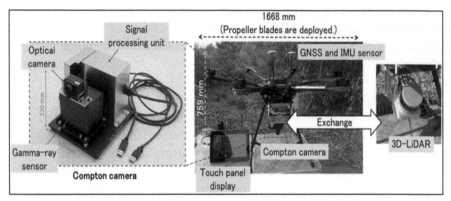

FIGURE 13.3   Drone with camera and LiDAR (center); zoom-in camera picture on left and zoom-in LiDAR (on right) [30].

The imaging results from the UAV device are shown in Figure 13.4. The distance is of 1 meter between the source of radioactive activity and the capturing camera. The radioactive intensity is reflected by the color bars.

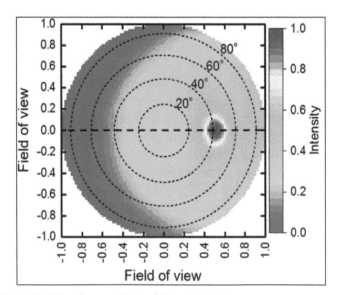

FIGURE 13.4   Radiation images obtained from the device.

This case study shows that an imaging system using drones is useful and safe to record the intensity of radioactive substances by observing gamma rays in the nuclear disaster area.

13.3   DRONES IN FLOOD DISASTER: CASE STUDY

In this section, the application of drones in the flood disaster has been explained taking an example of case study [31–35]. With respect to flood disaster, drones play vital role in three different timeframes. One, before the flood, Second, during the flood crisis and third, after the flood. The role before the flood is to take a survey over the water level in the rivers as a preventive measure. During floods, the drones act as 'eagle's eye' to provide the information. After disaster, drones help to map the affected area to quickly assess and provide relief services [36]. These services can be categorized as pre-flood management, flood management and post-flood management as shown in Figure 13.5.

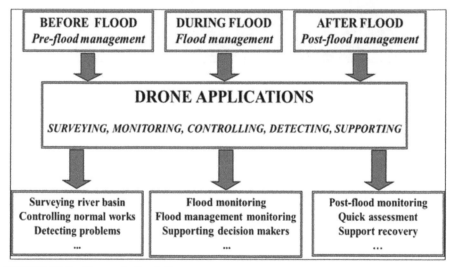

FIGURE 13.5   Drones in flood disaster.

The drones are mounted with high resolution cameras for surveying purposes by capturing high-definition images. It also remaps the river basin for better insight. During the flight, drones with mounted cameras take photos from which 3D models of the river basin can be generated. Using drones, dam status can be easily assessed, due to cracks in dams, most of the floods occur [37]. The high-definition photos captured using drones are used to evaluate secreted places. From the evaluation made by experts, it can be detected whether there are some cracks in dam or the hair lines occurred or steel parts corroded or else deformations resulted, as shown in Figure 13.6.

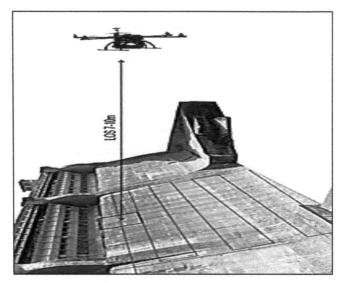

FIGURE 13.6   Dam survey using drones.

During the flood crisis, drones play a helpful tool to rescue team to map the affected areas and to find out the trapped people or animals. The rescue team utilizes the real-time information provided by the drones [37–43].

13.4   DRONES IN WILDFIRE DISASTER: CASE STUDY

Wildfires endangers the safety of property, human being, and the entire ecosystem. Whenever the wildfire occurs, a safe, in time, and least expensive method is desirable to quantify the damage caused by wildfire [44]. The expense in managing, monitoring, and extinguishing the wildfires is millions of dollars. As per the report [45], in the United States of America, $13 billion is spent for fire extinguish and $5 billion for managing the wildfires during the period from year 2006 to year 2015. Wildfires significantly impact the life of humans in an adverse manner financially and qualitatively. As the wildfire affects the ecosystem in a long-term, so the information and data regarding the damage caused to the vegetation is essential to be estimated effectively in a time-constrained way for recovery of current situation and possible predictions for the future accidents [46]. The agenda is to collect situational data in form of image captures and to derive classification information from the captured images as shown in Figure 13.7. The unique patterns of burnt vegetation have the features like severity of burn and the type of vegetation.

FIGURE 13.7   Detection of fire using drones.

The various mechanism the literature suggests for identifying these areas and remap on paper are as follows: (i) using a device mounted with global positioning system (GPS) sensor and rescue crew members traverse the burnt areas; (ii) using cameras and capture images of burnt area from the aircraft by ERT; (iii) using images captured by the satellites.

It is seen that using the sensors (multispectral), whether via satellites or human-pilot aircrafts are affective to estimate the recovery rate, to compute the level of severity over huge geographic areas but lacks in accuracy due to low power resolution [47]. From the study, it is clear that near-infrared bands are found to be effective in the evaluation, and estimation of wildfire impact [48]. The study reported the limitations of the usage of imaging spectrometer thermal emission and reflection radiometer to capture images of post-fire areas to estimate the severity of the disaster. The reason stated is that the technique used returns the images with poor resolution. They concluded that the images captured from the camera mounted in drones or UAS results in better resolution in comparison to the images captured using satellite and human-aircraft. Allison et al. [47] advocated the specific range of IR bands for the detection of wildfires. Aicardi et al. [48] showed the comparison between LiDAR and UAS-imagery. And concluded that UAS is more accurate for assessing the long-term impact and recovery rate. The use of UAS is two-fold beneficial: (i) high spatial resolution; and (ii) the low cost of operation in comparison to satellite-based system and the aircraft boarded with humans.

The measurement index is a ground reference along with a classification model. The map is produced reflecting the burn classification. Then, the comparison between the situation before the happening of disaster and the situation after the happening of the disaster over the criteria like area and volume of burnt vegetation is made [49–51].

A study [52] made an experimental comparison between the performance low-cost camera operating in multispectral range and the performance by a satellite. The authors favored the multispectral cameras mounted on UAS. In another study [53], the authors made a study and reported with positive results for sensor mounted on a UAS and operating in hyperspectral range. They also found that the range for bands which is most suitable for fire detection and estimation of fire-severity is between 400 nm and 900 nm. From the literature, the studies [54–56] proposed to use the UAS mounted with special and least expensive sensors to estimate the intensity of damage caused by fire and the rate of fire spreading over the specified area.

For this purpose, unmanned aerial system (UAS) or drone can be deployed with an embedded multi-spectral sensor. It allows to classify and estimate the damage caused by wildfire as shown in Figure 13.8. The agenda of deploying the drones is to gather imagery for both the situations-before the happening of wildfire and after the happening of wildfire, and to use the captured images to compute the pre-fire health, to estimate the damage caused by the wildfire, and to record the recovery of vegetation. For the revegetation computations [57–60], burnt area and volume are traced. The classification results for burnt area obtained from the drones are compared with those produced for unburnt vegetation area. The potential of using UAS is clear from the accuracy of classification, turnaround time, and high resolutions in spatial and temporal domains.

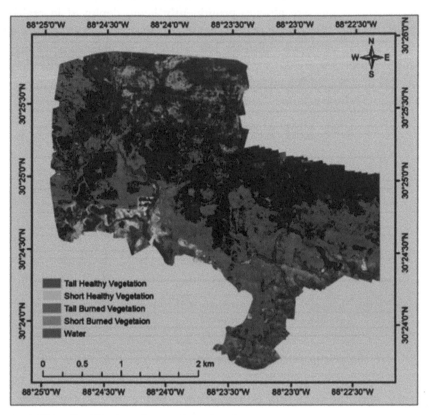

FIGURE 13.8 Classification of vegetation using drones.

13.5 CONCLUSION AND FUTURE SCOPE

This study shows that using Autonomous First Response Drones-based smart rescue system for critical situation management is a novel idea that can be used to save many lives by significantly reducing emergency response time [61, 62]. In addition, the paradigm is inexpensive, secure, and reliable compared to conventional response-time reduction methods. For the future scope of the reported study, I would like to work in two aspects. First, it is the fusion of drones or UAVs with other booming technologies. This work has bright future prospects as a project utilizing a number of drones mounted with sensors and interconnected with IoT systems. Then, applying some auto-classifiers and deep-learners over the images collected. The future agenda is to automate the reasoning of drones for fire extinguishing balls. The power of drones can be synergized with big data analytics and machine learning (ML) [63–68]. As these technologies are already prominently being deployed in real-life situations. Second, this work can be extended by applying UAVs in the catastrophic situation earthquake disaster [69, 70] over all three dimensions as shown in Figure 13.9. By the application of drones (UAS) in all three phases namely-pre disaster, during disaster and post disaster, the full proof safety can be ensured with high accuracy and low cost.

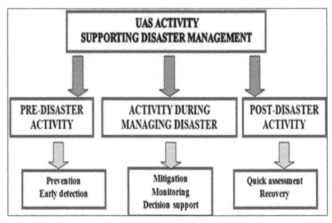

FIGURE 13.9   Possible application dimension for UAS.

In this way, the work will be extended in future for the application of drones in saving lives and property through natural disasters [71–75].

KEYWORDS

- **disaster management**
- **drone**
- **emergency response team (ERT)**
- **flood**
- **nuclear accident**
- **unmanned aerial vehicle (UAV)**
- **wildfire**

REFERENCES

1. Fennelly, L. J., & Perry, M. A., (2020). *Unmanned Aerial Vehicle (drone) Usage in the 21st Century,* 183–189. The Professional Protection Officer, Butterworth-Heinemann.

2. Erdelj, M., Natalizio, E., Chowdhury, K. R., & Akyildiz, I. F., (2017). Help from the sky: Leveraging UAVs for disaster management. *IEEE Pervasive Computing, 16*(1), 24–32.

3. Molina, P., Parés, M., Colomina, I., Vitoria, T., Silva, P., Skaloud, J., Kornus, W., Prades, R., & Aguilera, C. (2012). Drones to the Rescue! Unmanned aerial search missions based on thermal imaging and reliable navigation. *InsideGNSS, 7,* 36–47.

4. Chandhar, P., & Larsson, E. G., (2019). Massive MIMO for connectivity with drones: Case studies and future directions. *IEEE Access, 7,* 94676–94691.

5. Restas, A., (2015). Drone applications for supporting disaster management. *World Journal of Engineering and Technology, 3*(03), 316.

6. Apvrille, L., Tanzi, T., & Dugelay, J. L., (2014). Autonomous drones for assisting rescue services within the context of natural disasters. In: *2014 XXXIth URSI General Assembly and Scientific Symposium (URSI GASS)* (pp. 1–4). IEEE.

7. Glock, K., & Meyer, A., (2020). Mission planning for emergency rapid mapping with drones. *Transportation Science, 54*(2), 534–560.

8. Khayal, D., Pradhananga, R., Pokharel, S., & Mutlu, F., (2015). A model for planning locations of temporary distribution facilities for emergency response. *Socio-Econ. Plan. Sci. 52,* 22–30.

9. Kılcı, F., Kara, B. Y., & Bozkaya, B., (2015). Locating temporary shelter areas after an earthquake: A case for turkey. *Eur. J. Oper. Res., 243,* 323–332.

10. Langevin, A., Mbaraga, P., & Campbell, J. F., (1996). Continuous approximation models in freight distribution: An overview. *Transp. Res. Part B, 30*(3), 163–188.

11. Li, X., & Ouyang, Y., (2010). A continuous approximation approach to reliable facility location design under correlated probabilistic disruptions. *Transp. Res. Part B, 44,* 535–548.

12. Lin, Y. H., Batta, R., Rogerson, P. A., Blatt, A., & Flanigan, M., (2009). *Application of a Humanitarian Relief Logistics Model to an Earthquake Disaster.* Available from: http://www.acsu.buffalo.edu/~batta/TRB_Updated.pdf (accessed on 19 November 2021).

13. Gilman, D., (2014). In: Matthew, E., (ed.), *Unmanned Aerial Vehicles in Humanitarian Response.*

14. V, Béla., M, Golabi., A, Nedjati., Gümü¸sbu¨ga, F., & G, Izbirak., (2019). Top-down approach to design the relief system in a metropolitan city using UAV technology, part I: The first 48h. *Natural Hazards, 99*(1), 571–597.

15. Greenwood, F., (2015). *Above and Beyond: Humanitarian Uses of Drones.* Available from: http://www.worldpoliticsreview.com/articles/16750/above-and-beyondhumanitarian-uses-of-drones (accessed on 19 November 2021).

16. Hall, T., (2016). *Starting Your Own Gas Station or Convenience Store in Minnesota.* Available from: http://thompsonhall.com/starting-your-own-gas-stationminnesota-business-attorney/ (accessed on 19 November 2021).

17. Hamedi, M., Haghani, A., & Yang, S., (2012). Reliable transportation of humanitarian supplies in disaster response: Model and heuristic. *Procedia - Social. Behav. Sci., 54,* 1205–1219.

18. Ji, G., & Zhu, C., (2012). A study on emergency supply chain and risk based on urgent relief service in disasters. *Syst. Eng. Procedia, 5,* 313–325.

19. Jia, H., Ordóñez, F., & Dessouky, M. M., (2007). Solution approaches for facility location of medical supplies for large-scale emergencies. *Comput. Ind. Eng., 52*(2), 257–276.
20. Fadi, A., Lemayian, J., Alturjman, S., & Mostarda, L., (2019). Enhanced deployment strategy for the 5G drone-BS using artificial intelligence. *IEEE Access, 7*, 75999–76008.
21. Mietelski, J. W., & Povinec, J., (2014). Environmental radioactivity aspects of recent nuclear accidents associated with undeclared nuclear activities and suggestion for new monitoring strategies. *Journal of Environmental Radioactivity, 214*, 106–15.
22. Marturano, F., Chierici, A., et al., (2020). Enhancing radiation detection by drones through numerical fluid dynamics simulations. *Sensors, 20*(6), 1770.
23. Goyal, S., Bhatia, P. K., & Parashar, A., (2020). Cloud-assisted IoT-enabled smoke monitoring system (e-Nose) using machine learning techniques. In: Somani, A., Shekhawat, R., Mundra, A., Srivastava, S., & Verma, V., (eds.), *Smart Systems and IoT: Innovations in Computing. Smart Innovation, Systems and Technologies* (Vol. 141. pp. 743–754) Springer, Singapore. .https://doi.org/10.1007/978-981-13-8406-6_70.
24. Franco, C. D., & Buttazzo, G., (2015). Energy-aware coverage path planning of UAVs. *IEEE International Conference on Autonomous Robot Systems and Competitions.*
25. Ganeshan, R., (1999). Managing supply chain inventories: A multiple retailer, one warehouse, multiple supplier model. *Int. J. Prod. Econ., 59*(1–3), 341–354.
26. Geoffrion, A. M., (1976). The purpose of mathematical programming is insight not numbers. *Interfaces, 7*(1), 81–92.
27. Gonzales, D., Searcy, E. M., & Eksioglu, S. D., (2013). Cost analysis for high-volume and longhaul transportation of densified biomass feedstock. *Transp. Res. Part A, 49*, 48–61.
28. Graves, S. C., & Willems, S. P., (2003). Chapter 3: Supply chain design: Safety stock placement and supply chain configuration. In: De Kok, A. G., & Graves, S. C., (eds.), *Handbooks in OR and MS* (Vol. 11, pp. 95–132).
29. Li, X., Haipeng, Y., Jingjing, W., Xiaobin, X., Chunxiao, J., & Lajos, H., (2019). A near-optimal UAV-aided radio coverage strategy for dense urban areas. *IEEE Transactions on Vehicular Technology, 68*(9), 9098–9109.
30. Yuki, S., Shingo, O., Yuta, T., Kojiro, M., Satoshi, T., Kazutoshi, S., Makoto, N., & Tatsuo, T., (2020). "Remote detection of radioactive hotspot using a Compton camera mounted on a moving multi-copter drone above a contaminated area in Fukushioma. *Journal of Nuclear Science and Technology, 57*(6), 734–744. https://doi.org/10.1080/0 0223131.2020.1720845.
31. Dronethusiast, (2015). *EHang 184 is a Manned UAV You Will Never Get to Fly.* Available from: https://www.dronethusiast.com/ehang-184-is-a-manned-uav-you-will-never-get-to-fly/ (accessed on 19 November 2021).
32. Eksioglu, B., Vural, A. V., & Reisman, A., (2009). The vehicle routing problem: A taxonomic review. *Comput. Ind. Eng., 57*(4), 1472–1483.
33. Erlebacher, S. J., & Meller, R. D., (2000). The interaction of location and inventory in designing distribution systems. *IIE Trans., 32*(2), 155–166.
34. Erlenkotter, D., (1989). The general optimal market area model. *Ann. Oper. Res., 18*(1), 43–70.
35. Federal Aviation Administration, U.S. Department of Transportation (2016). *Unmanned Aircraft Systems.* Available from: https://www.faa.gov/uas/ (accessed on 19 November 2021).

36. Manalili, M. A., Schumann, G., Prades, L., Rosa, S., Reane, D., & Beleza, A. J., (2020). The value of drones for bespoke local flood risk assessment in the licungo basin. In: *EGU General Assembly Conference Abstracts* (p. 4473).

37. Mario, A. R. E., & Abrahim, N., (2019). The uses of unmanned aerial vehicles –UAV's- (or drones) in social logistic: Natural disasters response and humanitarian relief aid. *Procedia Computer Science, 149*, 375–383. https://doi.org/10.1016/j.procs.2019.01.151.

38. Center for Disaster Philanthropy, (2016). *The Disaster Life-Cycle*. Available from: http://disasterphilanthropy.org/the-disaster-life-cycle/ (accessed on 19 November 2021).

39. Chen, A., Yu, Y., & Ting-Yi, (2016). Network based temporary facility location for the emergency medical services considering the disaster induced demand and the transportation infrastructure in disaster response. *Transp. Res. Part B: Methodol., 91*, 408–423.

40. Cohen, R., (2014). *Humanitarian Aid Delivered by Drones: A New Frontier for NGOs?* Available from: https://nonprofitquarterly.org/nasa-experiments-with-drones-and-a-showy-delivery-to-a-rural-nonprofit/ (accessed on 19 November 2021).

41. Coxworthe, B., (2016). *Ehang 184 Drone Could Carry You Away One Day*. Available from: http://www.gizmag.com/ehang-184-aav-passenger-drone/41213/ (accessed on 19 November 2021).

42. Daganzo, C. F., (1996). *Logistics Systems Analysis*. Springer, Berlin.

43. Dasci, A., & Verter, V., (2001). A continuous model for production-distribution system design. *Eur. J. Oper. Res., 129*(2), 287–298.

44. Bowman, D. M. J. S., Williamson, G. J., Abatzoglou, J. T., Kolden, C. A., Cochrane, M. A., & Smith, A. M. S., (2017). Human exposure and sensitivity to globally extreme wildfire events. *Nature Ecol. Evol., 1*. 58.

45. Schoennagel, T., Balch, J. K., Brenkert-Smith, H., Dennison, P. E., Harvey, B. J., Krawchuk, M. A., Mietkiewicz, N., Morgan, P., Moritz, M. A., Rasker, R., et al., (2017). "Adapt to more wildfire in western North American forests as climate changes. *Proc. Natl. Acad. Sci. USA, 114*, 4582.

46. Carter, V. A., Power, M. J., Lundeen, Z. J., Morris, J. L., Petersen, K. L., Brunelle, A., Anderson, R. S., Shinker, J. J., Turney, L., Koll, R., et al., (2017). A 1,500-year synthesis of wildfire activity stratified by elevation from the U.S. Rocky Mountains. *Quatern. Int., 488*, 107–119.

47. Allison, R., Johnston, J., Craig, G., & Jennings, S., (2016). Airborne optical and thermal remote sensing for wildfire detection and monitoring. *Sensors, 16*, 1310.

48. Aicardi, I., Garbarino, M., Andrea, L., & Emanuele, L., (2016). Monitoring post-fire forest recovery using multi-temporal digital surface models generated from different platforms. In: *Proceedings of the EARSeL Symposium* (Vol. 15, pp. 1–8). Bonn, Germany.

49. Stow, D. A., Lippitt, C. D., Coulter, L. L., & Loerch, A. C., (2018). Towards an end-to-end airborne remote-sensing system for post-hazard assessment of damage to hyper-critical infrastructure: Research progress and needs. *Int. J. Remote Sens., 39*, 1441–1458.

50. Anonymous, (2017). Unmanned aerial vehicles for environmental applications. *Int. J. Remote Sens., 38*, 2029–2036.

51. Fernández-Guisuraga, M. J., Sanz-Ablanedo, E., Suárez-Seoane, S., & Calvo, L., (2017). Using unmanned aerial vehicles in postfire vegetation survey campaigns through large and heterogeneous areas: Opportunities and challenges. *Sensors, 18*, 586.

52. Hamilton, D., Bowerman, M., Colwell, J., Donohoe, G., & Myers, B., (2017). "Spectroscopic analysis for mapping wildland fire effects from remotely sensed imagery. *J. Unmanned Veh. Syst., 5*, 146–158.

53. McKenna, P., Erskine, P. D., Lechner, A. M., & Phinn, S., (2017). Measuring fire severity using UAV imagery in semi-arid central Queensland, Australia. *Int. J. Remote Sens., 38*, 4244–4264.

54. Cruz, H., Eckert, M., Meneses, J., & Martínez, J. F., (2016). Efficient forest fire detection Index for application in unmanned aerial systems (UASs). *Sensors, 16*, 893.

55. USGS. (2004). *USDA Burned Area Reflectance Classification (BARC)*. Available online: https://www.fs.fed.us/eng/rsac/baer/barc.html (accessed on 19 November 2021).

56. Patrick, R., (2020). *Ukraine says Wildfires Close to Chernobyl are Extinguished After Rain Falls.* ABC News.

57. Al-Tahir, R., Arthur, M., & Davis, D., (2011). *Low Cost Aerial Mapping Alternatives for Natural Disasters in the Caribbean.* Available from: https://www.fig.net/resources/proceedings/fig_proceedings/fig2011/papers/ts06b/ts06b_altahir_arthur_et_al_5153.pdf (accessed on 19 November 2021).

58. Balcik, B., & Beamon, B. M., (2008). Facility location in humanitarian relief. *Int. J. Logist. Res. Appl., 11*(2), 101–121.

59. Barnes, G., & Langworthy, P., (2003). *The Per-mile Costs of Operating Automobiles and Trucks.* Available from: https://www.lrrb.org/pdf/200319.pdf (accessed on 30 November 2021).

60. Ben-Tal, A., Chung, B. D., Mandala, S. R., & Yao, T., (2011). Robust optimization for emergency logistics planning: Risk mitigation in humanitarian relief supply chains. *Transp. Res. Part B, 45*(8), 1177–1189.

61. Blumenfeld, D. E., & Beckmann, M. J., (1985). Use of continuous space modeling to estimate freight distribution costs. *Transp. Res. Part A, 19*(2), 173–187.

62. Burns, L. D., Hall, R. W., Blumenfeld, D. E., & Dganzo, C. F., (1985). Distribution strategies that minimize transportation and inventory costs. *Oper. Res., 33*(3), 469–490.

63. Somya, G., & Pradeep, K. B., (2021). Empirical software measurements with machine learning. *Computational Intelligence Techniques and Their Applications to Software Engineering Problems, 1*, 49–64. Boca Raton: CRC Press. https://doi.org/10.1201/9781003079996.

64. Somya, G., Anubha, P., & Anita, S., (2018). Application of big data analytics in cloud computing via machine learning. *Data Intensive Computing Applications for Big Data, 29*, 236–266. IOS Press. https://doi.org/10.3233/978-1-61499-814-3-236.

65. Somya, G., & Pradeep, K. B. (2020). Comparison of machine learning techniques for software quality prediction. *International Journal of Knowledge and Systems Science (IJKSS), 11*(2), 21–40. IGI Global. doi: 10.4018/IJKSS.2020040102.

66. Abhijith Valsan, Parvathy, B., Vismaya Dev, G. H., Unnikrishnan, R. S., Praveen Kumar Reddy, & Vivek A. (2020). In: *2020 4ᵗʰ International Conference on Trends in Electronics and Informatics (ICOEI) (48184)* (pp. 684–687). IEEE.

67. Balmukund, M., Deepak, G., Pratik, N., & Vipul, M., (2020). Drone-surveillance for search and rescue in natural disaster. *Computer Communications, 156*, 1–10. https://doi.org/10.1016/j.comcom.2020.03.012.

68. Christen, M., Villano, M., Narvaez, D., Serrano, J., & Crowell, C., (2014). *Measuring the Moral Impact of Operating "Drones" on Pilots in Combat, Disaster Management*

and Surveillance. Available from: https://www.zora.uzh.ch/id/eprint/108498/1/2014_ECIS2014_drones.pdf (accessed on 30 November 2021).

69. Otto, A., Agatz, N., Campbell, J., Golden, B., & Pesch, E., (2018). Optimization approaches for civil applications of unmanned aerial vehicles (UAVs) or aerial drones: A survey. *Networks, 72*(4), 411–458.

70. Koubâa, A., Qureshi, B., Sriti, M. F., Javed, Y., & Tovar, E., (2017). A service-oriented Cloud-based management system for the Internet-of-Drones. In: *2017 IEEE International Conference on Autonomous Robot Systems and Competitions (ICARSC)* (pp. 329–335). IEEE.

71. Chowdhury, S., Emelogu, A., Marufuzzaman, M., Nurre, S. G., & Bian, L., (2017). Drones for disaster response and relief operations: A continuous approximation model. *International Journal of Production Economics, 188*, 167–184.

72. Alsamhi, S. H., Ma, O., Ansari, M. S., & Almalki, F. A., (2019). Survey on collaborative smart drones and internet of things for improving smartness of smart cities. *IEEE Access, 7*, 128125–128152.

73. Abounacer, R., Rekik, M., & Renaud, J., (2014). An exact solution approach for multiobjective location? Transportation problem for disaster response. *Comput. Oper. Res. 41*, 83–93.

74. Afshar, A., & Haghani, A., (2012). Modeling integrated supply chain logistics in real-time large-scale disaster relief operations. *Socio-Econ. Plan. Sci., 46*(4), 327–338.

75. Ahmadi, M., Seifi, A., & Tootooni, B., (2015). A humanitarian logistics model for disaster relief operation considering network failure and standard relief time: A case study on San Francisco district. *Transp. Res. Part E: Logist. Transp. Rev., 75*, 145–163.

Index

Milton Keynes UK
Ingram Content Group UK Ltd.
UKHW051534141024
449569UK00001B/33